D0343195

State and Local
Population Projections
Methodology and Analysis

The Plenum Series on Demographic Methods and Population Analysis

Series Editor: Kenneth C. Land, *Duke University, Durham, North Carolina*

ADVANCED TECHNIQUES OF POPULATION ANALYSIS
Shiva S. Halli and K. Vaninadhu Rao

ANALYTICAL THEORY OF BIOLOGICAL POPULATIONS
Alfred J. Lotka

CONTINUITIES IN SOCIOLOGICAL HUMAN ECOLOGY
Edited by Michael Micklin and Dudley L. Poston, Jr.

CURBING POPULATION GROWTH: An Insider's Perspective
on the Population Movement
Oscar Harkavy

THE DEMOGRAPHY OF HEALTH AND HEALTH CARE
Second Edition
Louis G. Pol and Richard K. Thomas

FORMAL DEMOGRAPHY
David P. Smith

HOUSEHOLD COMPOSITION IN LATIN AMERICA
Susan M. De Vos

HOUSEHOLD DEMOGRAPHY AND HOUSEHOLD MODELING
Edited by Evert van Imhoff, Anton Kuijsten, Pieter Hooimeijer,
and Leo J. G. van Wissen

MODELING MULTIGROUP POPULATIONS
Robert Schoen

POPULATION ISSUES: An Interdisciplinary Focus
Edited by Leo J. G. van Wissen and Pearl L. Dykstra

THE POPULATION OF MODERN CHINA
Edited by Dudley L. Poston, Jr., and David Yaukey

A PRIMER OF POPULATION DYNAMICS
Krishnan Namboodiri

STATE AND LOCAL POPULATION PROJECTIONS: Methodology and Analysis
Stanley K. Smith, Jeff Tayman, and David A. Swanson

State and Local Population Projections
Methodology and Analysis

Stanley K. Smith
University of Florida
Gainesville, Florida

Jeff Tayman
San Diego Association of Governments
San Diego, California

and

David A. Swanson
Helsinki School of Economics and Business Administration
Mikkeli, Finland

Kluwer Academic/Plenum Publishers
New York Boston Dordrecht London Moscow

Library of Congress Cataloging-in-Publication Data

Smith, Stanley K.
   State and local population projections: methodology and analysis/Stanley K. Smith,
Jeff Tayman, and David A. Swanson.
        p.   cm. — (The Plenum series on demographic methods and population analysis)
   Includes bibliographical references and index.
   ISBN 0-306-46492-6 (hc.)—ISBN 0-306-46493-4 (pbk.)
        1. Population forecasting—Methodology.   2. Population research—Methodology.   I.
Tayman, Jeff, 1951–   II. Swanson, David A. (David Arthur), 1946–   III. Title.   IV. Series.

   HB849.53 .S626 2000
   304.6′01′12—dc21   00-063818

ISBN 0-306-46492-6 (Hardbound)
ISBN 0-306-46493-4 (Paperback)

©2001 Kluwer Academic / Plenum Publishers
233 Spring Street, New York, N.Y. 10013

http://www.wkap.nl/

10 9 8 7 6 5 4 3 2 1

A C.I.P. record for this book is available from the Library of Congress

Printed in the United States of America

Preface

The initial plans for this book sprang from a late-afternoon conversation in a hotel bar. All three authors were attending the 1996 meeting of the Population Association of America in New Orleans. While nursing drinks and expounding on a variety of topics, we began talking about our current research projects. It so happened that all three of us had been entertaining the notion of writing a book on state and local population projections. Recognizing the enormity of the project for a single author, we quickly decided to collaborate. Had we not decided to work together, it is unlikely that this book ever would have been written.

The last comprehensive treatment of state and local population projections was Don Pittenger's excellent work *Projecting State and Local Populations* (1976). Many changes affecting the production of population projections have occurred since that time. Technological changes have led to vast increases in computing power, new data sources, the development of GIS, and the creation of the Internet. The procedures for applying a number of projection methods have changed considerably, and several completely new methods have been developed. A great deal of research has been done on the determinants of population growth and the construction, use, and evaluation of population projections. In addition, projections are now routinely made for very small geographic areas such as ZIP codes, census tracts, and even individual blocks; 25 years ago, this would have been considered akin to alchemy. Given the magnitude of these changes and the rapidly increasing use of population projections for planning, budgeting, and other decision-making purposes, we concluded that a new book on this topic was long overdue.

This volume addresses many of the changes that have occurred during the last quarter century. By covering both the methodology and analysis of state and local projections, it brings together two main branches of the projections literature, providing the reader with an understanding of the mechanics and underlying rationale of commonly used projection methods. We believe that both factors are essential for producing and evaluating population projections.

This book is intended as a practical guide for demographers, planners, and other analysts called upon to construct or evaluate state and local population projections. Using numerical examples and illustrations as well as equations and verbal descriptions, we discuss data sources and application techniques that can be used when implementing three major classes of population projection methods: trend extrapolation, cohort-component, and structural models. We also discuss the determinants of population growth, the formation of assumptions, the development of evaluation criteria, the determinants of forecast accuracy, and the strengths and weaknesses of each projection method. We close by looking ahead at some technological and methodological advances that are likely to influence the production and use of population projections.

We place particular emphasis on problems encountered in making projections for small areas (e.g., counties and subcounty areas). Although small-area projections are widely used, they are subject to several problems that are much less severe—or not encountered at all—in projections for larger areas. Despite their unique characteristics and widespread use, however, small-area projections have received relatively little attention in the scholarly research. We hope this book will help reverse that trend.

The chapters of this book follow a logical sequence, but do not have to be read sequentially. Rather than plowing straight through, many readers may find it more useful to browse, reading only a single chapter—or a section of a chapter—at any given time. To facilitate this approach, we have tried to make each chapter an independent reading capable of standing on its own. Although this creates some repetition from one chapter to another, it makes the volume more valuable as a reference tool.

This book was not easy to write. A number of methods can be used for state and local projections, each involving a variety of potential data sources and application techniques. Deciding which ones to include—and how to present them—was a difficult task. We wanted the book to be comprehensive but not long-winded, technically precise but not overly mathematical, clearly written but not simplistic. We wanted it to provide useful information to demographers yet be accessible to planners and other analysts who have little formal training in demography. We wanted it to reflect the complexity and lack of consensus regarding many demographic, statistical, and methodological issues, but also to offer some clear conclusions and operational guidelines. The reader must decide whether we have succeeded in accomplishing these often-conflicting goals.

Many people helped in writing this book. We would like to thank Dennis Ahlburg, Larry Long, Steve Murdock, Pam Perlich, Steve Putnam, Rick Rogers, Dennis Turner, and Carol Taylor West for reading individual chapters and offering many valuable comments and suggestions. Special thanks go to Ryan Burson, Ken Land, Peter Morrison, Don Pittenger, and John Weeks for reviewing the entire manuscript; their herculean efforts helped us tremendously. Eric Chiang, June Nogle, and Ed Schafer provided valuable assistance in assembling tables and figures and checking references. Wayne Losano provided many helpful suggestions regarding grammar, punctuation, and writing style. Any remaining errors, oversights, and other inadequacies, of course, are the full responsibility of the authors.

Above all, we express our gratitude to our wives—Rita, Melinda, and Phoenecia—for their love, patience, and understanding while we worked on this seemingly endless project.

Contents

CHAPTER 1

Introduction

People are fascinated by the future. Palm readers, astrologers, and crystal-ball gazers down through the ages have found eager customers for their predictions and views of the future. Modern-day analysts and forecasters—using computers and mathematical models instead of tea leaves and chicken entrails—continue to find willing audiences. The desire to see into the future is seemingly insatiable and apparently has not been diminished by the relatively low success rates achieved by visionaries and forecasters in the past (see Box 1.1).

The desire to see into the future extends to the course of the human population and reflects much more than idle curiosity. At the global level, many are concerned about the earth's ability to feed, clothe, and house the billions of people expected to be added to the world's population during the next century. Nations and states are concerned about the economic, social, political, and environmental consequences of population growth and demographic change. At the local level, many types of public- and private-sector planning—for schools, hospitals, shopping centers, housing developments, roads, and countless other projects—are strongly affected by expected population growth. Indeed, the success or failure of such plans often depends on the extent to which projected growth is realized over time. It is no wonder that population projections are of so much concern to so many people.

Despite this concern, the future is unknown and—in most respects—unknowable. Many factors influence population growth and demographic change, often in unpredictable ways. No matter how extensive our data and sophisticated our forecasting methods, we still cannot "see" into the future. One hundred years ago, who could have predicted the baby boom and bust, the tremendous increase in life expectancy, or the dramatic shifts in foreign immigration that occurred in the United States during the twentieth century? Who could have predicted microwave ovens, interstate highways, space exploration, or the development of the Internet? For that matter, who could have predicted the appearance of hula hoops,

---

Box 1.1

Blasts from the Past: Some Predictions That Missed the Boat

"This 'telephone' has too many shortcomings to be seriously considered as a means of communication. The device is inherently of no value to us."

Western Union internal memo, 1876

"Heavier-than-air flying machines are impossible."

LORD KELVIN, President of the Royal Society, 1895

"Everything that can be invented has been invented."

CHARLES DUELL, Commissioner, U.S. Office of Patents, 1899

"Who the hell wants to hear actors talk?"

H. M. WARNER, Warner Brothers, 1927

"Stocks have reached what looks like a permanently high plateau."

IRVING FISHER, Professor of Economics, Yale University, 1929

"I think there is a world market for maybe five computers."

THOMAS WATSON, Chairman of IBM, 1943

"We don't like their sound, and guitar music is on the way out."

DECCA RECORDING COMPANY, rejecting the Beatles, 1962

"640K ought to be enough for anybody."

BILL GATES, founder of Microsoft, 1981.

Sources: Automatic Forecasting Systems (www.autobox.com); D. Lambro, The Heritage Foundation (www.heritage.org.)

---

disco, beanie babies, or the Beatles? As Winston Churchill said, "the future is just one damn thing after another."

We are not completely lost, however. Although individual events may be unpredictable, clear patterns often emerge when the effects of individual events are combined. This is especially true in demography, where the momentum of demographic processes links the future with the past in clear and measurable ways. We can study demographic trends, collect historical data, and build projection models based on our knowledge of the past and our expectations for the future. Because the future is intimately tied to the past, these projections often provide reasonably accurate predictions of future population change. If constructed and interpreted properly, population projections—however imperfectly they predict the future—can be extremely useful tools for planning and analysis.

WHAT IS A POPULATION PROJECTION?

Projections, Forecasts, Estimates

A variety of terms can be used to describe calculations of past and future populations. Following demographic convention, we define a *population projection* as the numerical outcome of a particular set of assumptions regarding future population trends (e.g., Irwin, 1977; Isserman, 1984; Keyfitz, 1972; Pittenger, 1976; Shryock & Siegel, 1973). Some projections refer only to total population, but many make further distinctions by age, sex, race, and other characteristics. A variety of projection methods can be used. Some focus solely on changes in total population, whereas others distinguish among individual components of growth.

Strictly speaking, population projections are conditional statements about the future. They show what the population would be *if* particular assumptions were to hold true. However, they do not predict whether those assumptions actually *will* hold true. By definition, population projections are always "right," barring a mathematical error in calculating them. Because they make no predictions regarding the future, they can never be proven wrong by future events.

A *population forecast*, on the other hand, is the projection that the analyst (i.e., the person or agency that makes the projection) believes is most likely to provide an accurate prediction of the future population. Whereas projections are nonjudgmental, forecasts are explicitly judgmental. They are unconditional statements that reflect the analyst's views regarding the optimal combination of data sources, projection techniques, and methodological assumptions. Unlike projections, population forecasts *can* be proven right or wrong by future events (or, more precisely, it can be found that they approximate the future with relatively small or large errors).

Demographers have traditionally used the term *projection* to describe calculations of future population. There are several reasons for choosing this terminology. First, *projection* is a more inclusive term than *forecast*. A forecast is a particular type of projection; namely, the projection that the analyst believes is most likely to provide an accurate prediction of the future population. Given this distinction, all forecasts are projections, but not all projections are forecasts. Second—as we discuss later in this chapter—projections can serve other purposes besides predicting future population; we believe the term *projection* facilitates the discussion of these alternate roles. Finally, demographers have often intended their calculations of future population as merely illustrative rather than predictive; *projection* fits more closely with this intention than *forecast*.

We use both terms in this book, but use *projection* as the general term to describe calculations of future population. We use *forecast* and *forecasting* when the discussion focuses on predicting the most likely course of the future population. We believe the critical factor is not the term itself, but the purposes for which projections (or forecasts) are used (e.g., describing a hypothetical future scenario or selecting the most likely future outcome).

A distinction can also be made between population projections and *population estimates*. This distinction is based on both temporal and methodological considerations. The most fundamental difference is that projections refer to the future whereas estimates refer to the present or the past. In addition, estimates can often be based on data for corresponding points in time. For example, estimates for 1998 made in 2000 can be based on data series—such as births, deaths, building permits, school enrollments, automobile registrations, and Medicare enrollees—reflecting population growth through 1998. Projections for 2008 made in 2000, however, cannot use such data series because those series do not yet exist.

The distinction between estimates and projections is not always clear-cut. Sometimes, no data are available for constructing population estimates. In these circumstances, methods typically used for population projections can be used for population estimates. For example, calculations of a city's age-sex composition in 1998 made in 1999 may have to be based on the extrapolation of 1980–1990 trends because data series reflecting post-1990 changes in age-sex composition are not available. Should these calculations be called *estimates* or *projections*?

In this book, we refer to calculations that extend beyond the date of the last observed data point as *projections*; calculations for all prior dates are called *estimates*. For example, if we have population counts or symptomatic data through July 1, 2001, calculations for dates on or before that day are called *population estimates*, and calculations for dates after that day are called *population projections*.

Alternative Approaches to Projecting Population

Many approaches can be used to develop population projections. Some are subjective, others objective. Some are very simple, others extremely complex. Some are strictly extrapolative, others make use of structural models. Data requirements range from small to large. Levels of disaggregation vary from minimal to elaborate. There are also many ways to classify projection methods. Figure 1.1 shows the classification scheme we follow in this book.

A fundamental distinction in the general forecasting literature is between subjective and objective methods (Armstrong, 1985). Subjective methods are those that lack a clearly defined process for analyzing data and creating projections. Examples include projections based on general impressions, intuition, or analogy; sometimes they are simply wild guesses. Even when subjective methods are based partly on objective data and formal analyses, the exact nature of the projection process has not been clearly specified and cannot be replicated by other analysts.

Objective methods, on the other hand, are those in which the projection process has been clearly specified. Data sources, assumptions, and mathematical

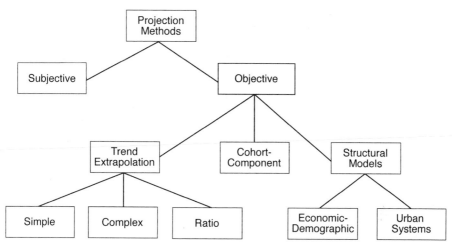

Figure 1.1. A typology of population projection methods.

relationships are defined in precise quantitative terms. Theoretically, the process can be specified so precisely that other analysts can replicate the method and obtain exactly the same results.

Although subjective methods are used frequently for some types of forecasting (e.g., technological change), they are not commonly used for population projections. In this book we focus primarily on objective methods. However, it is important to recognize that objective methods themselves contain many subjective elements. Even the most quantitative projection methods require subjective choices regarding variables, data sources, time periods, functional forms, and so forth. As we emphasize throughout this book, the application of every projection method involves using judgment. In general, the more complex the method, the greater the role of judgment in the projection process.

We describe three types of population projection methods: trend extrapolation, cohort-component, and structural models. *Trend extrapolation methods* are based on the continuation of observable historical trends. They can be simple (e.g., linear extrapolation) or complex (e.g., ARIMA time series models). They are often used for projecting total population but can also be used for projecting a particular population subgroup (e.g., a race or ethnic group) or for a particular component of population growth (e.g., births or birth rates). Trend extrapolation methods are generally applied to a single data series (e.g., total population) but can also be applied to data expressed as ratios (e.g., county shares of state population). The defining characteristic of trend extrapolation methods is that projected values for a particular variable are based solely on its historical values.

The *cohort-component method* accounts for births, deaths, and migration—the components of population growth. Most applications of this method divide the population into age-sex groups (i.e., cohorts) and project the components of growth for each cohort. The population can be further subdivided by race, ethnicity, and other demographic characteristics. There are a number of ways to construct cohort-component models and to project the components of growth. Because the cohort-component method is used more frequently than any other method of population projection, it is a major focus of this book.

The third approach to population projection focuses on the relationships between demographic and nondemographic variables. Models are developed in which projections of population change (or a particular component of population change) are based on changes in one or more independent variables. In many applications, the independent variables are economic in nature. For example, we might develop a model in which job growth in one area attracts migrants from another area. This model might specify how many migrants arrive each year for each 1,000 new jobs created. Such a model has economic and demographic components and translates projections of future job growth into projections of future migration.

We call these *structural models*. Their defining characteristic is that the projected values for a demographic variable are based not only on the historical values of that variable, but on other variables as well. Whereas trend extrapolation models tell us nothing about the causes of population change, structural models provide explanations as well as projections of future population change.

Many types of structural models can be developed. They range from simple recursive models involving only a few variables and a single equation to huge systems of simultaneous equations involving many variables and parameters. In this book, we discuss *economic-demographic* and *urban systems models*, the two types of structural models used most frequently for state and local population projections.

These three approaches—trend extrapolation, cohort-component, and structural modeling—are not mutually exclusive. Many projection models combine elements from several approaches. For example, cohort-component models might use structural models to project in- and out-migration or might simply extrapolate past migration trends. Although trend extrapolation methods can be applied independently of the other two approaches, cohort-component and structural models are often used in combination with another approach.

We discuss a wide variety of methods, models, and techniques in this book, including those most commonly used for state and local population projections. Those we do not discuss are not commonly used or are simply variants of one of the methods discussed. We believe that these methods reflect the current "state of the art" and provide the reader with an ample set of tools for producing and evaluating state and local population projections.

WHY MAKE POPULATION PROJECTIONS?

Roles of Projections

Predicting Future Population Change. Population projections can play a number of roles and serve several distinct purposes. The most fundamental is to predict future population change, giving a single point of reference to people planning for the future. Most data users (especially at the local level) rely on projections as guides for decision making and view them as forecasts regardless of the intentions of the analyst(s) who produced the projections. We give several examples of using projections for decision making in the next section. When projections are used in this manner, it is imperative to pay close attention to the plausibility of the underlying assumptions. If the assumptions are implausible, the projections will not provide viable forecasts. Chapter 13 discusses the forecast accuracy of population projections.

Analyzing Determinants of Population Change. Projections are often used as a tool to analyze the determinants of population change. This is the "what if" role of projections. What effect would a 20% increase in birth rates have on a state's population size and age composition? What would be the effect of eliminating all cancer deaths? What effect would the opening of a new factory employing 2,000 people have on a county's migration patterns? In this "simulation" role, projections are not necessarily meant to predict future population changes, but rather to illustrate the effects of specific changes in the model's underlying assumptions.

Presenting Alternative Scenarios. The third role of projections is closely related to the second. Projections can be used not only to analyze the determinants of population change, but also to give an indication of possible future scenarios. Because we cannot be sure what the future will bring, it is helpful to consider projections based on a variety of assumptions. Charting the implications of different combinations of assumptions gives some idea of the possible variation in future population values. These alternative scenarios are often based on expert judgment regarding the appropriate combination of assumptions but can also be based on formal techniques, giving a probabilistic description of the uncertainty inherent in a particular projection (i.e., a *prediction interval*). This possibility is discussed in Chapter 13.

Promoting Agendas and Sounding Warnings. Projections can also be used to support a particular political or economic agenda or to sound a warning. Here the emphasis shifts from what *will* happen to what *should* or *could* happen. For example, suppose that the growth of a county's population has been projected at 40% during the next decade. The county commission might use those projections of rapid growth as a marketing tool to attract new businesses to the area. Conversely, a citizen's group concerned about the environmental impact of rapid population growth might use those projections as a warning and a call for action to reduce the rate of growth. This role illustrates the political nature of population projections, a characteristic that may affect the way projections are made and how they are used. Chapter 12 discusses some of the political implications of population projections.

Providing a Base for Other Projections. Finally, population projections are often used as a base for constructing other types of projections used for decision making. For example, population projections can be used for school enrollment projections by applying enrollment rates to the appropriate age groups. They can be used for labor force projections by applying labor force participation rates to the appropriate age, sex, race, or ethnic groups. They can be used for projecting the number of births by applying birth rates to the projected female population. Analyses based directly or indirectly on population projections provide the basis for many types of decision making in both the public and private sectors. As a practical matter, this may be the most important use of population projections (including those used as forecasts).

Projections and Decision Making

Population projections are interesting to demographers because they reveal a great deal about the components of growth, the momentum of population change, and the implications of particular assumptions. To many nondemographers, however, these issues are—at best—only mildly interesting. The primary value of population projections to most people is how they can be used to improve decision making. The following examples illustrate some real-world uses of population projections.

Will Texas Run Out of Water? Texas is a rapidly growing state that draws its water from aquifers, lakes, and rivers. These sources are replenished over time by rainfall, but rainfall in Texas is wildly variable over time and space. Averages range from more than 40 inches per year in the eastern part of the state to fewer than 10 inches in the western part. Agriculture and industry are the major users of water, but households are important as well. Thus, population growth affects the demand for water directly through household formation and size and indirectly through its impact on the demand for agricultural and industrial output. In 1997, a state board published a 50-year state water plan (Texas Water Development Board, 1997). Population and household projections for counties and subcounty areas played an important role in the development of that plan.

How Much Will Income-Assistance Payments Cost the State of Washington? Income-assistance programs often account for a substantial chunk of a state's operating budget. Until recently, one of the major programs was Aid to Families with Dependent Children (AFDC), which accounted for more than 40% of total spending in the state of Washington during the mid-1990s (Opitz & Nelson, 1996). Washington uses a biennial budget cycle to plan for its revenues and expenditures, requiring a series of two-year projections. Given their impact on the budget, projections of future AFDC caseloads were crucial in developing the state's operating budget. Projections of population, family status, and household income made by the state's Office of Financial Management played a critical role in projecting those caseloads.

Where Should Encinitas Put Its New Fire Stations? Encinitas is a coastal city in southern California. Local officials were concerned about the city's ability to respond to future demands for fire protection and contacted the San Diego Association of Governments (SANDAG) for assistance in determining how many stations were needed and where they should be located. SANDAG developed a plan based on the projected number and location of households, road networks, travel times, access to "critical sites" (e.g., hospitals, schools, nursing homes), and land use plans. Population and household projections at the block level were essential for constructing this plan (Tayman, Parrott, & Carnevale, 1994).

Does Hillsboro Need a New School? Hillsboro is a suburb of Portland, Oregon. In the mid-1990s, the city had twenty elementary schools, four middle schools, and three high schools, containing a total of nearly 16,500 students (Swanson, Hough, Rodriguez, & Clemens, 1998). The school board faced a formidable planning task because of rapid population growth and the pending unification of separate elementary, middle, and high school districts. They contracted with a group of consultants to study how to combine the current districts to determine whether any new schools were needed and to develop attendance zones for each school. The consultants developed a plan based on population projections by age, sex, race, ethnicity, and income by traffic analysis zone. (Traffic analysis zones are small, user-designed areas developed for transportation planning; they are typically composed of one or more blocks.) These projections were then aggregated to form projections for each attendance zone within the newly formed school district.

Can Hospital X Support a New Obstetrical Unit? Hospital X is located in the central city of a medium-sized metropolitan area in the southeastern United States. Seeking to establish an obstetrical unit in a suburban satellite hospital, the administrators of Hospital X applied to the local health planning board for approval. The approval process required that Hospital X demonstrate the need for additional maternity services in the area. The application was opposed by a nearby suburban hospital that already offered maternity services. The opposing hospital retained a demographic consultant to determine if additional maternity services were actually needed. The critical factor in the case was the number of births projected for the satellite hospital's service area. The consultant used population projections by age and sex and assumptions regarding future birth rates to demonstrate that the projected increase in births was insufficient to justify establishing a new obstetrical unit (Thomas, 1994).

How Long Will GM's Retirees Live? Employers sponsor health benefits for the more than three-fourths of Americans who have health-care coverage. This is a major cost for many employers. General Motors, for example, spends more than $3 billion each year on health care for its current and former employees and their dependents. Retiree health-care costs are particularly important because health problems increase sharply in the older age groups. Concerned about the financial liabilities associated with rising health-care costs, GM managers directed a staff demographer and an outside consultant to analyze the expected longevity of its current and future retirees. These demographers developed a series of projections of GM retirees by age and sex and—based on those projections—made recommendations to the company's senior management, helping them improve corporate planning and reduce health-care costs (Kintner & Swanson, 1994).

Forecasting and Planning

Any plan for the future must be based—at least in part—on a projection (or forecast). Even the most mundane activities of everyday life illustrate this point. When one walks out the door in the morning and decides whether to take along an umbrella, one is forecasting the likelihood of rain that day. A household's saving decisions, a corporation's investment decisions, and a government's spending decisions all reflect plans based implicitly or explicitly on forecasts. The examples described before show how population projections affect plans developed by a variety of businesses and government agencies. Forecasting and planning are closely intertwined.

They are not synonymous, however. Each has its own goals and objectives. Forecasting attempts to *predict* the future, whereas planning seeks to *affect* it. Isserman (1984) described three types of forecasts. *Pure* forecasts describe the most likely future in the absence of intervention; *contingency* forecasts describe possible futures under different assumptions; and *normative* forecasts describe the desired future. These terms reflect the different roles played by projections. Because the objective of planning is to affect the future, successful planning may render a forecast obsolete and inaccurate; that is, active intervention may cause a deviation of future population trends from the paths they would have followed in the absence of intervention.

In addition, population projections *themselves* may affect future population growth. For example, areas projected to grow rapidly may do so in part because the predicted growth attracts job seekers and businesses wishing to relocate or expand. Areas projected to decline may do so in part because businesses and workers are driven away by their apparently poor prospects. In these circumstances, population forecasts influence the very trends they seek to predict, perhaps even becoming self-fulfilling prophecies (Isserman & Fisher, 1984; Moen, 1984). Thus planners may influence population growth through their forecasts as well as their plans.

Government planners use a variety of tools to influence the pace, distribution, and characteristics of population growth. Some local governments try to spur growth by actively courting new businesses, reducing tax rates, extending infrastructure into new areas, or improving the quality of government services. Others attempt to restrict growth by setting limits on the number of residential building permits issued, denying water and sewer services, instituting impact fees on new structures, banning certain types of industries, or limiting the number of unrelated persons living in one household. State governments influence population growth through policies that affect the economic environment, cost of living, and quality of services provided.

Forecasting and planning are distinct but closely related exercises. Successful planning not only reacts to projected population changes but may also seek to influence those changes. In some instances, then, it may be important that population projections account for potential changes in the local planning environment. This may be particularly important for very small areas, where planning decisions can have a substantial and immediate impact on population growth. There are some circumstances in which forecasting and planning cannot be separated.

HOW CAN THIS BOOK HELP?

A great deal has been written about population projections. One branch of the literature has addressed projection methodology (e.g., Davis, 1995; Irwin, 1977; Isserman, 1993; Pittenger, 1976; Shryock & Siegel, 1973). Another has focused on analysis and evaluation, with a particular emphasis on forecast accuracy (e.g., Keilman, 1990; Keyfitz, 1981; Long, 1995; Murdock, Hamm, Voss, Fannin, & Pecotte, 1991; Smith & Sincich, 1992; Stoto, 1983; Swanson & Tayman, 1995). In this book, we combine these two branches. We describe and illustrate the most commonly used projection methods, paying special attention to problems unique to applications for small areas. We also discuss determinants of population growth, quality of data sources, formation of assumptions, development of evaluation criteria, and determinants of forecast accuracy. Our aim is to provide the reader with an understanding not only of the mechanics of common projection methods but also of the complex analytical issues that make population forecasting an art as well as a science. Box 1.2 lists some of the major producers of state and local projections.

Objectives

We have three fundamental objectives in this book: to describe commonly used projection methods, to analyze their strengths and weaknesses, and to provide practical guidance to analysts called upon to produce or use population projections.

Describing Projection Methods.
First, we describe a variety of population projection methods. Some are quite simple, others far more complex. We use nontechnical language whenever possible and supplement the discussion with illustrations and step-by-step examples. We include equations where they illuminate the discussion but try not to overwhelm the reader with mathematics. For complex methods, of course, the discussion necessarily becomes more technical.

We emphasize methods that planners, demographers, and other analysts can use for small-area projections. For the cohort-component method and most of the trend extrapolation methods, most readers will be able to follow the procedures and illustrations to actually construct a set of population projections. Time series and structural models, however, may lie beyond the grasp of nontechnical readers. We do not provide step-by-step examples of the most complex methods, including the most data-intensive applications of the cohort-component method (e.g., parameterized multistate models). Interested readers may consult the references for further information on these methods.

Analyzing Strengths and Weaknesses.
Our second objective is to develop a set of criteria for evaluating the strengths and weaknesses of each projection method and to use those criteria to evaluate the methods described in this book. Each method has a unique combination of characteristics. Some have large data requirements, others have small data requirements. Some provide a great deal of demographic detail, others provide only a little. Some are time consuming and costly to apply, others are quick and inexpensive. Some require a high level of modeling skills, others require only simple skills. Some are easy to explain to data users, others are difficult to explain. For any given purpose, then, some methods are more useful than others. Only after considering the strengths and weaknesses of each method can the analyst choose the one(s) best suited to a particular project.

Box 1.2

Who Makes State and Local Population Projections?

Federal government. The Census Bureau makes state-level population projections several times each decade. The first state projections were made in the mid-1950s and provided breakdowns for several broad age groups. Recent projections have included breakdowns by age, sex, race, and Hispanic origin. The Census Bureau does not currently make any projections below the state level. The U.S. Bureau of Economic Analysis made several sets of population projections for states, economic areas, and metropolitan areas in the 1970s, 1980s, and 1990s, but no projections were made after 1995 because of budget cuts.

State demographers. Most state governments designate an agency to produce "official" population projections. These agencies are often part of the state government system; sometimes they are part of a state university. Most states make projections at the state and county levels; a few also make projections for selected subcounty areas. Most states make projections by age, sex, and race; a few also make projections by Hispanic origin. Representatives from each state and the Census Bureau formed the Federal-State Cooperative Program for Population Projections (FSCPPP) in 1981. This organization promotes the development and testing of projection methods; encourages the collection and exchange of reliable data; and provides a forum for sharing data, information, and ideas.

Local government agencies. Agencies such as city and county planning departments and associations (or councils) of governments often make population projections for subcounty areas such as census tracts, block groups, ZIP code areas, and traffic analysis zones. The level of geographic and demographic detail varies from place to place.

Private vendors. A number of private companies make population projections for states, counties, and subcounty areas such as census tracts, block groups, ZIP code areas, and a variety of customized geographic areas. These projections often contain a high level of geographic and demographic detail, are produced on a regular schedule, and are sold to data users in the public and private sectors.

Other private businesses. Some companies have staff members who produce population projections for use within the company. Demographic consultants also produce a variety of customized projections under contract with individual clients. These projections are often proprietary and are not generally available for public use.

Note: In 1999, the U.S. Bureau of the Census officially changed its name to the U.S. Census Bureau. For simplicity, we refer to this agency simply as the Census Bureau.

Because population projections are so widely used as forecasts of future population change, we examine forecast accuracy in detail, evaluating differences by projection method, population size, growth rate, length of base period, length of projection horizon, and launch year. We believe that this discussion will help the analyst choose the best method(s) for any particular purpose and provide a basis for judging the uncertainty inherent in population projections.

Providing Practical Guidance. Our most fundamental objective is to provide practical guidance to planners, demographers, and other analysts called upon to produce or use state and local population projections. Describing alternative projection methods and discussing their strengths and weaknesses is necessary but not sufficient for achieving this goal. One must also consider the context in which projections are made. What theoretical models and data sources are most appropriate for a particular type of projection? What social, economic, cultural, and political factors might affect the choice of assumptions? What special circumstances must be considered? Are any adjustments to the models or data necessary? If so, how can they be made? What potential pitfalls should the analyst watch out for? We provide answers to these questions throughout the book.

Geographic Focus

Much of the research on population projections has focused on the national level (e.g., Cohen, 1986; Keilman, 1990; Keyfitz, 1981; Land, 1986; Lee & Tuljapurkar, 1994; Pflaumer, 1992; Stoto, 1983). This research has been very valuable, but there are several reasons why conclusions based on studies of national projections may not be applicable to states and local areas.

First, there are substantial differences in data availability and reliability between national and subnational areas. Some data series are available only at the national level. Others are available with much greater frequency at the national level than at state and local levels. Due to reporting, coverage, and sampling errors, the quality of data is also likely to be better for nations than for subnational areas (especially for small areas). Consequently, certain projection methods that work well at the national level may not work well (or work at all) at the state and local levels.

Second, movements of population—both international and domestic— typically play a much greater role in population growth at the state and local levels than at the national level. For most countries, fertility and mortality are the major determinants of population growth and international migration has a relatively modest influence. For states and local areas, however, migration is often the major determinant of growth. Subnational migration rates are generally more variable across space and more volatile over time than either fertility or mortality rates, making them more difficult to forecast accurately. Consequently, subnational projections contain a major source of demographic uncertainty that is small or non-existent in most national projections.

Finally, population size itself plays a role. Individual activities such as the construction of a prison, the opening of a new highway, and the closing of a major employer can have a substantial impact on population growth in small areas. For large areas, however, the effects of these activities tend to cancel each other out. Consequently, population growth is much more volatile and unpredictable in small areas than in large areas. (By *small areas*, we mean counties and subcounty areas such as cities, townships, school districts, and traffic analysis zones. In some instances, we use the term expressly for very small areas such as census tracts and block groups.)

We believe an emphasis on population projections for states and local areas is justified by their unique characteristics, the relatively small amount of research that has been done on this topic, and the widespread use of small-area projections for decision making. This emphasis means that we pay special attention to problems of data availability and reliability, the role of migration, and the impact of special events and unique circumstances on population growth. However, except for urban systems models, all of the methods described in this book can be used for national projections as well as for state and local projections.

We focus on data sources and geographic areas in the United States—the country we know best—but we believe much of our discussion is relevant for subnational projections in other countries that have regularly conducted censuses, good vital registration systems, and a wide range of administrative records. For countries lacking these data resources, the discussion will provide a point of departure but not a complete road map to constructing population projections.

Coverage

This book is composed of four parts. Chapters 1 and 2 provide a general introduction to the book's topics and terminology. Chapter 1 discusses the reasons for making projections, defines some basic concepts, and describes the scope of the book. Chapter 2 is a primer on population analysis, covering basic demographic concepts. This chapter will be review material for readers who have a background in demography and can most likely be skipped. Box 1.3 defines some of basic terms used throughout the book.

Chapters 3 through 7 focus on the cohort-component method. Chapter 3 presents an overview of the method and provides a brief history of its development and use. Chapter 4 discusses mortality measures, sources of data, the construction of survival rates, alternative approaches to projecting survival rates, and some examples of mortality projections for states and local areas. Chapter 5 discusses fertility measures, sources of data, a description of the cohort and period perspectives, alternative approaches to projecting fertility rates, and some examples

---

Box 1.3

Some Basic Terminology

Projection. The numerical outcome of a particular set of assumptions regarding future values of a variable (e.g., population).

Forecast. The projection selected as the one most likely to provide an accurate prediction of the future value of a variable (e.g., population).

Estimate. A calculation of a current or past value of a variable (e.g., population), typically based on symptomatic indicators of change in that variable.

Base year. The year of the earliest data used to make a projection.

Launch year. The year of the most recent data used to make a projection.

Target year. The year for which a variable is projected.

Base period. The interval between the base year and launch year of a projection.

Projection horizon. The interval between the launch year and target year of a projection.

Projection interval. The increments in which projections are made.

For example, if data from 1990 through 2000 were used to project the population in 2005 and 2010, then 1990 would be the base year; 2000, the launch year; 2005 and 2010, the target years; 1990–2000, the base period; and 2000–2005 and 2000–2010, the projection horizons. These projections would be made in five-year intervals.

---

of fertility projections for states and local areas. Chapter 6 discusses migration definitions and measures, sources of data, the determinants of migration, alternative approaches to projecting migration, and some critical issues related to migration projections for states and local areas. Chapter 7 provides several step-by-step examples that show how to apply the cohort-component method and discusses its strengths and weaknesses.

Chapters 8 through 11 discuss other approaches for projecting state and local populations. Chapter 8 covers trend extrapolation methods, including a brief history of their application, a description of the most commonly used methods, examples of each method, and an assessment of their strengths and weaknesses. Chapters 9 and 10 cover economic-demographic and urban systems models, the two main types of structural models used for state and local projections. These chapters discuss some of the theory underlying these models, give examples of the way they can be used, and evaluate their strengths and weaknesses. Chapter 11 discusses special circumstances that may be encountered when making population projections and describes several adjustments and refinements to make projections more useful.

Chapters 12 through 15 focus on evaluating and analyzing population projections. Chapter 12 discusses criteria for evaluating projections and evaluates the methods discussed in this book. Chapter 13 analyzes forecast accuracy and bias, covering alternative error measures, factors that affect accuracy and bias, the potential for combining projections, and ways to account for uncertainty. Chapter 14 presents a practical guide to small-area projections and details each step that must be taken to construct a set of projections. This chapter summarizes many points made throughout the book. Chapter 15 discusses recent innovations and possible new directions in population projection research.

Target Audience

This book is aimed primarily at two groups of readers. The first consists of analysts who work for state and local governments, private businesses, universities, and nonprofit organizations and are responsible for making population projections for states and local areas. This group includes demographers, land use planners, transportation planners, environmental planners, school district administrators, market analysts, personnel managers, retirement benefits administrators, and sales forecasters. We believe that this book gives practitioners the information they need to decide which data sources to use, which methods to apply, and what problems to watch out for. It also provides guidance for evaluating the validity of population projections.

The second group is comprised of students in courses dealing with demographic methods or state, regional, and local planning. We believe this book will be useful as the primary textbook in courses focusing on population projections. It also will be useful as supplementary reading or resource material for courses in which population projections are covered in a short module. This book is not highly mathematical, but it assumes that readers have at least a basic knowledge of mathematics and statistics. We believe that it is suitable for both graduate students and upper level undergraduate students.

CHAPTER 2

Fundamentals of Population Analysis

Demography is defined as the scientific study of population (Shryock & Siegel, 1973, p. 2). It typically concentrates on the human population, although many of its concepts, measures, and techniques can be extended to nonhuman populations as well. It focuses on five basic topics:

1. The size of the population.
2. Its distribution across geographic areas.
3. Its composition (e.g., age, sex, race, and other characteristics).
4. Changes in population size, distribution, and composition over time.
5. The determinants and consequences of population growth.

In this chapter, we focus on the first four of these topics. We describe a number of basic demographic concepts, define some commonly used terms, and discuss several sources of demographic data. Our objective is to give a brief introduction to the reader who has little training or experience in formal demography. More comprehensive discussions can be found in Newell (1988), Shryock and Siegel (1973), Smith (1992), and Weeks (1999).

DEMOGRAPHIC CONCEPTS

Size

The most basic demographic concept is population size. Typically, *population size* refers to the number of people residing in a specific area at a specific time. According to estimates prepared by the Census Bureau, California had a population of 32,666,550 on July 1, 1998, whereas Wyoming had a population of

Table 2.1. Population Change for States, 1990–1998

	1990	1998	Absolute	Percent
			Change	
Northeast				
Connecticut	3,287,116	3,274,069	−13,047	−0.4
Maine	1,227,928	1,244,250	16,322	1.3
Massachusetts	6,016,425	6,147,132	130,707	2.2
New Hampshire	1,109,252	1,185,048	75,796	6.8
New Jersey	7,747,750	8,115,011	367,261	4.7
New York	17,990,778	18,175,301	184,523	1.0
Pennsylvania	11,882,842	12,001,451	118,609	1.0
Rhode Island	1,003,464	988,480	−14,984	−1.5
Vermont	562,758	590,883	28,125	5.0
Midwest				
Illinois	11,430,602	12,045,326	614,724	5.4
Indiana	5,544,156	5,899,195	355,039	6.4
Iowa	2,776,831	2,862,447	85,616	3.1
Kansas	2,447,588	2,629,067	151,479	6.1
Michigan	9,295,287	9,817,242	521,955	5.6
Minnesota	4,375,665	4,725,419	349,754	8.0
Missouri	5,116,901	5,438,559	321,658	6.3
Nebraska	1,578,417	1,662,719	84,302	5.3
North Dakota	638,800	638,244	−556	−0.1
Ohio	10,847,115	11,209,493	362,378	3.3
South Dakota	696,004	738,171	42,167	6.1
Wisconsin	4,891,769	5,223,500	331,731	6.8
South				
Alabama	4,040,389	4,351,999	311,610	7.7
Arkansas	2,350,624	2,538,303	187,679	8.0
Delaware	666,168	743,603	77,435	11.6
District of Columbia	606,900	523,124	−83,776	−13.8
Florida	12,938,071	14,915,980	1,977,909	15.3
Georgia	6,478,149	7,642,207	1,164,058	18.0
Kentucky	3,686,892	3,936,499	249,607	6.8
Louisiana	4,221,826	4,368,967	147,141	3.5
Maryland	4,780,753	5,134,808	354,055	7.4
Mississippi	2,575,475	2,752,092	176,617	6.9
North Carolina	6,632,448	7,546,493	914,045	13.8
Oklahoma	3,145,576	3,346,713	201,137	6.4
South Carolina	3,486,310	3,835,962	349,652	10.0
Tennessee	4,877,203	5,430,621	553,418	11.3
Texas	16,986,335	19,759,614	2,773,279	16.3
Virginia	6,189,197	6,791,345	602,148	9.7
West Virginia	1,793,477	1,811,156	17,679	1.0

Table 2.1. (*Continued*)

	1990	1998	Change Absolute	Percent
West				
Alaska	550,043	614,010	63,967	11.6
Arizona	3,665,339	4,668,631	1,003,292	27.4
California	29,785,857	32,666,550	2,880,693	9.7
Colorado	3,294,473	3,970,971	676,498	20.5
Hawaii	1,108,229	1,193,001	84,772	7.6
Idaho	1,006,734	1,228,684	221,950	22.0
Montana	799,065	880,453	81,388	10.2
Nevada	1,201,675	1,746,898	545,223	45.4
New Mexico	1,515,069	1,736,931	221,862	14.6
Oregon	2,842,337	3,281,974	439,637	15.5
Utah	1,722,850	2,099,758	376,908	21.9
Washington	4,866,669	5,689,263	822,584	16.9
Wyoming	453,589	480,907	27,318	6.0

Source: U.S. Census Bureau, Population Division, Population Estimates Branch, Internet Release Date: December 31, 1998.

only 480,907 (Table 2.1). These were the largest and smallest states in the United States in terms of population size.

Population size is seemingly a simple concept, but there is some ambiguity in the way it is applied. Americans are a very mobile population, and many people spend part of their time in one place and part in another (e.g., at home, at work, on vacation, or on a business trip). Where should these people be counted when a census is conducted?

There are two basic approaches to answering this question. The de facto approach counts people where they were physically located on census day, regardless of how much time they spent at that location. Under this approach, all tourists, business travelers, and seasonal residents present in Phoenix on census day would be counted as Phoenix residents, along with the usual residents of Phoenix who were in town that day. Usual residents who were out of town, however, would not be counted as Phoenix residents. The de jure approach counts people at their usual (or permanent) place of residence, regardless of where they were physically located on census day. Under this approach, tourists, business travelers, and other visitors temporarily present in Phoenix on census day would be counted as residents of Chicago, Omaha, or any other place in which they normally resided. The first approach is used in many countries that lack well-developed statistical systems. The second approach is used in the United States, Canada, and most other industrialized countries, and is the approach we follow in this book.

The de jure approach means that many people present in an area at one time or another are omitted from population counts, estimates, and projections. These omissions may be substantial for some places. For example, Florida has huge numbers of seasonal residents who spend three to six months in the state during the winter; these people are not included in Florida's population estimates and projections. The Census Bureau has established guidelines for determining place of residence for population subgroups such as military personnel, college students, snowbirds, prison inmates, business travelers, and the homeless. Despite these guidelines, there is still some ambiguity concerning who should or should not be included in official population statistics. Chapter 6 provides a more detailed discussion of these issues.

Populations need not refer to specific *geographic* areas. For example, a population could refer to all of the employees who work for a company or all of the enrollees in a health-care plan. Demographic analyses can be performed for populations defined for these entities as well as for populations defined for geographic areas. In this book, however, we focus primarily on populations defined for geographic areas.

Distribution

The *distribution* of a population refers to its geographic location. There are several different ways to define geographic areas. The most basic is the set of geographic areas developed specifically for statistical purposes, such as collecting and reporting data in the decennial census (U.S. Bureau of the Census, 1992). The boundaries for these areas are determined by the Census Bureau in consultation with local government agencies. These statistical areas form the building blocks used in constructing data sets for other types of geographic areas.

The most important geographic areas defined strictly for statistical purposes are census tracts, block numbering areas, block groups, and individual blocks. *Census tracts* (CTs) are small, relatively permanent areas that do not cross county boundaries and generally contain between 2,500 and 8,000 persons. They are designed to be relatively homogeneous in population characteristics, living conditions, and economic status. Census tracts were designated for all counties in the 2000 Census. In previous censuses, they were designated only for metropolitan areas and other densely populated counties. In the 1990 Census, *block numbering areas* (BNAs) were used instead of census tracts in more sparsely populated counties; they have the same general characteristics as census tracts. *Block groups* are clusters of blocks that generally contain 250 to 550 housing units; they do not cross CT or BNA boundaries. *Blocks* are small areas bounded on all sides by visible features such as streets or railroad tracks or by invisible boundaries such as city or township limits; they are the smallest geographic units for which census data are tabulated.

Geographic areas can also be defined according to administrative or political criteria. Examples include states, counties, cities, townships, congressional districts, school districts, and water management districts. For many purposes, these are the most important types of geographic areas that can be defined. They play an important role in planning, budgeting, and political representation and are often used for analyzing population growth and demographic change. However, geographic areas defined according to administrative or political criteria also have several limitations. Their boundaries are somewhat arbitrary and do not always account for important economic, cultural, social, or geographic factors. In addition, their boundaries often change over time, making it difficult (or impossible) to conduct time series analyses.

Geographic boundaries can also be defined according to other criteria. The U.S. Postal Service defines ZIP code areas for delivering the mail. Businesses define service areas to identify the locations of their customers or clients. Local planners define traffic analysis zones for developing transportation plans. Population data pertaining to these geographic areas are typically based on data collected or estimated for the statistical and administrative/political areas described before. Clearly, a wide variety of geographic areas can be used when analyzing the distribution of the population.

Composition

Composition refers to the characteristics of the population. For population projections, the most commonly used characteristics are age, sex, race, and ethnicity. These are the characteristics to which we refer most frequently in this book.

Age is the most important demographic characteristic because it has such a large impact on so many aspects of life, for individuals as well as for society as a whole. The age structure of a population affects its birth rate, death rate, and crime rate. It affects the demand for public education, health care, and nursing home care. It affects the housing market, the labor market, and the marriage market. It has tremendous implications for Social Security and Medicare. No other characteristic is more valuable for a wide variety of planning and analytical purposes than the age composition of the population.

Sex composition is also important for many purposes. In fact, these two characteristics are often combined to reflect the age-sex structure of the population (e.g., Weeks, 1999). We use the term *sex* to refer to the strictly biological differences between males and females; *gender* refers to nonbiological differences resulting from social, cultural, political, and economic factors.

The age-sex structure is often illustrated by using *population pyramids*. Population pyramids are graphic representations that show the number (or proportion) of the population in each age-sex group. For a country, the shape of a pyramid is determined primarily by its fertility history. For states and local areas, however, migration plays an important role as well. A pyramid that has a wide base reflects a young population.

Population pyramids tell a great deal about a population. Consider Figure 2.1, which shows pyramids for the Hispanic and non-Hispanic populations of the United States in 1999. Both populations have more males than females in the youngest age groups. This reflects the larger number of male births, a worldwide phenomenon (typically, there are about 105 male births for every 100 female births). In the middle and older age groups, however, both populations have more females than males. This reflects lower mortality rates for females than males, also a widespread phenomenon. The impact of the baby boom—people born between 1946 and 1964—is clearly evident in the non-Hispanic population: the age groups between 35 and 49 have the largest populations. The large numbers in the youngest age groups for the Hispanic population reflect their high birth rates.

Race and ethnicity are two other widely used demographic characteristics. During the last several decades, the Census Bureau has used four broadly defined race categories: white; black; American Indian, Eskimo, or Aleut; and Asian or Pacific Islander (U.S. Bureau of the Census, 1992). More detailed categories based on ethnicity and national origin were also used (e.g., Chinese, Filipino, and Samoan). In addition, the population was classified as Hispanic or non-Hispanic. It should be noted that "Hispanic" is an ethnic category, not a race category; consequently, people are classified both by race and by Hispanic origin.

The 2000 Census incorporated several changes in collecting race data. The number of basic categories was expanded from four to five: white; black; American Indian or Alaska Native; Asian; and Native Hawaiian or Other Pacific Islander. More importantly, the Census Bureau for the first time allowed respondents to list themselves as belonging to more than one race category; prior to that time, respondents were asked to list only a single category. The 2000 Census thus included a large number of potential multiracial combinations, in addition to the five single-race categories. This change created a great deal of uncertainty regarding the interpretation and use of race data and the consistency of those data over time.

Composition also refers to characteristics such as marital status, household relationship, employment status, income, education, and occupation. These characteristics are useful for various types of demographic analyses. They are sometimes included in population projections, but the specific methodological issues related to projecting these characteristics are not covered in this book.

Change

Population *change* is measured as the difference in population size between two points in time (i.e., two specific dates). A time point can correspond to the date of a census or a population estimate or projection. Because censuses are typically more accurate than estimates or projections, measures of change based on censuses will generally be more accurate than measures based on estimates or projections.

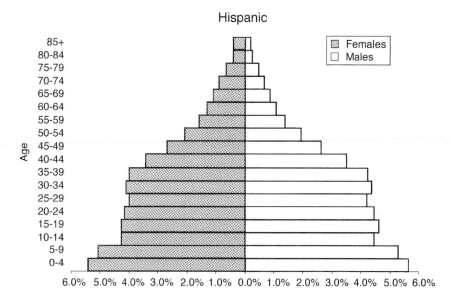

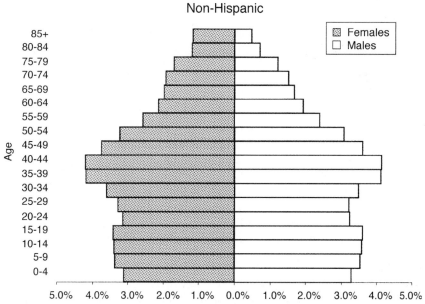

Figure 2.1. Population pyramids, United States, 1999. (*Source*: U.S. Census Bureau, Population Division, Population Estimates Branch. Internet Release Date, October 29, 1999)

Population change can be expressed in either absolute or percentage terms, as shown in Table 2.1. Absolute change is computed by subtracting the population at the earlier date from the population at the later date. A negative sign indicates a population loss. Percent change is computed by dividing the absolute change by the population at the earlier date and multiplying by 100. For example, Alabama had a population of 4,351,999 on July 1, 1998 and a population of 4,040,389 on April 1, 1990, yielding

Absolute change: 4,351,999 − 4,040,389 = 311,610

Percent change: (311,610 / 4,040,389)100 = 7.7%

Population change can also be expressed in terms of an average annual change. The average annual absolute change (AAAC) can be computed simply by dividing the total change by the number of years between the two dates:

$$\text{AAAC} = (P_l - P_b)/y$$

where P_l is the population at the later date, P_b is the population at the earlier date, and y is the number of years between the two. In the example for Alabama, there are 8.25 years between April 1, 1990 and July 1, 1998, yielding

$$\text{AAAC} = (4,351,999 - 4,040,389)/8.25 = 37,771$$

For some purposes, it is helpful to view annual population change in relative rather than absolute terms, or as percent changes (i.e., growth rates) rather than as absolute changes. Average annual growth rates can be calculated in two slightly different ways. The first is based on a geometric model:

$$r = (P_l/P_b)^{(1/y)} - 1$$

where r is the average annual geometric growth rate and the other terms are defined as before. Again, using Alabama as an example,

$$r = (4,351,999/4,040,389)^{(1/8.25)} - 1 = 0.00905, \text{ or } 0.905\% \text{ per year}$$

The geometric growth rate calculated in this manner is based on compounding in discrete intervals (i.e., at specific dates). In this example, growth is compounded once a year. Because populations grow continuously, it is often useful to use an exponential model based on continuous compounding:

$$r = [\ln(P_l/P_b)]/y$$

where r is the average annual exponential growth rate and ln is the natural logarithm. For Alabama, the average annual exponential growth rate is calculated as

$$r = [\ln(4,351,999/4,040,389)]/8.25 = 0.00901, \text{ or } 0.901\% \text{ per year}$$

Geometric and exponential growth rates are generally about the same. Geometric rates are always slightly larger than exponential rates because they are calculated at discrete intervals rather than continuously. The more rapidly growing the area, the greater the difference between geometric and exponential growth rates.

These measures of population change are simple and straightforward. However, they are not always easy to implement properly because of changes in geographic boundaries, changes in the accuracy of the base data, and changes in definitions.

The geographic boundaries of states have been constant for a long time. The same has been true for most counties, at least for the last several decades. Many other geographic areas, however, have experienced sudden (and sometimes large) boundary changes. For example, the boundaries of many cities, metropolitan areas, block groups, voting districts, and ZIP code areas have changed substantially in recent years. Although the Census Bureau attempts to hold census tract boundaries constant from one decennial census to another (except for subdivisions into coterminous sets of smaller tracts), changes at other levels of census geography occur frequently. Analysts must be aware of changes in geographic boundaries and make adjustments when necessary. Consistent measures of population change are possible only if geographic boundaries are held constant over time.

Changes in the accuracy of the base data also affect the measurement of population change. For example, suppose that a city's population was counted as 10,000 in 1990 and 11,000 in 2000, but that it was later discovered that the 2000 Census had missed an apartment complex with 1,000 residents. The population change based on the corrected numbers (2,000) would be twice as large as the change based on the uncorrected numbers (1,000). Because population estimates are typically less accurate than population counts, they introduce an additional source of error. It is difficult to uncover and correct errors in the underlying data, but in some places these errors have a substantial impact on calculations of population change.

Changes in the definition or interpretation of demographic concepts can also affect the measurement of change. Take race, for example. Because respondents were allowed to list only one race category in the 1990 Census but could list multiple categories in 2000, apparent changes in the number of persons in each race category between 1990 and 2000 may have been caused by changes in reporting practices, as well as by changes in the actual population. Problems of definition and interpretation are particularly significant in censuses based on self-enumeration, such as those used in the United States and Canada. Although guidelines for answering questions are provided, they are not followed in the same way by all respondents. Problems of definition and interpretation are particularly acute for measurement of race and ethnicity, but potentially affect every type of data collected in a census or survey, from the number of persons in a household to their detailed personal characteristics.

Measures of population change always refer to a specific population and a specific period of time; in most instances, they refer to a specific geographic area as well. Table 2.1, for example, refers to changes in total population for states between April 1, 1990 and July 1, 1998. Population change can also be measured for various subgroups of the population (e.g., females, Asians, or teenagers), different geographic areas (e.g., counties, cities), and different time periods (e.g., 1980–1990). In other words, *population change* can refer to changes in size, distribution, or composition, or to any combination of the three.

COMPONENTS OF CHANGE

There are only three components of population change: births, deaths, and migration. A population grows through the addition of births and in-migrants and declines through the subtraction of deaths and out-migrants. Understanding these three demographic processes is essential to understanding the nature and causes of population change.

Fertility

Fertility is the occurrence of a live birth (or births). It is determined by a variety of biological, social, economic, psychological, and cultural factors. Biological factors determine the physiological capacity to reproduce and the other factors determine perceptions regarding the costs and benefits of children. The availability and effectiveness of contraceptives also play a role, affecting the ability to control the number and timing of births.

Fertility rates have declined dramatically over the last two centuries in Europe, North America, and other high-income countries. Causes of this decline have included higher costs and lower economic benefits of children, lower rates of infant and child mortality, changes in female roles in the home and society, and improvements in contraceptive effectiveness. Fertility rates have also declined significantly in recent decades in many countries in Asia and Latin America. In many African and Middle Eastern countries, however, they remain very high.

Although fertility rates are low (sometimes very low) in high-income countries, there is often a substantial degree of variation from place to place and from one race, ethnic, or socioeconomic group to another within a given country. This is especially true in the United States, which has a very heterogeneous population and covers a vast geographic area. Chapter 5 discusses the fertility component of population growth.

Mortality

Mortality is the occurrence of deaths in a population. Changes in mortality rates are determined primarily by changes in a population's standard of living and advances in medicine, public health, and science. Low-income countries typically have higher mortality rates than high-income countries; within countries, low-income people typically have higher mortality rates than high-income people. Education also has a substantial impact on mortality rates, even when differences in income are accounted for.

Mortality rates have declined tremendously during the last two centuries in Europe, North America, and other high-income countries. They have also declined dramatically in many low- and middle-income countries, especially during the last 50 to 60 years. However, there is much more variability in mortality rates among low- and middle-income countries than among high-income countries. Although there is not a great deal of variability in mortality rates among different geographic areas within high-income countries, there are still some differences among race, ethnic, and socioeconomic groups. Chapter 4 discusses the mortality component of population growth.

Migration

Migration is the change in one's place of residence from one political or administrative area to another. It refers solely to changes in place of usual residence, thereby excluding all short-term or temporary movements such as commuting to work, visiting friends or relatives, going away on vacation, and taking business trips. Migration is often distinguished from *local mobility*, which refers to a change of address within a particular political or administrative area. At the aggregate level, factors that affect migration include area-specific characteristics such as wage rates, unemployment rates, costs of living, and amenities (e.g., climate, recreational opportunities). At the individual level, migration is affected by these factors plus a host of personal characteristics such as age, education, occupation, family connections, and marital status.

The migration literature uses a number of descriptive terms. *Gross migration* refers to the total number of migrants into or out of an area (e.g., 600 in-migrants and 400 out-migrants). *Net migration* refers to the difference between the two (e.g., a net influx of 200). *Internal* (or *domestic*) *migration* refers to changes in residence from one place to another within the same country. *International* (or *foreign*) *migration* refers to changes in residence from one country to another.

International migration is a minor component of population growth for most countries but is substantial in the United States. Migration attains its greatest importance, however, at the subnational level. Migration levels vary tremendously from one place to another in the United States and—for any given place—are subject to large, sudden changes. Migration affects not only the total population of an area, but its age, sex, race, income, education, and other characteristics as well. Chapter 6 discusses the migration component of population growth.

Demographic Balancing Equation

The overall growth or decline of a population is determined by its mortality, fertility, and migration, as formalized in the demographic balancing equation:

$$P_l - P_b = B - D + IM - OM$$

where P_l is the population at the end of the time period; P_b is the population at the beginning of the time period; and B, D, IM, and OM are the number of births, deaths, in-migrants, and out-migrants during the time period, respectively.

The difference between the number of births and the number of deaths is called *natural increase* $(B - D)$; it represents population change coming from within the population itself. It may be either positive or negative, depending on whether births exceed deaths or deaths exceed births. (If it is negative, it is called *natural decrease*.) As noted before, the difference between the number of in-migrants and the number of out-migrants is called *net migration* $(IM - OM)$; it represents population growth coming from the movement of people into and out of the area. It may be either positive or negative, depending on whether in-migrants exceed out-migrants or out-migrants exceed in-migrants.

For the demographic balancing equation to be exactly correct, it must apply to a geographic area with boundaries that do not change over time; in addition, there must be no measurement errors in any of the equation's variables. Because there will virtually always be errors in one or more of the variables, an error term is sometimes added to the right-hand side of the equation. This error term is often called the *residual error* or the *error of closure* (Shryock & Siegel, 1973, p. 6). Because the error term is difficult to measure precisely, it is often lumped with net migration.

The demographic balancing equation is one of the most basic formulas in demography and has a number of uses. For example, if we have a good population count in a census year (P_b) and good data on births, deaths, and in- and out-migration, we can estimate the population in a later year (P_l) as

$$P_l = P_b + B - D + IM - OM$$

Another common use of the demographic balancing equation occurs when there are good data from two consecutive censuses and good data on births and deaths, but no migration data. In this case, net migration $(IM - OM)$ can be calculated as a residual by subtracting natural increase $(B - D)$ from total population change $(P_l - P_b)$:

$$(IM - OM) = (P_l - P_b) - (B - D)$$

Table 2.2 shows natural increase and net migration for states for 1980–1990 and 1990–1998. There is a tremendous amount of variability among states, especially for net migration. All states had positive natural increase in both time periods, but not all had positive net migration. Seventeen states had positive net migration in both time periods, seven had negative net migration in both time periods, and 27 changed signs between the two time periods. It should be noted that these estimates of net migration include the effects of errors in population counts, estimates, and vital statistics data.

STATISTICAL MEASURES

Demographic analysis requires the use of statistical measures. Two types can be identified. *Absolute* measures focus on single numbers such as population size, births, deaths, natural increase, or net migration. *Relative* measures focus on the relationship between two numbers; they are typically expressed as ratios, proportions, percentages, rates, or probabilities. All of the relative measures are similar to each other, but each has a distinct meaning.

A *ratio* is simply one number divided by another. These could be any two numbers; they do not need to have any particular relationship to each other. For example, one could calculate the ratio of dogs to cats at the animal pound, the ratio of cars to bicycles at an intersection, or the ratio of desserts to casseroles at a potluck. To be useful, of course, the comparison of the two numbers should provide some type of meaningful information.

A commonly used ratio in demography is the sex ratio, which is the number of males per 100 females. In the United States in 1999, there were 133,352,000 males and 139,526,000 females, yielding a sex ratio of

$$(133,352,000 / 139,526,000) 100 = (0.956) 100 = 95.6$$

That is, there were 95.6 males for every 100 females in the United States. Sex ratios can also be calculated for different subgroups of the population. For example, for the U.S. population aged 85 and older in 1999, the sex ratio was

$$(1,242,000 / 2,942,000) 100 = (0.422) 100 = 42.2$$

That is, there were only 42.2 males aged 85+ for every 100 females aged 85+. This low ratio reflects the cumulative impact of differential mortality: In the United States (and most other countries), males have higher mortality rates than females at every age.

A *proportion* is a special type of ratio in which the numerator is a subset of the denominator. For example, we might calculate females, Hispanics, blondes, left-handers, or right-brained people as a proportion of the total population. In the United States in 1999, there were 34,578,000 people aged 65 and older and a total population of 272,878,000. The proportion aged 65+ can thus be calculated as

$$34,578,000 / 272,878,000 = 0.127$$

Table 2.2. Components of Population Change for States, 1980–1990 and 1990–1998

	1980–1990			1990–1998		
	Population change	Natural increase	Net migration[a]	Population change	Natural increase	Net migration[a]
Northeast						
Connecticut	179,539	161,285	18,254	−13,047	141,396	−154,443
Maine	103,269	58,121	45,148	16,322	28,493	−12,171
Massachusetts	279,388	259,886	19,502	130,707	245,099	−114,392
New Hampshire	188,642	71,870	116,772	75,796	53,527	22,269
New Jersey	365,365	360,063	5,302	367,261	369,607	2,346
New York	432,383	899,311	−466,928	184,523	917,479	−732,956
Pennsylvania	17,748	386,112	−368,364	118,609	255,771	−137,162
Rhode Island	56,310	36,552	19,758	−14,984	33,588	−48,572
Vermont	51,302	34,737	16,565	28,125	20,542	7,583
Midwest						
Illinois	4,082	804,590	−800,508	614,724	687,447	−72,723
Indiana	53,934	336,612	−282,678	355,039	264,978	90,061
Iowa	−137,053	143,375	−280,428	85,616	83,865	1,751
Kansas	113,896	176,152	−62,256	151,479	118,034	33,445
Michigan	33,219	618,127	−584,908	521,955	478,355	43,600
Minnesota	299,130	327,884	−28,754	349,754	237,576	112,178
Missouri	200,387	264,686	−64,299	321,658	185,368	136,290
Nebraska	8,561	107,511	−98,950	84,302	69,520	14,782
North Dakota	−13,916	57,894	−71,810	−556	22,983	−23,539
Ohio	49,485	634,794	−585,309	362,378	452,070	−89,692
South Dakota	5,237	54,910	−49,673	42,167	31,864	10,303
Wisconsin	186,002	311,957	−125,955	331,731	205,488	122,243
South						
Alabama	146,699	234,792	−88,093	311,610	164,228	147,382
Arkansas	64,290	115,245	−50,955	187,679	76,709	110,970
Delaware	71,828	42,993	28,835	77,435	36,031	41,404
Dist. of Columbia	−31,434	28,874	−60,308	−83,776	25,762	−109,538
Florida	3,191,602	417,975	2,773,627	1,977,909	376,097	1,601,812
Georgia	1,015,111	491,068	524,043	1,164,065	465,685	698,373
Kentucky	24,520	195,833	−171,313	249,607	137,209	112,398
Louisiana	14,073	426,418	−412,345	147,141	241,711	−94,570
Maryland	564,492	320,806	243,686	354,055	282,400	71,655
Mississippi	52,579	195,980	−143,401	176,617	128,339	48,278
North Carolina	746,872	372,311	374,561	914,045	337,377	576,688
Oklahoma	120,296	227,108	−106,812	201,137	123,348	77,789
South Carolina	364,883	254,088	110,795	349,652	178,360	171,292
Tennessee	286,067	248,446	37,621	553,418	200,268	353,150
Texas	2,757,320	1,811,997	945,323	2,773,279	1,572,789	1,200,490
Virginia	840,540	409,708	430,832	602,148	355,324	246,824
West Virginia	−156,167	52,023	−208,190	17,679	11,298	6,381

Table 2.2. (*Continued*)

	1980–1990			1990–1998		
	Population change	Natural increase	Net migration[a]	Population change	Natural increase	Net migration[a]
West						
Alaska	148,193	96,308	51,885	63,967	63,350	−5,383
Arizona	947,014	337,287	609,727	1,003,292	313,640	689,652
California	6,092,119	2,737,680	3,354,439	2,880,693	2,898,196	−17,503
Colorado	404,429	333,371	71,058	676,498	254,712	421,786
Hawaii	143,539	128,877	13,662	84,772	96,584	−11,812
Idaho	62,814	104,260	−41,446	221,950	77,883	144,067
Montana	12,375	65,320	−52,945	81,388	31,414	49,974
Nevada	401,341	86,634	314,707	545,223	103,023	442,200
New Mexico	212,176	175,206	36,970	221,862	128,222	93,106
Oregon	209,216	171,644	37,572	439,637	128,756	311,415
Utah	261,812	293,509	−31,697	376,908	236,931	139,977
Washington	734,535	364,061	370,474	822,594	319,780	502,814
Wyoming	−15,968	59,106	−75,074	27,318	25,155	2,163

[a]Net migration is population change minus natural increase. The net migration shown here includes international and domestic migration, census enumeration errors, and a statistical residual that results when the Census Bureau adjusts the state population estimates to the U.S. population.
Source: U.S. Census Bureau, Population Division, Population Estimates Branch, Internet Release Dates: September, 1995, December 31, 1998.

If we multiply a proportion by 100, we get a *percentage*. For example, people aged 65 and older accounted for 12.7% of the U.S. population in 1999.

A *rate* is also a special type of ratio. Strictly speaking, a rate is the number of events that occur during a given time period divided by the population at risk of the occurrence of those events. For example, the death rate is the number of deaths divided by the population exposed to the risk of dying; the birth rate is the number of births divided by the population exposed to the risk of giving birth. In demography, rates are generally based on a period of one year.

Although the concept of a rate is clear, it is often difficult or impossible to develop an exact measure of the population at risk of the occurrence of a particular event. For example, only females in their childbearing years are subject to the risk of giving birth; in addition, some die during the year and—for any given area—some move away and others move in. How can the population exposed to the risk of giving birth be measured?

This problem is usually solved by using the midyear population as an approximation of the population at risk. This solution is based on the assumption that births, deaths, and migration occur evenly throughout the year, so that the midyear population is a measure of the average population during the year. For example, the *crude birth rate* (CBR) is calculated by dividing the number of births during the year by the midyear population. It is often multiplied by 1,000 to express the CBR as the number of births per 1,000 persons:

$$CBR = (B/P)\,1{,}000$$

where B is the number of births during the year and P is the midyear population. This is called a "crude" birth rate because the denominator—which includes men, children, and older people—is only a crude approximation of the population at risk of giving birth.

The *crude death rate* (CDR) is defined similarly:

$$CDR = (D/P)\,1{,}000$$

where D is the number of deaths during the year and P is the midyear population. This is also a "crude" rate because not everyone represented in the denominator has an equal risk of dying. For example, males have a greater risk of dying than females and older people have a greater risk of dying than younger people.

In both the CBR and the CDR, the denominator is only a rough approximation of the population at risk of the occurrence of an event. A commonly used strategy for refining crude rates is to develop rates for specific age groups:

$$_nR_x = {}_nE_x/{}_nP_x$$

where x is the youngest age in the age interval, n is the number of years in the age interval, $_nR_x$ is the age-specific rate (ASR), $_nE_x$ is the number of events, and $_nP_x$ is the mid-year population. For example, if $x = 20$ and $n = 5$, the ASR would be based on data for the population aged 20–24. We will give a number of examples of age-specific rates in Chapters 4–6, along with a variety of other demographic rates.

In addition to the distinction between crude and age-specific rates, a distinction can also be made between *central rates* and *probabilities* (Shryock & Siegel, 1973, p. 8). In a central rate, the denominator is the population at the *midpoint* of the time period (e.g., the middle of the year). It is meant to represent the average population during the time period. The CBR, CDR, and ASR defined above are all central rates. In a probability, the denominator is the population at the *beginning* of the time period, which is thought to correspond more closely to the population at risk of the occurrence of an event during the period.

In reality, the distinction between central rates and probabilities is somewhat fuzzy because of the movement of migrants into and out of the area. Consider age-specific death rates, for example. A true one-year probability can be calculated by using the population of the area at the beginning of the year and the number of deaths that occurred during the year to members of the beginning population, for each age cohort. However, some deaths will be missed by the death registration system because they occur to people who moved out of the area during the year. In addition, some deaths will be improperly included because they occur to people who moved into the area during the year. Consequently, it is difficult (if not

---

Box 2.1

Web Site Addresses of Major Pubic-Sector Data Providers

U.S. Census Bureau. http://www.census.gov
U.S. Bureau of Economic Analysis. http://www.bea.doc.gov
U.S. Bureau of Labor Statistics. http://www.bls.gov
U.S. Immigration and Naturalization Service. http://www.ins.usdoj.gov/
graphics/index.htm
U.S. National Center for Health Statistics. http://www.cdc.gov/nchs/default/htm
U.S. Geological Survey. http://www.usgs.gov
U.S. State Census Data Centers. http://www.census.gov/sdc/www
U.S. Bureau of Transportation Statistics. http://www.bts.gov
Federal-State Cooperative Program for Population Estimates. http://www.
census.gov/population/www.coop/fscpe.html
Statistics Canada. http://www.statcan.ca

---

impossible) to construct true death probabilities. Central rates are widely used to approximate probabilities for a variety of demographic measures.

The term *rate* is used very loosely in demography, as it is elsewhere (Newell, 1988). Many measures called *rates* are really *ratios*. A growth rate, for example, is a ratio of population change over a time period to the population at the beginning of the period. It is not a rate in the strictest sense because an area's growth comes not only from the population of the area itself, but from other populations as well. Another commonly used measure in demography is the *infant mortality rate*, which is the ratio of deaths of children less than age one to the number of births during the year. We follow common demographic practices in this book, using *rate* to refer to changes or events in relation to some reference population. However, we give frequent reminders that these are not always rates in the strictest sense of the word.

SOURCES OF DATA

Demographic data are collected, produced, and distributed by a variety of federal, state, and local government agencies and private companies. Data from primary sources are available as printed publications, unpublished reports, and electronic data files. In recent years, more and more data have become directly available on the Internet (see Box 2.1 for some commonly used public-sector web sites). Primary data are often replicated in secondary sources such as professional journals, textbooks, and statistical abstracts. In this section we briefly discuss the most important sources of demographic data in the United States.

Decennial Census

The decennial census is by far the most important source of demographic data in the United States. Every 10 years, the federal government attempts to count the entire population of the country. The results determine each state's representation in Congress and are used by state legislatures and local governments to redraw electoral boundaries. They form the basis for the distribution of billions of dollars in federal and state funds each year through a variety of revenue-sharing and grant-in-aid programs. Businesses and government agencies use the results for planning, budgeting, marketing, and policy making. Scholars and the media use them to analyze social, economic, and political issues. They form the basis for population estimates and projections made during the following decade. No other data source is nearly as comprehensive or used for as many purposes. Excellent discussions of the decennial census can be found in Anderson (1988), Anderson and Fienberg (1999), Choldin (1994), and Edmonston and Schultze (1995).

In 1790, the United States became the first country in the world to institute a regular periodic census of the national population (Anderson, 1988). Since then, a census has been conducted every 10 years without interruption. The first census compiled a list of household heads and counted people in five demographic categories. More and more questions were added over the following decades, covering social, economic, and housing characteristics, as well as demographic characteristics. The practice of collecting a limited amount of data from all households (sometimes called *short-form data*) and a larger amount from a sample of households (sometimes called *long-form data*) was begun in 1940 and has continued since that time. Nationally, about five in every six households receive the short form of the census and one in six receives the long form. Sampling rates vary from one in two households in small areas to one in eight in large areas. The long form asks the same questions as the short form, plus a number of others. Table 2.3 describes the types of data collected on the short and long forms of the 2000 Census.

The starting point for the census enumeration is an address list developed by the Census Bureau. This list, called the *Master Address File* (MAF), is based on previous censuses and delivery records from the U.S. Postal Service. In an attempt to make the address list as accurate as possible, the Census Bureau gives local officials an opportunity to review the list and note additions, deletions, and any other changes that may be needed. Addresses are allocated to specific geographic areas using the Topologically Integrated Geographic Encoding and Referencing (TIGER) system. Developed during the 1980s and first used in the 1990 Census,

Table 2.3. Types of Data Collected in the 2000 Census of Population and Housing

All households (short-form data)

Population: Name, relationship to householder, sex, age, date of birth, race, and Hispanic origin.

Housing: Number of people in household, telephone number, and tenure (ownership status).

Sample of households (long-form data)

Population: Same as short form, plus marital status, school enrollment, educational attainment, ethnic origin (ancestry), language spoken at home, place of birth, citizenship status, year of entry into the United States, place of residence five years ago, disability status, living with grandchildren, military service, employment status, employment history, place of work, transportation to work, occupation, industry, and income.

Housing: Same as short form, plus type of housing unit, year built, length of residence in current unit, number of rooms, number of bedrooms, plumbing facilities, kitchen facilities, telephone in unit, type of heating fuel, number of motor vehicles, size of lot, presence of home business, annual costs of utilities, monthly rent or mortgage payment, second mortgage, real estate taxes, property insurance, and value of property.

this system includes coordinates (latitude and longitude) for the boundaries of all geographic areas in the United States and contains information on a variety of natural and man-made features. When combined with geographic information system (GIS) software, the TIGER system offers unprecedented opportunities to use census data in combination with other types of geographically referenced information (Murdock & Ellis, 1991).

The decennial census in the United States is based on self-enumeration. Most households are mailed census forms in late March and are asked to fill them out and return them by mail; in some rural areas the forms are delivered by a census enumerator. The Census Bureau follows a number of procedures designed to maximize response rates and to collect information from nonresponding households. Despite these procedures, the data collected are incomplete and sometimes incorrect. Post-enumeration surveys and demographic analyses are used to measure the extent and nature of census errors and to develop estimates of the net undercount (or, in some instances, the net overcount).

The Census Bureau tabulates aggregate census results for a variety of geographic areas, including states, metropolitan areas, counties, county subdivisions, cities, census designated places, census tracts, block groups, and blocks. Not all types of data are tabulated for all levels of geography, however. For example, long-form data are tabulated only down to the block group level. Public Use Microdata Sample (PUMS) files are compiled for areas that have 100,000 or more

residents; these files provide individual records (stripped of identifying information) for use in more detailed analyses.

Ever since George Washington complained about an undercount in the first
census, government officials and other interested parties have been concerned
about the accuracy of census results (Anderson & Fienberg, 1999). Census errors
may be caused by missed households, refusal to respond, recording errors, sampling errors, geographic assignment errors, duplication errors, coding and data-
processing errors, and the incorrect imputation of missing data. Although these
errors can make census counts either too high or too low for any given geographic
area or population subgroup, in most instances they lead to net undercounts of the
"true" population. Nationally, the net census undercount was estimated as 5.4%
in 1940, 4.1% in 1950, 3.1% in 1960, 2.7% in 1970, 1.2% in 1980, and 1.8% in 1990
(Edmonston & Schultze, 1995).

The net undercount is not the same for all geographic areas and population
subgroups. For example, the net undercount is much greater for blacks, Hispanics,
and American Indians than for non-Hispanic whites. Because of the increase in the
net undercount between 1980 and 1990 and the large differences found among
population subgroups, many concerns about the accuracy of the decennial census
were voiced during the 1990s. The Census Bureau responded to these concerns by
developing plans to use statistical sampling to account for nonresponses and to
adjust for the net undercount, but those plans encountered strong political opposition in Congress. A Supreme Court decision in 1999 prohibited the use of adjustments based on sampling for the reapportionment of Congress after the 2000
Census, but left unresolved several broader issues related to the use of statistical
adjustments. Sampling and statistical adjustment are hot political (and academic)
topics that will continue to generate a great deal of discussion during the coming
years.

Vital Statistics

Data on events such as births, deaths, marriages, and divorces are called *vital
statistics*. In the United States, the collection of these data is the responsibility of
individual states, not the federal government. As early as 1639, the Massachusetts
Bay Colony began reporting births, deaths, and marriages as part of its administrative/
legal system (Shryock & Siegel, 1973). Other states gradually began doing the
same thing, and today all states keep records of births, deaths, and other vital
events. The federal government sets standards for collecting and reporting the
data, compiles summaries from data collected by each state, and publishes a
variety of reports (e.g., National Center for Health Statistics, 1999). The quality of
vital statistics data is generally very good in the United States and other high-
income countries.

Before 1945, vital statistics reports were published by the Census Bureau.

Beginning in 1945, this task was taken over by the U.S. Public Health Service, National Office of Vital Statistics. In 1960, this office was reorganized and became part of the National Center for Health Statistics (NCHS), which today is a branch of the Centers for Disease Control (CDC). Annual and monthly reports on births, deaths, marriages, and divorces are available from the NCHS. It should be noted that some of the concepts and definitions used by the NCHS do not precisely match those used by the Census Bureau (Hahn, Mulinare, & Teutsch, 1992). Consequently, adjustments may have to be made when combining population data from the Census Bureau with vital statistics data from the NCHS (e.g., Sink, 1997).

Data from the NCHS are available only at the national and state levels; vital statistics data for local areas must be obtained elsewhere. Most states tabulate data at the county (or county-equivalent) level, but few go beyond that to develop regular data series for subcounty areas (Bogue, 1998). Although individual records generally contain the information needed to allocate them to different types of subcounty areas (e.g., cities, census tracts), actually doing so requires a substantial effort. In addition, there are often errors in geocoding birth and death records at the subcounty level (Flotow & Burson, 1996). Analysts who need vital statistics data for subcounty areas may have to develop those data themselves.

Sample Surveys

The decennial census and vital statistics reports are valuable sources of demographic data. However, the census is conducted only once every 10 years, and vital statistics data cover only a small portion of the variables of interest to demographers, planners, and other analysts. Sample surveys can be used to collect data on a variety of topics between censuses.

One of the most important sample surveys in the United States is the Current Population Survey (CPS), a monthly survey of about 50,000 households conducted by the Census Bureau for the U.S. Bureau of Labor Statistics. Started in the early 1940s, this survey originally focused on collecting labor force and unemployment data. It has since been expanded to include a variety of topics including occupation, industry, education, income, veteran status, marital status, living arrangements, fertility, and migration, as well as demographic data on age, sex, race, and ethnicity. Data from the CPS are currently tabulated at the national, regional, and state levels and for large metropolitan areas, but small sample sizes lead to erratic trends for many states and metropolitan areas.

The Census Bureau also conducts the American Housing Survey and the Survey of Income and Program Participation. Both of these surveys provide useful information on a variety of demographic, socioeconomic, and housing characteristics. However, the small sample size and the nature of the sample universe limits the usefulness of these surveys for applications that involve states and local areas.

GRIN & BEAR IT   **By Wagner**

"Well then, how about a rough estimate, madam?"

Reprinted with special permission of King Features Syndicate.

The American Community Survey (ACS) is a relatively new survey conducted by the Census Bureau that has a great deal of potential as a demographic data source (U.S. Census Bureau, 1999a). It was started in four sites in 1996 and has been expanded every year since that time. If carried out as planned, it will be fully implemented by 2003 with some three million households—drawn from all counties or county equivalents in the United States—contacted each year. Starting in 2004, the ACS will provide annual estimates of demographic, housing, social, and economic characteristics for every state, as well as for all cities, counties, metropolitan areas, and places of 65,000 or more. For places smaller than 65,000, it will take two to five years to accumulate a large enough sample to produce reasonably accurate estimates. These estimates will be available annually but will be based on a multiyear average. According to current plans, estimates down to the block-group level will be available everywhere in the country by 2008.

The ACS will be conducted similarly to the decennial census. The sample will be drawn from an updated MAF, and a larger proportion of addresses will be sampled for small governmental units than for large units. The survey will rely on self-enumeration using a mail-out/mail-back methodology. Nonrespondents will be contacted by telephone or by personal visits from Census Bureau field representatives. If successful, the ACS is expected to replace the long form of the decennial census in 2010.

Administrative Records

Administrative records are records kept by agencies of federal, state, and local governments for purposes of registration, licensing, and program administration. Although not always designed explicitly to do so, these records provide valuable information on specific demographic events or subgroups of the population. We have already discussed vital statistics, one type of administrative record that is very valuable for demographic analysis. Other types include Social Security, Medicare, Internal Revenue Service (IRS), Immigration and Naturalization Services (INS), Aid for Families with Dependent Children, food stamps, drivers licenses, building permits, school enrollment, voter registration, and property tax records. All of these data sources can be used for various types of demographic analyses, including the production of population estimates and projections. We discuss several of these data sources—and how they can be used for population projections—later in this book.

Population Estimates

A final source of demographic data is represented by population estimates produced by public and private sector agencies. Population estimates are not primary data in the same sense as the data sources discussed before; rather, they are derived from (or based on) those data sources. They play an important role in supplementing and updating data from the other data sources.

The Census Bureau's first annual report in 1903 described its plans for making population estimates (Shryock & Siegel, 1973). The plans called for issuing estimates as of the first of June for each year after 1900, covering the nation, each state, cities of 10,000 or more, and the urban and rural balances of each state. These plans have changed over time and the Census Bureau currently produces annual estimates at the national, state, and county levels by age, sex, race, and Hispanic origin. Estimates of total population for places (i.e., cities, towns, and townships) are made roughly every other year (e.g., U.S. Census Bureau, 1999b).

Population estimates are also produced by a variety of state and local government agencies. Many state agencies participate with the Census Bureau in the Federal-State Cooperative Program for Population Estimates (FSCPPE); this program serves as a clearinghouse for demographic data and as a forum for discussing methods and exchanging ideas related to population estimation. Some states produce independent population estimates at the state, county, and/or city level. Some city and county governments—and Councils of Governments for large metropolitan areas—also produce population estimates, often for small areas such as census tracts and traffic analysis zones. Finally, several private companies produce estimates at the county and subcounty levels. Further information on the production of population estimates can be found in Murdock and Ellis (1991); Rives, Serow, Lee, Goldsmith, and Voss (1995); and Raymondo (1992).

CHAPTER 3

Overview of the Cohort-Component Method

We begin our discussion of population projection methods with the cohort-component method, which has a longstanding tradition in demography (e.g., Bowley, 1924; Cannan, 1895; Whelpton, 1928). The Census Bureau began using this method for national projections in the 1940s and for state projections in the 1950s and has used some version of the method ever since (Long & McMillen, 1987). A recent survey conducted by the Federal-State Cooperative Program for Population Projections found that 89% of states that make state-level projections of total population use some form of the cohort-component method; for states that make projections by age, sex, and race, 95% use the cohort-component method (Judson, 1997). It is also widely used for projections at the county and subcounty level. Although current applications of the cohort-component method are more detailed and sophisticated than the earliest applications, the basic framework is much like it was 100 years ago.

The cohort-component method is widely used because it provides a flexible and powerful approach to population projection. It can take the form of a purely atheoretical accounting procedure or can incorporate insights from a variety of theoretical models. It can incorporate many application techniques, types of data, and assumptions regarding future population change. It can be used at any level of geography from nations down to states, counties, and subcounty areas. Perhaps most important, it provides projections of total population, demographic composition, and individual components of growth. The cohort-component method provides a good starting point for the study of state and local population projections.

CONCEPTS AND TERMINOLOGY

A *cohort* is defined as a group of people who experience the same demographic event during a particular period of time and who may be identified at later dates on the basis of this common experience (Shryock & Siegel, 1973, p. 712). For example, all babies born during the 1990s comprise the birth cohort for that decade; all persons married in 2000 form the marriage cohort for that year; and all immigrants who entered the United States between 1996 and 1998 make up the immigration cohort for that period. Cohorts can also be defined for other significant events, such as graduation from college or entry into the labor force.

Dividing the population into age cohorts was an important methodological innovation. It permits the analyst to account for differences in mortality, fertility, and migration rates among different age groups and to consider how those rates change over time. For example, fertility rates are higher for women in their 20s than for women in their 30s, but the rates for both groups rose during the 1940s and 1950s and declined during the 1960s and 1970s. Mortality rates for infants less than age one are higher than mortality rates for teenagers, but declined much more rapidly during the twentieth century.

Age cohorts are typically split between males and females and are often subdivided by race, ethnicity, and other characteristics. These divisions allow the analyst to account for additional types of demographic variation and permit the construction of more finely detailed projections.

The components of population change (births, deaths, and migration) were discussed in Chapter 2. It is useful to distinguish among these components when producing population projections for several reasons. First, such distinctions enable us to account separately for the demographic causes of population change. Is an area changing primarily because of natural increase or net migration? Is the birth rate unusually high or the death rate unusually low? Are in-migrants coming mostly from other parts of the same state, from other states, or from abroad? If a population is aging rapidly, is it because many older people are moving in (as in Florida) or because many younger people are moving out (as in West Virginia)? Making these distinctions is the first step in gaining insight into why some areas grow more rapidly than others and why growth rates and demographic composition change over time.

Second, each component of change typically responds differently to changes in economic, social, political, cultural, medical, environmental, and other factors that affect population change. For example, medical advances lead to greater life expectancies but have little impact on migration, whereas changing employment conditions have a substantial impact on migration but little impact on life expectancies. Developing an understanding of nondemographic causes of population change requires breaking down population change into its individual components.

Finally, the behavior of each component varies among places and follows different trends over time. The number of births in one area may be increasing and the number of deaths declining, while in another area the opposite is occurring. In-migrants may exceed out-migrants in one area while out-migrants exceed in-migrants in another. Separating the components of change enables the analyst to account for these differences when developing assumptions about future population trends.

Information on the components of population change is important for many types of population analysis. Information on the demographic composition of the population is also important because overall birth, death, and migration patterns are strongly affected by demographic characteristics. For example, births occur primarily to women between the ages of 15 and 44, death rates are much higher for older persons than younger persons, and migration rates are typically highest for people in their twenties and decline steadily thereafter. Differences in demographic composition can have a major impact on a population's birth, death, and migration patterns.

Demographic composition differs considerably among states, counties, and subcounty areas. Figure 3.1 shows the 1998 age structures for Utah and Florida. Geographically, these states are on opposite sides of the continent; demographically, their age characteristics are at opposite ends of the spectrum as well. Utah's population is relatively young and Florida's is relatively old. Differences in age structure have a substantial impact on the components of growth in these two states. Natural increase accounted for 63% of the population growth in Utah between 1990 and 1998; in Florida, it accounted for only 19%. Age structure was not the entire cause of these differences, but it was an important contributing factor.

Differences in demographic composition are even wider among counties and subcounty areas than among states. For counties in Florida, for example, 34% of the population in Charlotte County was aged 65 and older in 1990, compared to only 7% in Union County. Almost 58% of the population in Gadsden County was black, compared to only 2% in Citrus County. More than 49% of the population in Miami-Dade County was of Hispanic origin, compared to less than 1% in Wakulla County. The median age in the town of South Bay in southeast Florida was 24 in 1990; in nearby Century Village, it was 78. Local variations like these are found throughout the nation. The cohort-component method allows the analyst to account for these variations when developing population projections.

BRIEF DESCRIPTION OF PROCEDURES

By today's standards the earliest applications of the cohort-component method were somewhat crude (e.g., Bowley, 1924; Cannan, 1895). Although they projected the components of population change separately, early models did not fully account for differences in demographic composition. This soon changed, and birth and death rates were calculated separately for each age-sex cohort in the

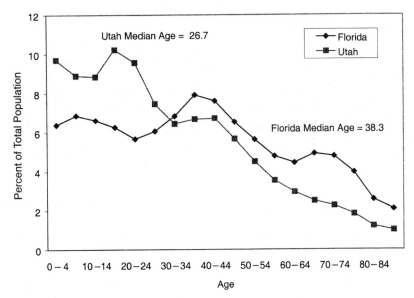

Figure 3.1. Age distribution, Florida and Utah, 1998. (*Source*: U.S. Census Bureau, Population Division, Population Estimates Branch. Internet Release Date, June 15, 1999)

population (e.g., Whelpton, 1928). Cohort-component models have since been extended to cover differences by race and ethnicity as well (e.g., Campbell, 1996; Day, 1996a). Figure 3.2 provides an overview of the steps involved in applying the cohort-component method.

The starting point is the launch-year population (i.e., the population at the beginning of the projection period) divided into age-sex cohorts. Age cohorts can be defined in a number of ways, but one- and five-year groups are the most commonly used. As we show later, the construction of projections is simpler if the number of years in the projection interval is equal to or exactly divisible by the number of years in the age cohort (e.g., five-year age cohorts for projections made in five- or 10-year intervals, but not for projections made in one-year intervals). The oldest cohort is typically age 75+ or 85+, but age 100+ is sometimes used. Many applications of the cohort-component method further subdivide the population by race and ethnicity. This adds to the data requirements and the number of calculations, but the logic and procedures remain the same.

The first step in the projection process is to calculate the number of persons who survive to the end of the projection interval. This is accomplished by applying age-sex-specific survival rates to each age-sex cohort in the initial population. These survival rates indicate the probability of surviving throughout the projec-

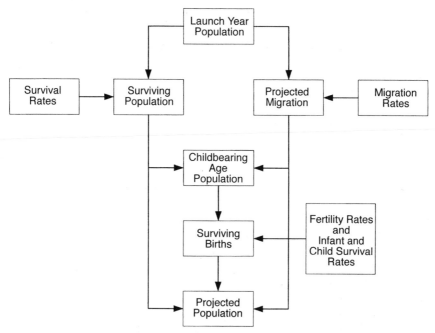

Figure 3.2. Overview of the cohort-component method.

tion interval; they are commonly based on life tables derived from compilations of recent mortality data. Future survival rates can be based on the extrapolation of historical trends, structural models, simulation techniques, or rates found in other areas. Chapter 4 describes a number of approaches used for projecting survival rates, the mortality component of population growth.

The second step is to project migration during the projection interval. Migration rates are calculated for each age-sex cohort in the population. These rates can be based on either gross migration data (i.e., separate calculations for in-migrants and out-migrants) or net migration data (i.e., one calculation reflecting the net change due to migration). Projections of future rates can be based on recent values, trend extrapolations, simulated values, model schedules, or structural models. The application of projected migration rates provides a projection of the number of persons in each age-sex cohort who move into or out of an area during the projection interval (or, for models using net migration data, the net change due to migration). These numbers are added to or subtracted from the surviving population to provide a projection of persons born before the launch date (e.g., persons aged five and older for a five-year projection interval). Chapter 6 discusses sources of migration data and various approaches to projecting migration rates.

The third step is to project the number of births that will occur in the projection interval. This is accomplished by applying age-specific birth rates to the female population in each age cohort. Projected birth rates can be based on recent values, trend extrapolations, simulated values, model schedules, or structural models. Chapter 5 describes sources of fertility data and several approaches to projecting birth rates.

The final step in the process is to add the number of births (distinguishing between males and females and adjusting for migration and mortality) to the rest of the population. This provides a projection of the total population by age and sex at the end of the projection interval. This population serves as the base for projections for the following interval. The process is repeated until the final target year in the projection horizon has been reached. Chapter 7 gives several step-by-step examples of the entire process and discusses the strengths and weaknesses of the cohort-component method.

The logic underlying the cohort-component method is simple and straightforward. Collecting the data and developing the assumptions and procedures needed to apply the method, however, is much more complicated. The next three chapters provide a detailed description of the data sources, statistical measures, theoretical perspectives, and projection techniques that can be used for projecting the three components of population change. We start with mortality, the simplest of the three to forecast accurately. Then, we consider fertility, which presents more challenges to the forecaster. We conclude with migration, the most difficult component to forecast accurately at the state and local level.

CHAPTER 4

Mortality

Survival has been a central preoccupation of humankind since the origin of the species. For most of history, however, human beings were not particularly successful at fending off deprivation, disease, destruction, and death. As recently as 200 years ago, life expectancy at birth was only 30–40 years, even in the richest countries. These levels were not much higher than they had been thousands of years earlier. In many low-income countries, life expectancies remained at very low levels well into the twentieth century.

Substantial increases in life expectancies have occurred during the last century or two. Driven by improved standards of living and scientific, medical, and public health advances, life expectancies at birth have risen to 75–80 in most high-income countries and 60–75 in most middle-income countries. Even the world's poorest countries experienced substantial gains in life expectancy during the second half of the twentieth century (although the spread of AIDS has recently wiped out many of those gains in some countries, especially in Sub-Saharan Africa). Changes in life expectancy have had a dramatic impact on the size and composition of the human population.

Life expectancy at birth in the United States rose from age 47 to 77 during the twentieth century. It has been estimated that more than 68 million Americans alive in 2000 (about 25% of the total population) would have died without the improvements in survival rates that occurred since 1900; another 25% would never have been born because their ancestors would have died before they had children (White & Preston, 1996). Given the magnitude of past changes and the likelihood of future changes, mortality clearly plays a central role in the production of cohort-component population projections.

We start this chapter with a description of several basic measures of mortality. Then, we discuss survival, the converse of mortality: A person either lives from one time to another or dies during that interval. We consider two types of survival rates, focusing primarily on the type used most frequently for projections in the United States. We discuss data sources and techniques used in constructing survival rates and consider several perspectives regarding future mortality trends. We pay particular attention to the special problems of projecting survival rates for small areas. We close with an assessment of the impact of mortality assumptions on population projections.

MORTALITY MEASURES

The most commonly used mortality measures relate the number of deaths during a particular period (usually a year) to the "at-risk" population (usually approximated by the midyear population). These measures are typically defined as mortality rates but are not rates in a true probabilistic sense (see Chapter 2 for an explanation). In the United States, the mortality data used in constructing mortality rates are collected by the vital statistics agencies of each state and are compiled by the National Center for Health Statistics (NCHS). The population data are taken from either censuses or estimates, depending on the year(s) for which the rates are to be constructed.

Crude Death Rate

The simplest mortality measure is the crude death rate (CDR), which is calculated by dividing the number of deaths during a year by the midyear population. It is generally multiplied by 1,000 to reflect the number of deaths per 1,000 persons:

$$CDR = (D/P)\,1,000$$

where D is the number of deaths during the year and P is the midyear population. For example, there were 2,338,070 deaths in the United States in 1998, and the midyear population was estimated as 270,298,524, yielding a CDR of

$$(2,338,070/270,298,524)\,1,000 = 8.6$$

This means there were 8.6 deaths for every 1,000 residents of the United States in 1998. Crude death rates can be calculated for race, ethnic, occupational, educational, and other subgroups of the population. For example, the CDR in the United States in 1998 was 9.1 for whites and 8.0 for blacks:

Whites: $(2,020,230/223,000,729)\,1,000 = 9.1$

Blacks: $(275,469/34,430,569)\,1,000 = 8.0$

Crude death rates can also be calculated for different geographic regions. In 1998, CDRs for states ranged from 4.2 in Alaska to 11.5 in West Virginia (National Center for Health Statistics, 1999). For nations, CDRs in 1999 ranged from about

2.0 in Kuwait, Qatar, and the United Arab Emirates to around 24.0 in Malawi and Niger (Population Reference Bureau, 1999).

The CDR provides an indication of the incidence of deaths relative to the overall size of a population. For many purposes, however, the usefulness of the CDR is limited because it does not account for one of the major determinants of mortality, the age structure of the population. A young age structure is the primary reason that the CDR for blacks is lower than the CDR for whites in the United States and that the CDR for Alaska is lower than the CDR for West Virginia. A second mortality measure deals with this problem by focusing on deaths within each age group.

Age-Specific Death Rate

An age-specific death rate (ASDR) shows the proportion of persons in each age group (x to $x+n$) that dies during a year:

$$_n\text{ASDR}_x = {_nD_x} / {_nP_x}$$

where x is the youngest age in the age interval, n is the number of years in the age interval, $_nD_x$ is the number of deaths of persons between the ages of x and $x+n$ during the year, and $_nP_x$ is the midyear population of persons between the ages of x and $x+n$. For example, there were 91,014 deaths to males aged 45–54 in the United States in 1998 and a midyear population of 16,900,000 males aged 45–54, yielding an ASDR of

$$_{10}\text{ASDR}_{45} = 91{,}014 / 16{,}900{,}000 = 0.0054$$

ASDRs are typically calculated for one-, five-, or 10-year age groups. They are called *central death rates* (denoted as $_nm_x$) because they are based on the average population during the year, typically represented by the midyear population. To control for short-run fluctuations, ASDRs are often based on a three-year average rather than a single year of mortality data; such adjustments are particularly important for places that have small populations. ASDRs are generally calculated separately for males and females because of their well-known differences in longevity. They can also be calculated separately for different races, ethnic groups, and other demographic categories.

ASDRs are often expressed in terms of deaths per 100,000 persons. This conversion is made by multiplying the ASDR by 100,000. In the example shown above, the ASDR of 0.0054 can be expressed as 540 deaths per 100,000 males aged 45–54. Figure 4.1 shows ASDRs for males in the United States in 1998. The J-shaped pattern reflects the relatively high death rates for newborn babies, the considerably lower rates for young children, the slowly increasing rates at the middle ages, and the rapidly increasing rates at the older ages. This general pattern is found for virtually every population and population subgroup throughout the

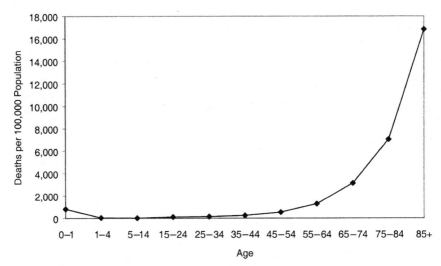

Figure 4.1. Age-specific death rates for males, United States, 1998. (*Source*: National Center for Health Statistics, *National Vital Statistics Reports*, Vol. 47, No. 25. Hyattsville, Maryland, 1999)

world. It should be noted, however, that the rate of increase in mortality rates sometimes slows down in the very oldest age groups (e.g., Horiuchi & Wilmoth, 1998).

SURVIVAL RATES

Survival rates show the probability of surviving from one age (or age group) to another. There are two main approaches to constructing survival rates. One is based on life tables, which are statistical tables that summarize a population's mortality characteristics. The other is based on a comparison of age cohorts in two consecutive censuses. The first is used far more frequently than the second in countries that have good vital statistics data.

Life Table Survival Rates

Constructing Life Tables. Life tables have a long history. In fact, the origins of formal demography are often traced to Englishman John Graunt, who analyzed mortality records for London in the 1660s and developed a precursor to the modern life table (Weeks, 1999). Other inexact life tables based on limited data were prepared for several places in Europe during the seventeenth and eighteenth centuries. Joshua Milne constructed the first scientifically correct life table for several parishes in England in 1815.

The first official life tables for the United States were prepared for 1900–1902 and have been produced at least once each decade ever since. Annual life tables have been prepared for every year since 1945. Life tables for individual states have been produced in conjunction with every decennial census since 1940 (Shryock & Siegel, 1973).

The principal focus of a life table is the probability of dying between one exact age and another. The empirical foundation of a life table for a particular year is a complete set of ASDRs for that year (the $_nm_x$ values described before). ASDRs, however, are based on midyear population estimates. They do not provide exact measures of the risk of dying because some people die before midyear. To be useful for constructing life tables, the denominators of the ASBRs must be adjusted to account for deaths occurring during the year. This is often done by assuming that deaths are evenly distributed throughout the year. Age-specific death rates for life tables ($_nq_x$) can thus be calculated as

$$_nq_x = {_nD_x} / ({_nP_x} + 0.5{_nD_x})$$

If we divide both the numerator and denominator by $_nP_x$, we can express the life table death rate ($_nq_x$) as a function of the central death rate ($_nm_x$):

$$_nq_x = {_nm_x} / (1 + 0.5{_nm_x})$$

This transformation of $_nm_x$ to $_nq_x$ is an approximation; its accuracy depends on the degree to which deaths are evenly distributed throughout the year. This assumption will be valid for most age groups, but not for the very young or (to a lesser extent) the very old. The validity of this assumption will also be reduced for areas in which the population is growing (or declining) very rapidly.

The elements of a life table are defined as follows:

1. *Proportion dying* ($_nq_x$). The proportion of persons who are alive at exact age x but die before reaching exact age $x+n$. An *exact age* refers to a birthday. For example, $_5q_{30}$ refers to the proportion of persons alive on their thirtieth birthdays who die before reaching their thirty-fifth birthdays.
2. *Number surviving* (l_x). The number of persons who survive to exact age x out of a beginning cohort of 100,000 live births (called the *radix*).
3. *Number dying* ($_nd_x$). The number of deaths between exact ages x and $x+n$, out of the number of persons alive at the beginning of that interval.
4. *Person-years lived during an age interval* ($_nL_x$). The summed total of person-years lived between exact ages x and $x+n$, based on each person's record of survival during that age interval. For example, a person who lived from age 60 to 65 would count as five person-years lived during this five-year interval; a person who died at exact age 64 would count as four person-years lived.

5. *Total person-years yet to be lived* (T_x). The summed total of person-years lived during this and all following age intervals.
6. *Life expectancy* (e_x). The average number of years of life remaining to persons alive at exact age *x*.

There are two types of life tables. A *period life table* is based on the ASDRs calculated for a particular period of time (usually one, two, or three years). For example, recent life tables prepared by the National Center for Health Statistics (NCHS) were based on average annual mortality data from 1989–1991 and population data from the 1990 Census (National Center for Health Statistics, 1997). Period life tables may be interpreted as showing the lifetime mortality patterns that would be experienced by a cohort of newborn babies if the age-specific death rates observed at the time of their births continued unchanged throughout their lifetimes.

A *cohort life table*, on the other hand, is based on the mortality patterns *actually experienced* by members of a particular birth cohort (e.g., all persons born in 1900) during their lifetimes. Age-specific death rates are calculated at each age as the cohort moves from infancy through old age. Obviously, cohort life tables require many more years of data than period life tables; consequently, they can be constructed only for cohorts born long ago. Although they are valuable for analyzing mortality trends over time, they are not very useful for producing population projections. In this book, we consider only period life tables. Cohort life tables for the United States may be found in Bell, Wade, and Goss (1992).

Life tables can be classified as complete or abridged. *Complete* (unabridged) life tables provide data by single year of age; *abridged* life tables provide data by age group (usually five-year groups, with the youngest group subdivided at age one). Both types can be used for population projections. Complete life tables are particularly useful for making projections by single year of age and for single-year projection intervals, whereas abridged life tables are particularly useful for making projections for five- or 10-year age groups and five- or 10-year intervals. Table 4.1 shows an abridged period life table for females in the United States in 1996.

Life tables are useful for many purposes besides mortality projections. The same techniques used to measure mortality can be used to measure the duration of variables such as marriage, employment, education, and the housing stock. When two or more variables are combined into one life table, it is called a *multiple-decrement* table. For example, mortality and divorce can be combined into one life table for married persons, where changes in marital status are caused by either death or divorce. An extension of this technique that is particularly relevant to many planners and policy makers combines mortality rates with disability rates, providing an indication of changing needs for medical care, living assistance, and institutionalization (e.g., Crimmins, Saito, & Ingegneri, 1997). Life tables also play an important role in a highly mathematical subarea of demography known as

Table 4.1. Abridged Life Table for Females, United States, 1996

| Period of life between exact ages x and x + n | Proportion of persons alive at beginning of age interval dying during interval $_nq_x$ | Of 100,000 born alive | | Stationary population | | Life expectancy^a e_x^o |
		Number alive at beginning of age interval l_x	Number who die during age interval $_nd_x$	In the age interval $_nL_x$	In this and all subsequent age intervals T_x	
0–1	0.00659	100,000	659	99,435	7,907,507	79.1
1–5	0.00135	99,341	134	397,043	7,808,072	78.6
5–10	0.00083	99,207	82	495,812	7,411,029	74.7
10–15	0.00093	99,125	92	495,426	6,915,217	69.8
15–20	0.00220	99,033	218	494,654	6,419,791	64.8
20–25	0.00242	98,815	239	493,488	5,925,137	60.0
25–30	0.00311	98,576	307	492,128	5,431,649	55.1
30–35	0.00430	98,269	423	490,336	4,939,521	50.3
35–40	0.00608	97,846	595	487,848	4,449,185	45.5
40–45	0.00858	97,251	834	484,325	3,961,337	40.7
45–50	0.01269	96,417	1,224	479,247	3,477,012	36.1
50–55	0.02036	95,193	1,938	471,421	2,997,765	31.5
55–60	0.03150	93,255	2,938	459,363	2,526,344	27.1
60–65	0.05068	90,317	4,577	440,808	2,066,981	22.9
65–70	0.07484	85,740	6,417	413,497	1,626,173	19.0
70–75	0.11607	79,323	9,207	374,780	1,212,676	15.3
75–80	0.17495	70,116	12,267	321,360	837,896	12.0
80–84	0.27721	57,849	16,036	250,275	516,536	8.9
85+	1.00000	41,813	41,813	266,261	266,261	6.4

^{a}Average number of years of life remaining at the beginning of the age interval.

Source: National Center for Health Statistics, National Vital Statistics Reports, *United States Abridged Life Tables, 1996*, Vol. 47, No. 13, Hyattsville, Maryland, 1998.

stationary population analysis. When used in this manner some elements of the life table take on a different interpretation than those given before (e.g., Shryock & Siegel, 1973).

The idea behind a period life table is clear: It summarizes the mortality (and survival) probabilities observed in a particular population during a particular period of time. The actual construction of a life table is not as simple as it might appear, however. Problems include adjusting for the accuracy of the underlying data; smoothing out data fluctuations over time; adjusting for the digit preference often found in age data reported by censuses and surveys; matching deaths during a calendar year with changes in age during the year; transforming observed age-specific death rates into survival probabilities; and developing techniques for converting unabridged to abridged life tables (and vice versa). For further reading see Bell et al. (1992), Shryock and Siegel (1973), and Smith (1992).

Constructing Life Table Survival Rates. In countries that have good vital statistics data, life tables provide the most frequently used source of data for calculating survival rates. For population projections, survival rates are often based on five-year time horizons and five-year age groups and are calculated as

$$_5S_x = {_5L_{x+5}} / {_5L_x}$$

where $_5S_x$ is the survival rate, $_5L_{x+5}$ is the number of person-years lived between ages $x + 5$ and $x + 10$, and $_5L_x$ is the number of person-years lived between ages x and $x + 5$. For U.S. females aged 20–24 in 1996, for example, the five-year survival rate is

$$_5L_{25} / {_5L_{20}} = 492,128 / 493,488 = 0.99724$$

In other words, given the survival rates observed in 1996, only about three out of 1,000 women aged 20–24 would be expected to die during the next five years.

Survival rates can be calculated for different time horizons and different age groups by changing the subscripts in the equation shown above. For example, a 10-year survival rate for a five-year age group can be calculated as

$$_5S_x = {_5L_{x+10}} / {_5L_x}$$

Using the data in Table 4.1, a 10-year survival rate for females aged 20–24 is

$$_5L_{30} / {_5L_{20}} = 490,336 / 493,488 = 0.99361$$

Due to the peculiar nature of mortality patterns in the first year of life, the 0–4 age cohort is often split into two groups: less than 1 and 1–4. Then, survival rates are calculated separately for each group. Rates for children aged 1–4 are often calculated in the manner described before, but rates for infants less than age 1 are based on procedures that account for the high mortality rates in the first days and weeks of life (see Shryock & Siegel, 1973, pp. 435–438 for details). Making this distinction may be important when the size of the newborn cohort has been changing rapidly or when the projections are used for detailed analyses of mortality.

Calculating survival rates for one-year age groups requires an unabridged life table, but the approach is the same. For example, a five-year survival rate for a one-year age group can be calculated as

$$S_x = L_{x+5}/L_x$$

Using data from an unabridged life table for the United States in 1990 (Bell et al., 1992), we can calculate the five-year survival rate for 50-year-old males as

$$L_{55}/L_{50} = 86,000/89,559 = 0.96026$$

The procedure for calculating survival rates for the oldest age group is slightly different because it is an open-ended group. For this age group, T values rather than L values are used. Suppose that 85+ is the oldest age group to be projected. The five-year survival rate for this age group is calculated as

$$S_{80} = T_{85}/T_{80}$$

where T_{85} and T_{80} are the total person-years lived after ages 85 and 80, respectively. For example, the five-year survival rate for females aged 80+ in 1996 is calculated as

$$T_{85}/T_{80} = 266,261/516,536 = 0.51547$$

In other words, given the continuation of 1996 mortality rates, only 51.6% of females aged 80+ would be expected to live for at least another five years. This stands in sharp contrast to the 99.7% survival rate noted previously for females aged 20–24.

Survival rates are typically calculated separately for males and females and are often further subdivided by race and ethnicity. The reason for drawing these distinctions is that mortality rates vary from one demographic subgroup to another. Table 4.2 shows that mortality rates in the United States are highest for black males and lowest for white females in every age group. Rates for white males and black females fall somewhere in between. The use of separate rates is particularly important when several population subgroups account for a substantial proportion of the total population and when they have survival rates that differ considerably from each other.

Census Survival Rates

The second approach to constructing survival rates does not require age-specific mortality data, making it particularly useful for countries (or regions) that lack vital statistics data. This approach is based on the formation of ratios between

Table 4.2. Life Table Mortality Rates ($_nq_x$) by Age, Sex, and Race, United States, 1996

Age	White females	White males	Black females	Black males
0–1	0.00544	0.00667	0.01325	0.01602
1–5	0.00113	0.00147	0.00245	0.00275
5–10	0.00076	0.00100	0.00129	0.00173
10–15	0.00087	0.00131	0.00130	0.00207
15–20	0.00213	0.00489	0.00266	0.00923
20–25	0.00213	0.00650	0.00408	0.01440
25–30	0.00265	0.00666	0.00614	0.01556
30–35	0.00356	0.00865	0.00913	0.02023
35–40	0.00502	0.01116	0.01336	0.02627
40–45	0.00725	0.01492	0.01853	0.03667
45–50	0.01115	0.02072	0.02541	0.05050
50–55	0.01859	0.03163	0.03654	0.06842
55–60	0.02940	0.04906	0.05164	0.09418
60–65	0.04817	0.07965	0.07734	0.13552
65–70	0.07234	0.11902	0.10490	0.16448
70–75	0.11269	0.17863	0.16305	0.24710
75–80	0.17280	0.25667	0.21280	0.31195
80–84	0.27543	0.38180	0.31464	0.42807
85+	1.00000	1.00000	1.00000	1.00000

Source: National Center for Health Statistics, National Vital Statistics Reports, *United States Abridged Life Tables, 1996*, Vol. 47, No. 13, Hyattsville, Maryland, 1998.

between age cohorts in two consecutive censuses. These ratios are called *census survival rates* and are calculated as

$$_nS_x = {_nP_{x+t,c+t}} / {_nP_{x,c}}$$

where $_nS_x$ is the survival rate for age group x to $x+n$, P is population size, c is the year of the second most recent census, and t is the number of years between the two most recent censuses. In the United States, t will usually be 10 because there are 10 years between complete national censuses. For example, 11,669,408 residents aged 40–44 were counted in the 1980 Census and 11,350,513 residents aged 50–54 in the 1990 Census. The census survival rate for this age cohort is calculated as

$$_5P_{50,1990} / {_5P_{40,1980}} = 11,350,513 / 11,669,408 = 0.97267$$

Census survival rates are typically constructed separately for males and females and can be further differentiated by race, ethnicity, and other demographic characteristics as well.

This approach to constructing survival rates has several problems. The size of a cohort changes because of deaths and people moving into and out of an area. Consequently, census survival rates mix the effects of mortality and migration and give misleading estimates of survival probabilities for places that have even moderate levels of in- or out-migration.

A second problem is that census survival rates are affected by changes in coverage from one decennial census to the next. As mentioned in Chapter 2, no census enumeration is perfect. Some people are missed, others are counted twice, and still others are counted in the wrong place. If coverage rates were the same in every census and in every demographic subgroup, enumeration errors would create no major problems for constructing survival rates. However, because coverage rates differ from one subgroup to another and change over time, changes in coverage rates introduce additional errors into the estimation of census survival rates.

Because of these problems, census survival rates are seldom used for mortality projections in countries that have good vital statistics data. As Chapter 6 shows, however, they can be used as a measure of the *joint* effect of mortality and migration. This measure is very useful for constructing population projections for small areas that lack migration data. For the remainder of this chapter, we focus strictly on life table survival rates.

APPROACHES TO PROJECTING MORTALITY RATES

Mortality rates in the United States declined considerably during the twentieth century. Figure 4.2 shows female mortality rates for selected age groups from 1900–1990. Mortality rates declined substantially for females in each age group (especially the youngest). Similar declines would be found for males and for

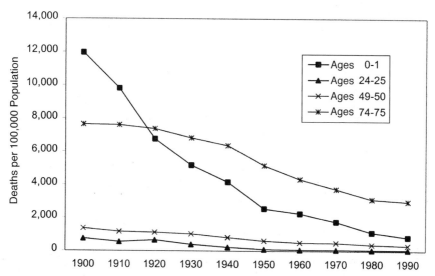

Figure 4.2. One-year life table mortality rates ($_nq_x$) for selected female age groups, United States, 1900–1990. (*Source:* F. Bell, A. Wade, & S. Goss [1992]. Life tables for the United States social security area 1900–2080. Actuarial Study No. 107. Washington, DC: Social Security Administration.)

females in other age groups. Will mortality rates continue declining in the future? If so, how rapidly will they decline? How can one go about projecting future mortality and survival rates?

Constant Rates

The simplest approach is to assume that the most recent mortality rates will continue unchanged. This assumption will generally be reasonable for short projection horizons of five or 10 years because mortality rates in more developed countries change relatively slowly over time, especially for the younger and middle age groups. For longer horizons, however, the no-change assumption is likely to diverge farther and farther from actual mortality trends. Assuming that mortality rates will remain unchanged for the next 20, 30, or 40 years would be unrealistic and would likely lead to population projections that were too low, especially for the older age groups.

A better approach is to develop assumptions that account for the likelihood that mortality rates will continue to fall. Although a war, epidemic, or some other calamity could cause U.S. mortality rates to rise, most observers believe the chances of further declines are much greater than the chances of a sudden increase. The only questions are: "How much farther will mortality rates fall?" and "At what pace will they fall?" Following Olshansky (1988), we identify several basic approaches to projecting mortality rates. One entails extrapolating past trends, another is based on the movement of current rates toward those observed in a target population, and a third is based on the reduction or elimination of particular causes of death. We also discuss an approach in which future changes in mortality rates for one area are based on changes projected for another. Each approach requires applying professional judgment.

Trend Extrapolation

One common approach to projecting mortality rates is to extrapolate past trends. Perhaps the foremost practitioner of extrapolation techniques is the Office of the Actuary in the Social Security Administration (SSA), which has been projecting U.S. survival rates since the 1930s. The SSA has used some type of extrapolation technique in every set of projections since 1952 (Olshansky, 1988). Initially, SSA projections accounted only for differences in mortality rates by age and sex; since the 1950s, they have accounted for differences in the cause of death as well.

In its most recent set of projections (Bell et al., 1992), the SSA first calculated average annual rates of change in mortality rates by age, sex, and cause of death from 1968 to 1988. It was assumed that the rates of decline in mortality rates would become smaller over time, and that no further changes in rates of decline would occur after the year 2016. Assumptions regarding the ultimate rates of decline were based on expectations regarding the development of new diagnostic and surgical techniques, increases in environmental pollutants, changes in lifestyle (e.g., nutrition and exercise), and similar factors. Thus, SSA mortality projections were based on extrapolation techniques informed by expert judgment.

Until recently, the Census Bureau used SSA mortality rates to construct population projections for the United States and for each state. Now, the Census Bureau develops its own mortality projections by age, sex, race, and ethnicity (Hispanic and non-Hispanic). These projections are also based on extrapolation techniques. The Census Bureau used three alternative assumptions in a recent set of national projections: (1) low—no change in 1990 survival rates, except for a decrease due to the impact of AIDS; (2) medium—1980–1990 trends in survival rates were extrapolated into the future, and an additional adjustment was made to account for the impact of AIDS; and (3) high—1970–1980 trends in survival rates were extrapolated into the future. Several constraints based on expert judgment were also imposed (Day, 1996a).

Extrapolation techniques assume that the future will mirror the past in certain important ways. This is not always a valid assumption. U.S. mortality rates have declined continuously throughout the twentieth century, but the pace of those declines has varied considerably. Rates of decline were fairly moderate from 1900 to the mid-1930s, much larger from the mid-1930s to the mid-1950s, considerably smaller from the mid-1950s to the late 1960s, larger again from the late 1960s to the early 1980s, and smaller during the rest of the 1980s (Bell et al., 1992). Simple extrapolation techniques cannot pick up the timing and magnitude of changes such as these.

A more sophisticated type of extrapolation technique attempts to capture some of these changes through the use of ARIMA time series models (e.g., Lee & Carter, 1992; Lee & Tuljapurkar, 1994; McNown & Rogers, 1989, 1992). Time series models account for changing trends over time and provide probabilistic prediction intervals for each projection. However, they require a high level of modeling expertise and have limited applicability to small areas due to the lack of reliable data and the high degree of variability in small-area mortality rates. In addition, projected changes tend to level off or converge to constant oscillations within a relatively short time (Land, 1986).

Targeting

The targeting approach is based on the assumption that mortality rates in the population to be projected will gradually converge toward those observed in another population (i.e., the target). A target population is chosen that provides a set of mortality rates believed to be realistic for the population to be projected. This choice is based on similarities in socioeconomic, cultural, and behavioral characteristics; levels of medical technology; and primary causes of death. Statistical techniques ranging from simple percentage reductions to complex curve-fitting procedures can be used to guide the convergence of one set of rates to another (Olshansky, 1988).

The targeting approach has been around almost as long as the cohort-component method itself. Whelpton (1928) used both extrapolation and targeting techniques for projections of the U.S. population; mortality rates in New Zealand were used as target rates toward which U.S. rates would move during a 50-year period. Siegel and Davidson (1984) compared U.S. mortality rates with those found in Hawaii, the state that has the greatest life expectancy at birth and at age 65; they concluded that the movement of U.S. rates toward Hawaiian rates would have only a moderate impact on U.S. mortality rates. Nakosteen (1989) projected that county mortality rates in Massachusetts would gradually move toward national rates, reaching equality by 2020. A variation of the targeting approach was applied by the Social Security Administration, using targeted rates of change in age-, sex-, and cause-specific mortality rates rather than targeted levels (Bell et al., 1992).

Cause-Delay

Cause-delay models focus on the implications of delaying (or completely eliminating) the occurrence of one or more causes of death (Manton, Patrick, & Stallard, 1980; Olshansky, 1987). The basic premise behind this approach is that changes in lifestyle and medical technology have delayed the occurrence of various types of deaths until progressively older ages. Consequently, as time goes by, each cohort faces smaller mortality risks at each age than did the previous cohort.

Cause-delay models are often operationalized by assuming that cause-specific mortality rates for one age group in a population will gradually move toward those currently found in a younger age group in the same population (e.g., rates for persons aged 60–64 will eventually become the same as those currently found for persons aged 55–59). The impact of such changes can be substantial. One application of a cause-delay model, for example, found that delaying cancer mortality rates by 10 years raised life expectancy at birth by 1.3 years for white males. The complete elimination of cancer as a cause of death raised life expectancy at birth by only one additional year (Manton et al., 1980).

Cause-delay models are similar to targeting models in that one set of mortality rates gradually converges toward another. In cause-delay models, however, the target population is a younger cohort in the same population rather than the same cohort in a different population. An advantage of cause-delay models compared to targeting models is that by staying within the same population, cause-delay models control for many factors that cause mortality rates to differ from one population to another. The basic problems with using cause-delay models are deciding what causes of death to focus on, how rapidly mortality rates for one cohort will move toward those currently found in a younger cohort, and how long the whole process will continue. To our knowledge, cause-delay models have not been used to prepare official population projections in the United States.

Synthetic Projection

In the final approach, synthetic mortality or survival rates are created by linking changes in rates in one area to changes projected for a different area. This approach is similar to targeting, but it adopts the *rates of change* in mortality or survival rates from the model population rather than the rates themselves. This is a simple, straightforward approach that is used frequently for state and local mortality projections (e.g., Campbell, 1996; Department of Rural Sociology, 1998; Smith & Nogle, 2000; Treadway, 1997).

Suppose that the middle series of the Census Bureau's national projections is accepted as a reasonable model of future mortality changes in New Jersey. Then, projected changes in national rates can be used to guide projected changes in New Jersey. For example, if an increase of 2% in survival rates for females aged 70–74 in the national population were projected for the next 10 years, the same percentage increase could be applied to survival rates for females aged 70–74 in New Jersey. This procedure can be carried out for every subgroup of the population and will provide a complete set of projected survival rates.

The synthetic approach can be applied by using either survival rates or mortality rates. However, to ensure that projected survival rates do not take on values greater than 1.0, survival rates are often converted into mortality rates before the adjustments are made (Shryock & Siegel, 1973, p. 453). The procedure is as follows:

1. Mortality rates are calculated by subtracting survival rates from 1.0.
2. Adjustments to the mortality rates are made based on the changes projected for the model population.
3. Mortality rates are converted back into survival rates by subtracting the adjusted rates from 1.0.

Regardless of whether mortality or survival rates are used, the analyst must make sure that the projected rates are reasonable. If survival/mortality rates for the model population are similar to those for the region being projected, the synthetic approach generally produces reasonable results (at least for short- and medium-range projection horizons). If the rates are considerably different, however, changes in the model population may overstate or understate likely future changes in the population being projected. In these instances, further adjustments must be made or a different approach used.

IMPLEMENTING THE MORTALITY COMPONENT

We have described the construction of survival rates and several techniques that can be used to project those rates into the future. Where can the analyst find a set of survival rates for a particular area? Which technique (or techniques) should be used to project those rates into the future? How can those techniques best be applied? What can be done if no current rates can be found? We offer the following suggestions for implementing the mortality component of the cohort-component method.

Sources of Data

Life tables for the United States and each state are readily available and can be used as a base for mortality projections. National life tables are published annually by the NCHS and several times each decade by the SSA (e.g., Bell et al., 1992; National Center for Health Statistics, 1997). State life tables are typically constructed once every 10 years, when decennial census data become available to serve as denominators for the mortality rates; they are prepared by the NCHS and by the vital statistics agencies in many states.

For substate areas, however, life tables are constructed only infrequently. Although the requisite data are generally available at the county level, small population sizes and variations in data quality exacerbate the problems of life table construction discussed earlier. For subcounty areas, reliable data are rarely available, which makes constructing life tables even more difficult (if not impossible). Life tables for some cities and counties have been constructed on an ad hoc basis, but only for a few places. Must the analyst construct a new set of life tables or use census survival rates if life table survival rates are not available for the areas to be projected?

Happily, the answer to this question is "No." Large regional differences in mortality rates have mostly disappeared in the United States. Table 4.3 shows life expectancy at birth (e_0) by race and sex for each state in 1990. This is a handy way to summarize a population's age-specific mortality patterns. Although some differences in e_0 values can be seen, they are generally small, especially when compared to state-to-state differences in fertility and migration levels.

Because survival rates for many areas in the United States are similar, proxy rates from a different area can generally be used for areas for which no life tables have been constructed (e.g., state life tables can be used for county projections).

However, it is important to choose a model population whose characteristics are similar to those of the area to be projected. If the population has large numbers of racial or ethnic minorities that have substantially different survival rates, it is advisable to make separate projections for each race/ethnic group or to construct survival rates weighted by race and ethnicity. Applying proxy survival rates to the region's population in a recent year and comparing the resulting number of deaths with the number actually occurring provides a test of the validity of the proxy rates.

Life tables based on county-specific population and mortality data could be constructed, of course. However, we believe that the costs of constructing those tables—combined with problems of data reliability—generally outweigh the benefits. Furthermore, other types of errors typically swamp errors caused by using proxy survival rates. We believe that scarce resources can be better spent elsewhere. (One exception may be counties that have unique population or mortality characteristics, for which no reliable proxy rates can be found).

Age-specific survival rates for counties and other small areas are sometimes estimated as $1 - {}_nASDR_x$ (e.g., Isserman, 1993). However, rates calculated in this manner understate the probability of surviving from one age to another because they do not account for deaths that occur during the year (London, 1988). Although this approach is acceptable for short horizons of five or 10 years, it is problematic for longer horizons because of the cumulative impact of systematically underestimating the number of survivors.

Views of the Future

Once a set of base survival rates has been chosen, the next step is to decide how to project those rates into the future. Given the wide range of methods that can be used, how can the analyst choose the one(s) that will be most appropriate for a particular set of projections? This choice will be determined partly by the availability of relevant data for the area(s) to be projected, but will also be affected by the analyst's views of future mortality trends.

Demographers, scientists, physicians, and health care experts are sharply divided in their views of future mortality trends. One group of researchers believes that there are biological limits to the human life span and that it is unlikely that life expectancy at birth will increase a great deal beyond the levels currently found in low-mortality countries (e.g., Fries, 1980, 1989; Olshansky, Carnes, & Cassel, 1990; R. Rogers, 1995). They point to the finite number of cell doublings in the life span of a species; the steady loss of organ capacity that begins around age 30 in human beings; the much smaller increases in life expectancy at older ages than younger ages during the twentieth century; the failure of many people to adopt lifestyle habits known to increase health and longevity; and the relatively modest increases in life expectancy that would be implied even by the total elimination of

Table 4.3. Life Expectancy at Birth by Sex and Race,
United States and Each State, 1989–1991

	White females	White males	Black females	Black males
Northeast				
Connecticut	80.4	74.3	75.4	66.0
Maine	79.6	73,0	*	*
Massachusetts	80.0	73.5	76.5	68.2
New Hampshire	79.7	73.5	*	*
New Jersey	79.3	73.4	72.9	63.9
New York	79.0	72.8	74.4	63.9
Pennsylvania	79.3	72.0	73.0	63.3
Rhode Island	80.0	73.3	*	*
Vermont	79.7	73.3	*	*
Midwest				
Illinois	79.3	72.8	72.4	62.4
Indiana	79.0	72.4	73.6	65.9
Iowa	80.6	74.0	*	*
Kansas	80.3	73.7	75.0	67.5
Michigan	79.1	73.1	73.2	63.7
Minnesota	81.0	74.8	*	*
Missouri	79.5	72.4	73.5	63.9
Nebraska	80.4	73.9	*	*
North Dakota	81.3	74.7	*	*
Ohio	79.0	72.7	74.3	65.8
South Dakota	81.6	74.3	*	*
Wisconsin	80.3	74.0	75.3	66.4
South				
Alabama	78.9	71.1	73.8	64.4
Arkansas	78.9	71.5	73.6	64.0
Delaware	78.6	72.8	72.9	65.5
District of Columbia	81.1	71.4	71.6	57.5
Florida	80.5	73.2	73.3	64.3
Georgia	78.9	71.5	73.7	64.0
Kentucky	78.2	71.0	74.1	66.1
Louisiana	78.5	71.2	73.2	63.8
Maryland	79.2	73.2	74.3	64.7
Mississippi	78.8	70.7	73.8	64.4
North Carolina	79.4	72.2	74.2	64.4
Oklahoma	78.6	71.8	74.5	67.1
South Carolina	79.0	71.6	73.4	64.1
Tennessee	79.1	71.4	73.2	64.4
Texas	79.4	72.1	74.2	65.4
Virginia	79.5	73.0	74.4	65.8
West Virginia	78.0	70.7	74.4	65.0

Table 4.3. (*Continued*)

	White females	White males	Black females	Black males
West				
Alaska	79.4	72.8	*	*
Arizona	79.8	73.0	74.9	67.2
California	79.3	72.6	74.1	65.4
Colorado	80.1	73.9	75.9	69.0
Hawaii	81.1	75.1	*	*
Idaho	79.9	73.9	*	*
Montana	79.9	73.6	*	*
Nevada	78.0	71.3	*	*
New Mexico	79.5	72.7	*	*
Oregon	79.7	73.3	*	*
Utah	80.4	75.0	*	*
Washington	79.8	74.0	75.6	67.9
Wyoming	79.5	73.3	*	*
United States	79.5	72.7	73.7	64.5

Note: An asterisk (*) indicates that the figure does not meet standards of reliability or precision (based on fewer than 20 cases).
Source: National Center for Health Statistics, *U.S. Decennial Life Tables for 1989–1991*, Vol. 1, No. 3, Hyattsville, Maryland, 1999.

several leading causes of death. Many researchers believe that life expectancies at birth are likely to level off at around age 85.

Another group of researchers believes that considerably larger gains in life expectancy are possible (e.g., Ahlburg & Vaupel, 1990; Fogel & Costa, 1997; Manton, Stallard, & Tolley, 1991). They point to the technological and biomedical advances that have already occurred and to the likelihood of further break-throughs; the increased awareness of the benefits of healthy lifestyles (e.g., re-duced smoking, improved nutrition, increased exercise); the very high life expec-tancies currently found in several population subgroups that practice healthy lifestyles and have access to good medical care; the persistence of mortality declines throughout the twentieth century; and the tendency for past forecasts to understate future increases in longevity. These researchers see life expectancies at birth rising to age 95, 100, or even higher by the end of the twenty-first century.

The Census Bureau and the Social Security Administration are the two main sources of national population projections in the United States. Each has a major impact on the development of public policy. Historically, projections from both agencies have been based on relatively conservative views of future mortality trends. The Census Bureau (Day, 1996a) recently projected that life expectancy at birth would reach 82.0 by 2050 in its middle series; the Social Security Adminis-tration (Bell et al., 1992) projected 79.7 for the same year. Some analysts believe that even these relatively conservative projections may turn out to be too high, especially for the older age groups (e.g., Bennett & Olshansky, 1996).

Ahlburg and Vaupel (1990) are considerably more optimistic about future declines in mortality rates. They presented two scenarios, one based on an extrapolation of 2% annual reductions in mortality rates at each age and the other based on an extrapolation of 1% annual reductions. These reductions are consistent with those that occurred in the United States at various times during the twentieth century. The first scenario produced life expectancies at birth of 100 for females and 96 for males in 2080; the second produced life expectancies of 89 and 84, respectively. The second scenario is similar to the ARIMA time series forecasts produced by Lee and Tuljapurkar (1994), which showed a life expectancy at birth (both sexes) of 86.1 in 2065, with a predicted range of 81–90.

When the purpose of projecting is to *forecast* the future population, the analyst who anticipates only modest improvements in longevity may want to apply assumptions similar to those used by the Census Bureau or the Social Security Administration. Conversely, one who anticipates larger increases in life expectancy might favor assumptions similar to those used by Ahlburg and Vaupel (1990). When the purpose of projecting is simply to *illustrate* or *analyze*, these contrasting projections provide an opportunity to explore the implications of alternative mortality scenarios.

Examples

The following examples illustrate two of the methods that can be used to project future mortality rates. One is from Florida, where demographers used a synthetic approach that tied projected changes in state survival rates to projected changes in national survival rates (Smith & Nogle, 2000). The starting point in these projections was a set of five-year survival rates by age and sex, based on Florida life tables for 1990. These survival rates were adjusted upward at 10-year intervals, based on adjustment factors derived from projected changes in U.S. survival rates through 2050 (Day, 1996a, including unpublished data available on the Census Bureau web site).

Adjustment factors ($_5A_x$) were calculated for each age-sex group by forming ratios of survival rates in projected year $t+10$ to survival rates in year t:

$$_5A_x = {_5S_{x,t+10}} / {_5S_{x,t}}$$

where $_5S_{x,t+10}$ is the five-year U.S. survival rate for age group x to $x + 5$ in year $t + 10$ and $_5S_{x,t}$ is the five-year survival rate for age group x to $x + 5$ in year t. For example, the five-year national survival rate for males aged 50–54 was 0.95474 in 1990 and was projected at 0.95985 in 2000. The adjustment factor for males aged 50–54 is calculated as

$$_5S_{50,2000} / {_5S_{50,1990}} = 0.95985 / 0.95474 = 1.00535$$

Table 4.4. Projected 2000 Survival Rates for Florida Males,
Based on Projected U.S. Survival Rates

Age	1990 survival rate[a]	U.S. adjustment factor[b]	2000 survival rate[c]
0–1	0.98836	1.00138	0.98972
1–4	0.99734	1.00020	0.99754
5–9	0.99825	1.00015	0.99840
10–14	0.99579	1.00030	0.99609
15–19	0.99246	1.00055	0.99301
20–24	0.99026	1.00054	0.99079
25–29	0.98808	1.00020	0.98828
30–34	0.98481	0.99977	0.98458
35–39	0.98110	0.99973	0.98084
40–44	0.97666	1.00055	0.97720
45–49	0.96670	1.00267	0.96928
50–54	0.94920	1.00535	0.95428
55–59	0.92715	1.00837	0.93491
60–64	0.90136	1.01226	0.91241
65–69	0.86270	1.01889	0.87900
70–74	0.80417	1.02861	0.82718
75–79	0.71500	1.03940	0.74317
80+	0.50268	1.03435	0.51995

[a]Florida Office of Vital Statistics, *Abridged Life Tables, 1990*, Department of Health and Rehabilitative Services, Jacksonville, 1993.
[b]Unpublished tables that accompany J. Day (1996a). Population Projections of the United States by Age, Sex, Race, and Hispanic Origin: 1995–2050. U.S. Bureau of the Census, *Current Population Reports*, P25-1130, Washington, DC.
[c]2000 survival rate = 1990 survival rate × U.S. adjustment factor.

Similar adjustment factors were calculated for males and females in each five-year age group through age 80+. These adjustment factors were then multiplied by the 1990 Florida survival rates to give projected survival rates in 2000. The mortality assumption used in Florida, then, was that the state's age-sex-specific survival rates would change at the same rate as the corresponding rates for the nation as a whole.

Table 4.4 shows 1990 survival rates, adjustment factors, and projected 2000 survival rates for males in Florida. Projections were also made for 2010 and 2020 (not shown here). It is noteworthy that the survival rates for ages 30–34 and 35–39 shown in Table 4.4 were adjusted downward rather than upward. This was caused by the impact of AIDS on mortality rates during the 1980s. Although long-term survival trends in the United States have been steadily upward, the reversal for these age groups points to the importance of considering contextual factors when projecting mortality rates.

The same Florida data can be used to illustrate another projection method.

Table 4.5. Projected 2000 Survival Rates for Florida Males,
Based on 1% Annual Declines in Mortality Rates[a]

Age	1990 survival rate[b]	1990 mortality rate[c]	2000 mortality rate[d]	2000 survival rate[e]
0–1	0.98836	0.01164	0.01053	0.98947
1–4	0.99734	0.00266	0.00241	0.99759
5–9	0.99825	0.00175	0.00158	0.99842
10–14	0.99579	0.00421	0.00381	0.99619
15–19	0.99246	0.00754	0.00682	0.99318
20–24	0.99026	0.00974	0.00881	0.99119
25–29	0.98808	0.01192	0.01078	0.98922
30–34	0.98481	0.01519	0.01374	0.98626
35–39	0.98110	0.01890	0.01709	0.98291
40–44	0.97666	0.02334	0.02111	0.97889
45–49	0.96670	0.03330	0.03012	0.96988
50–54	0.94920	0.05080	0.04594	0.95406
55–59	0.92715	0.07285	0.06588	0.93412
60–64	0.90136	0.09864	0.08921	0.91079
65–69	0.86270	0.13730	0.12417	0.87583
70–74	0.80417	0.19583	0.17710	0.82290
75–79	0.71500	0.28500	0.25775	0.74225
80+	0.50268	0.49732	0.44977	0.55023

[a]A 1% annual decline in mortality for 10 years is represented by an adjustment factor of 0.90438
(i.e., 0.99^{10}).
[b]Florida Office of Vital Statistics, *Abridged Life Tables, Florida, 1990*, Department of Health
and Rehabilitative Services, Jacksonville, 1993.
[c]1990 mortality rate = 1 − 1990 survival rate.
[d]2000 mortality rate = 1990 mortality rate × 0.90438.
[e]2000 survival rate = 1 − 2000 mortality rate.

Following Ahlburg and Vaupel (1990), we applied 1% annual reductions to 1990 mortality rates for males in Florida to create projected rates for 2000 (Table 4.5). The first column shows 1990 survival rates; these are the same rates shown in the first column of Table 4.4. The second column shows the mortality rates obtained by subtracting each survival rate from 1.0. The third column shows the adjusted mortality rates for year 2000 after 1% annual reductions have been applied to the mortality rates shown in column 2. The fourth column shows the new survival rates implied by the adjusted mortality rates.

It is interesting to note that the 2000 survival rates shown in Table 4.5 are very similar to those shown in Table 4.4 (except for the oldest age group). Differences in projection methodology generally will not lead to significant differences in survival rates unless the projections are based on dramatically different assumptions or extend well into the future.

CONCLUSIONS

We have looked at a number of ways to project survival rates based on different techniques, assumptions, and perspectives regarding future mortality trends. How sensitive are population projections to these differences? That is, just how important is the mortality component in the production of cohort-component population projections?

In terms of its impact on total population size, the choice of mortality rates is not very important, especially for short- and medium-range projections (i.e., less than 20 years). Long (1989) reported that for 20-year national projections, the population projected under the high mortality assumption was only 1.0% below that projected under the medium assumption; the population projected under the low mortality assumption was only 1.3% above the medium projection. Even for a 50-year horizon, the total population projected under the high mortality assumption was only 3% smaller than that projected under the medium assumption; the low mortality assumption led to a population only 4% larger.

The impact of differences in mortality assumptions on projections of total population may be somewhat larger at the state and local level than at the national level. States and local areas exhibit some variability in mortality rates, even after accounting for differences in race (Isserman, 1993). Because of their socio-economic and demographic differences, they also have the potential for larger changes in future mortality rates than would be expected at the national level. However, a study of 30-year projections for 55 counties in West Virginia found that using state mortality rates instead of county-specific rates led to differences in total population that averaged only 1% (Isserman, 1993). In general, reasonable differences in mortality assumptions will have relatively little impact on state and local projections of total population, even for fairly long-range projections.

For projections of the older population, however, differences in mortality rates can have a substantial impact. Mortality rates for young and middle-aged persons in the United States are already so low that there is little room left for further improvements. Indeed, eliminating *all* deaths below age 50 would increase life expectancy at birth by only 3.5 years (Olshansky et al., 1990). Among older age groups, however, mortality rates are considerably higher, which leaves more room for improvement and creates more possibilities for differences in projection assumptions.

This can be illustrated by comparing two sets of projections from the Social Security Administration. Projections for the year 2000 made in 1974 showed only 31.0 million persons aged 65+, whereas projections made in 1984 showed 36.2 million. This increase of more than 5 million was caused by changes in mortality assumptions for the elderly population; it accounted for about 80% of the 6.4 million increase projected for the total population (Olshansky, 1988). Differences in projected mortality rates clearly have much more impact on the elderly population than on the population as a whole.

Projections of the older population take on great practical importance through their impact on health care, housing, transportation, and entitlement programs like Social Security and Medicare (e.g., Bennett & Olshansky, 1996; Fogel & Costa, 1997). This impact is magnified when projections of the size of the elderly population are combined with projections of health status. Does greater longevity lead to more years of healthy living or simply to longer periods of illness, disability, and institutionalization? What are the connections between changes in mortality rates and changes in health status? These are hotly debated questions that have tremendous social, economic, and ethical implications (e.g., Crimmins et al., 1997; Fries, 1989; Olshansky, 1988; R. Rogers, 1995). Although decisions regarding mortality assumptions have only a modest impact on projections of total population in the United States, they are very important for other reasons.

CHAPTER 5

Fertility

The term *fertility* refers to the occurrence of a live birth (or births) to an individual, a group, or an entire population. It is determined by a combination of biological, social, psychological, economic, and cultural factors. Biological factors affect fecundity (i.e., the physiological capacity to reproduce), whereas social, psychological, economic, and cultural factors affect choices regarding whether to have children, how many to have, and when to have them. Although biological factors set an upper limit on a woman's lifetime fertility, most women bear children at levels far below that limit. The broad array of factors that affect personal choice is thus paramount in the study of fertility.

Fertility rates vary considerably among individuals and populations. Some women have no children, others have one or two, and a few have 10 or more. At current rates, American women will average just over two births during their lifetimes. This is considerably higher than the rates of 1.2 births per women found in Russia, Spain, Italy, and several other European countries, but much lower than the rates of 6.0 or more found in Niger, Ethiopia, Oman, and many other African and Middle Eastern countries (Population Reference Bureau, 1999). The most prolific mother ever reported was a Russian woman in the eighteenth century who was alleged to have borne 69 children through 27 pregnancies (Weeks, 1999).

Fertility rates also change over time. Two hundred years ago, women in the United States averaged around 8.0 births during their lifetimes (Weeks, 1999). These very high fertility rates fell rapidly during the nineteenth and early twentieth centuries, reaching levels below 2.3 during the 1930s. The postwar baby boom brought fertility rates up to 3.7 by the late 1950s, but the baby bust dropped them below 1.8 by the mid-1970s, their lowest levels ever. Rates have since increased by more than 10%. Figure 5.1 shows total fertility rates in the United States from 1917 to 1998.

Fertility rates followed a pattern very different from mortality rates during the twentieth century. Whereas mortality rates fell slowly and steadily, fertility

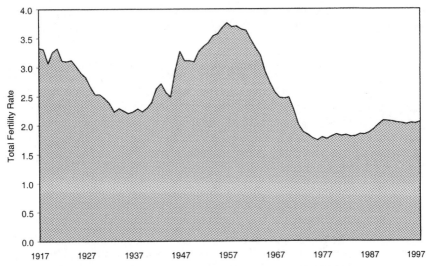

Figure 5.1. Total fertility rates, United States, 1917–1998. (*Sources*: National Center for Health Statistics, *Fertility Tables for Birth Cohorts by Color: United States, 1917–73*, DHEW Publication No. (HRA) 76-1152, Rockville, Maryland, 1976; U.S. Bureau of the Census, *Statistical Abstract of the United States, 1998*, Washington, DC, 1998; National Center for Health Statistics, *National Vital Statistics Report*, Vol. 47, No. 25, Hyattsville, Maryland, 1999)

rates fluctuated markedly, often within a relatively short time. These wide swings make it more difficult to develop accurate forecasts of fertility rates than mortality rates. In this chapter we describe several measures of fertility and discuss various approaches to formulating assumptions for projecting fertility rates. Again, we pay special attention to the problems of constructing projections for small areas. We close with an assessment of the impact of fertility assumptions on population projections.

FERTILITY MEASURES

A number of measures have been developed to reflect the fertility behavior of a population. We will describe several of the most commonly used. All are based on two types of data: the number of live births (hereinafter referred to simply as births) that occur in a geographic area during a particular period and the population of that area. These measures are called *birth* (or *fertility*) *rates* because they relate the number of births (the numerator) to the population at risk of giving birth (the denominator). However, as discussed in Chapter 2, they are not rates in a true probabilistic sense.

Fertility measures typically refer to a calendar year. The numerator is the number of births that occurred during the year and the denominator is the number of persons alive at midyear. Sometimes a three-year average of births is used to smooth out the effects of annual fluctuations. Some fertility measures use data on the total population; others use population data broken down by age, sex, race, or other characteristics. Birth data in the United States are collected by the vital statistics agencies of each state and are compiled nationally by the National Center for Health Statistics (NCHS). Population data are based on census counts or population estimates, depending on the year(s) for which rates are constructed.

Crude Birth Rate

The simplest fertility measure is the crude birth rate (CBR), which is calculated by dividing the number of births during a year by the midyear population. It is generally multiplied by 1,000 to reflect the number of births per 1,000 persons:

$$CBR = (B/P)\,1,000$$

where B is the number of births during the year and P is the midyear population. For example, there were 3,944,046 births in the United States in 1998, and the midyear population was 270,298,524, yielding a CBR of

$$(3,944,046\,/\,270,298,524)\,1,000 = 14.6$$

This means there were 14.6 births for every 1,000 residents of the United States in 1998. Crude birth rates can also be calculated for different race or ethnic groups and for different geographic regions. For example, the 1998 CBR was 14.0 for whites and 17.7 for blacks:

$$\text{Whites: } (3,122,391\,/\,223,000,729)\,1,000 = 14.0$$
$$\text{Blacks: } (610,203\,/\,34,430,569)\,1,000 = 17.7$$

For states, CBRs in 1998 ranged from 11.0 in Maine to 21.5 in Utah (National Center for Health Statistics, 1999).

The usefulness of the CBR as a measure of fertility is limited because it does not account for differences in demographic characteristics. Births occur only to females, primarily those between 15 and 44. Thus, the age-sex structure of a population has a major impact on its fertility behavior. Other fertility measures have been developed to account for differences in age and sex characteristics.

General Fertility Rate

The general fertility rate (GFR) relates the number of births to the number of females in their prime childbearing years. It is calculated by dividing the number of births by the number of females aged 15–44. It is typically expressed in terms of births per 1,000:

$$GFR = (B \,/\, FP_{15-44})\, 1,000$$

where FP_{15-44} is the midyear population of females aged 15–44. For example, there were 3,944,046 births in the United States in 1998 and 60,112,000 women aged 15–44, yielding a general fertility rate of

$$(3,944,046 \,/\, 60,112,000)\, 1,000 = 65.6$$

GFRs can also be calculated for different race and ethnic groups and for different geographic regions. For example, the GFR for Hispanic women in the United States in 1998 was 101.1, compared to 57.8 for non-Hispanic white women. For states, GFRs varied from 49.1 in Vermont to 91.4 in Utah (National Center for Health Statistics, 1999).

The GFR (sometimes simply called the *fertility rate*) provides a more refined measure than the CBR because it relates the number of births to the population most likely to give birth. It has several shortcomings, however. Some births occur to women younger than age 15 or older than age 44. More important, the age distribution of persons *within* the 15–44 age group differs from one population to another and changes over time. A third measure accounts for these differences by focusing on birth rates for each individual age group.

Age-Specific Birth Rate

The age-specific birth rate (ASBR) is calculated by dividing the number of births to females in a given age group by the number of females in that age group. It is often multiplied by 1,000 to reflect the number of births per 1,000 females:

$$_{n}ASBR_{x} = (_{n}B_{x} \,/\, _{n}FP_{x})\, 1,000$$

where x is the youngest age in the age interval, n is the number of years in the age interval, $_{n}B_{x}$ is the number of births to females between the ages of x and $x+n$, and $_{n}FP_{x}$ is the number of females between the ages of x and $x+n$ at midyear. Age groups are typically expressed in five-year intervals. For example, there were 965,414 births to females aged 20–24 in the United States in 1998 and a midyear population of 8,678,000 females aged 20–24, yielding an ASBR of

$$_{5}ASBR_{20} = (965,414 \,/\, 8,678,000)\, 1,000 = 111.2$$

Table 5.1 shows ASBRs for the United States in 1998. Several things stand out in this table. First, ASBRs were very low for women younger than age 15 and older than age 44. In fact, women in these two age groups accounted for less than 1% of all births in 1998, making ages 15–44 a reasonable choice for the denominator in the GFR. Second, ASBRs were higher for women in their 20s than for

Table 5.1. Age-Specific Birth Rates, United States, 1998[a]

Age	Births	Female population[a]	Age-specific birth rates[b]
10–14	9,481	9,387,000	1.0
15–19	484,976	9,494,000	51.1
20–24	965,414	8,678,000	111.2
25–29	1,083,894	9,341,000	116.0
30–34	890,336	10,179,000	87.5
35–39	425,194	11,370,000	37.4
40–44	80,982	11,049,000	7.3
45–49	3,769	9,607,000	0.4
Total	3,944,046	79,105,000	

[a]July 1, 1998 estimate.
[b]ASBR = (births / female population) × 1,000.
Sources: National Center for Health Statistics, *National Vital Statistics Reports*, Vol. 47, No. 25, Hyattsville, Maryland, 1999; U.S. Census Bureau, Population Division, Population Estimates Branch, Internet Release Date: October 29, 1999.

any other age group. This is a pattern found in many countries throughout the world.

Because birth rates vary so much by age, focusing on ASBRs yields a great deal of useful information. However, all of this detail makes it difficult to evaluate changes in fertility behavior over time and to compare differences among regions. The next measure summarizes the entire array of ASBRs and facilitates such comparisons.

Total Fertility Rate

The total fertility rate (TFR) is the sum of all of the individual ASBRs. When ASBRs are computed for one-year age groups, the TFR is calculated as

$$TFR = \sum ASBR_x$$

When age groups are defined in five-year intervals—which is more common—the TFR is calculated by multiplying the sum of the ASBRs by five (to account for the fact that females spend five years in each age group):

$$TFR = 5 \sum {}_5ASBR_x$$

Both approaches lead to similar results; TFRs based on five-year age groups are generally about the same as those based on one-year age groups (Shryock & Siegel, 1973).

The TFR can be interpreted as the number of children that a hypothetical cohort of 1,000 women would have during their lifetimes if none died and if their fertility behavior at each age conformed to a given set of ASBRs. Using the data from Table 5.1, we can calculate the TFR for the United States in 1998 as

$$\text{TFR} = 5\,(1.0 + 51.1 + 111.2 + 116.0 + 87.5 + 37.4 + 7.3 + 0.4) = 2,060$$

TFRs are often expressed as the average number of births per woman, rather than as the total number of births per 1,000 women. Using the example shown above, women just entering their childbearing years in 1998 would have an average of 2.06 births by the time they stopped having children, if none died and if 1998 ASBRs remained constant.

The TFR is similar to life expectancy at birth (e_0) in that both measures use hypothetical cohorts and both assume that a given set of age-specific rates will continue indefinitely. One measure shows the average number of children a cohort of women would have if a given set of ASBRs persisted throughout their lifetimes. The other shows the average life span that a cohort of newborn babies would have if a given set of ASDRs persisted throughout their lifetimes. Because they have clear intuitive meanings and are unaffected by the age-sex structure of a population, both measures are widely used for making comparisons among regions and over time.

Further refinements to these measures could be made. The gross reproduction rate (GRR) is similar to the TFR but focuses on female births rather than total births. The net reproduction rate (NRR) adjusts the GRR to account for survival rates at each age. These measures are useful for many analytical purposes but are not commonly used for population projections. Discussions of these and other fertility measures can be found in Newell (1988), Shryock and Siegel (1973), and Smith (1992).

Child–Woman Ratio

One additional measure is sometimes used in population projections, the child–woman ratio (CWR):

$$\text{CWR} = (P_{0-4} \,/\, FP_{15-44})\,1,000$$

where P_{0-4} is the number of children aged 0–4 and FP_{15-44} is the number of women aged 15–44. For example, there were 85,722 children aged 0–4 in Maine in 1990, and 286,473 women aged 15–44, yielding a CWR of 85,722 / 286,473 = 0.2992, or 299 children for every 1,000 women aged 15–44. In Utah, the CWR was 169,633 / 394,331 = 0.4302, or 430 children for every 1,000 women aged 15–44.

The CWR is neither a rate nor a true fertility measure. It is simply a ratio of one population subgroup to another. It incorporates the effects of past mortality and migration patterns, as well as past fertility behavior. In contrast to the other fertility measures discussed before, it does not require any data specifically related to births. This would be a shortcoming for many analytical purposes but can be very useful for geographic areas that lack vital statistics data. The CWR ratio is often used in demographic analyses of less developed countries. As we show in Chapter 7, it can also be used in more developed countries for projections of small areas that lack birth data.

TWO PERSPECTIVES: PERIOD AND COHORT

Fertility can be viewed from either of two perspectives; each is useful for particular purposes. The period perspective is cross-sectional, focusing on births during a particular period (e.g., one year). All of the fertility measures discussed above are period measures: CBR, GFR, ASBR, and TFR are based on the number of births that occurred during a year (or an average of several years) and the size of the midyear population. The cohort perspective, on the other hand, is longitudinal, focusing on the cumulative fertility behavior of a particular cohort of women (e.g., those born in 1950) as they pass through their childbearing years. Each perspective has its advantages and disadvantages, and each has its proponents and critics. It is important to recognize the differences and similarities between these two perspectives before considering techniques for projecting fertility rates.

Defining the Relationship

Tables 5.2 and 5.3 illustrate the relationship between period and cohort fertility. Table 5.2 shows annual ASBRs and TFRs at five-year intervals from 1940

Table 5.2. Age-Specific Birth Rates and Total Fertility Rates, United States, 1940–1995

Year	Age of mother								TFR
	10–14	15–19	20–24	25–29	30–34	35–39	40–44	45–49	
1940	0.7	54.1	135.6	122.8	83.4	46.3	15.6	1.9	2.30
1945	0.8	51.1	138.9	132.2	100.2	56.9	16.6	1.6	2.49
1950	1.0	81.6	196.6	166.1	103.7	52.9	15.1	1.2	3.09
1955	0.9	90.5	242.0	190.5	116.2	58.7	16.1	1.0	3.58
1960	0.8	89.1	258.1	197.4	112.7	56.2	15.5	0.9	3.65
1965	0.8	70.5	195.3	161.6	94.4	46.2	12.8	0.8	2.91
1970	1.2	68.3	167.8	145.1	73.3	31.7	8.1	0.5	2.48
1975	1.3	55.6	113.0	108.2	52.3	19.5	4.6	0.3	1.77
1980	1.1	53.0	115.1	112.9	61.9	19.8	3.9	0.2	1.84
1985	1.2	51.0	108.3	111.0	69.1	24.0	4.0	0.2	1.84
1990	1.4	59.9	116.5	120.2	80.8	31.7	5.5	0.2	2.08
1995	1.3	56.8	109.8	112.2	82.5	34.3	6.6	0.3	2.02

Source: U.S. Bureau of the Census, *Statistical Abstract of the United States: 1961, 1985, and 1998*, Washington, DC, 1961, 1985, and 1998.

Table 5.3. Age-Specific Birth Rates and Cohort Fertility Rates,
United States, Birth Cohorts 1920–1924 to 1950–1954

	Age of mother						
Year of birth	15–19	20–24	25–29	30–34	35–39	40–44	CFR
1920–24	54.1	138.9	166.1	116.2	56.2	12.8	2.72
1925–29	51.1	196.6	190.5	112.7	46.2	8.1	3.03
1930–34	81.6	242.0	197.4	94.4	31.7	4.6	3.26
1935–39	90.5	258.1	161.6	73.3	19.5	3.9	3.03
1940–44	89.1	195.3	145.1	52.3	19.8	4.0	2.53
1945–49	70.5	167.8	108.2	61.9	24.0	5.5	2.19
1950–54	68.3	113.0	112.9	69.1	31.7	6.6	2.01

Source: U.S. Bureau of the Census, *Statistical Abstract of the United States: 1961, 1985, and 1998*,
Washington, DC, 1961, 1985, and 1998.

to 1995 for women in the United States. The rows summarize age-specific fertility behavior for each year; the columns show the changes in ASBRs and TFRs over time. Looking across each row shows the typical inverted U-shaped relationship between ASBRs and age: rates increase from the teens to the twenties and decline thereafter. Looking down each column shows the large increases and declines in fertility rates that occurred during the baby boom and bust. Thus, each row provides a snapshot of fertility behavior at a particular point, and comparing rows shows how those snapshots have changed over time.

A different picture emerges if we look at Table 5.2 diagonally instead of by rows and columns. Consider females born in 1925–1929, for example. They were aged 10–14 in 1940, 15–19 in 1945, 20–24 in 1950, and so forth. The ASBRs for this cohort were 0.7 at age 10–14, 51.1 at 15–19, 196.6 at 20–24, and so forth. If we add up all these ASBRs, multiply by 5, and divide by 1,000, we get the cohort fertility rate (CFR):

$$CFR = 5 (0.7 + 51.1 + 196.6 + 190.5 + 112.7 + 46.2 + 8.1 + 0.3)/1,000 = 3.03$$

The CFR calculated in this manner is sometimes called the *cumulative fertility rate* or *completed family size*. In contrast to the TFR defined earlier, the CFR is a measure of the actual fertility behavior of a real cohort of women over their lifetimes. Table 5.3 shows the ASBRs taken from the diagonals of Table 5.2 and the associated CFRs. (Ages 10–14 and 45–49 were excluded to increase the number of cohorts for which CFRs could be calculated.) The impact of the baby boom and bust is clearly evident, but changes in CFRs over time are considerably smaller than the changes in TFRs shown in Table 5.2.

It is interesting to note that, whereas TFRs increased between 1975 and 1995, CFRs continued their post-baby boom decline. This suggests that the very low period fertility rates observed during the 1970s were caused partly by timing factors; that is, by the decisions of many women to delay childbearing until older ages (e.g., Bongaarts & Feeney, 1998; Lesthaeghe & Willems, 1999). One indication of delayed childbearing is that the ASBR for women aged 25–29 is now higher than the ASBR for women aged 20–24, after being much lower during the 1950s and 1960s. It is also noteworthy that the TFR for 1995 and the CFR for the 1950–1954 birth cohort—which had completed its childbearing by 1995—were virtually the same: 2.02 and 2.01, respectively. We will discuss the implications of these observations later in this chapter.

Assessing the Issues

Which is better for studying fertility, the period perspective or the cohort perspective? It has been argued that the cohort perspective is superior for many purposes because it more accurately describes the sequential nature of childbearing (e.g., Ryder, 1965, 1986, 1990). Cohorts are socially and demographically distinct, and their distinctiveness—including fertility attitudes and behavior—tends to persist over time. Most theories of fertility focus on completed family size rather than on the timing of births. In addition, cohort measures change more smoothly over time than period measures; this is generally considered an advantage. By focusing on the ASBRs of a particular cohort of women as they pass through their childbearing years, the cohort perspective picks up age, period, and cohort influences on fertility.

From the period perspective, it is not clear whether year-to-year changes in fertility rates reflect changes in long-term fertility behavior or a shift in the timing of births. An example from Japan illustrates this problem. Many Japanese believed that a girl born during the Year of the Fiery Horse (1966) would grow up to be ill-tempered and no one would want to marry her (Weeks, 1999). The TFR dropped from 2.2 in 1965 to 1.6 in 1966 but rose to 2.1 in 1967. The dramatic drop in 1966 clearly reflected changes in the timing of births, not their ultimate number. Wars, economic recessions, changes in government policies, and other events may cause similar (if less dramatic) shifts in the timing of births. These shifts may have a substantial impact on period fertility measures but little or no impact on cohort measures.

Proponents of the cohort perspective view cohorts as the vehicles of causation, whereas periods simply reflect the consequences of changes in cohort behavior. In this view, each birth cohort is unique, developing its own ideas, values, sentiments, vocabulary, and style as its members age together and experience the same events, institutions, economic conditions, and social norms at various stages of their lives. As an Arab proverb puts it, "Men resemble the times more than they do their fathers" (quoted in Ryder, 1965, p. 853). Attitudes regarding sexuality, contraception, and ideal family size are strongly affected by "the times."

Many studies of fertility behavior have been based on a cohort perspective (e.g., Bloom & Trussell, 1984; Evans, 1986; Lesthaeghe & Surkyn, 1988; Ryder, 1986, 1990). Additional refinements can be made to permit more detailed analyses. One uses parity-progression ratios, which are the proportions of women at each level of childbearing (e.g., no children, one child, two children, etc.) who go on to have at least one more child. Another incorporates the length of the time interval between births. A third focuses on marital fertility by combining marriage probabilities with parity-progression ratios and birth intervals for married women. These refinements lead to insights such as changes in the incidence of childlessness or of very large families that might not otherwise be apparent.

Studies based on a cohort perspective are not without problems, of course. First, cohort analyses require a great deal of birth and population data. Such data may be difficult to obtain for states and are often unobtainable for counties and subcounty areas. Even when available, data for small areas may be unreliable due to small population sizes. When marriage rates, parity-progression ratios, and birth intervals are incorporated, cohort analyses become even more data-intensive. Period analyses require much less data.

Second, complete cohort fertility data become available only after women reach age 45 or 50. Data for younger cohorts must either be excluded from the analysis or estimated indirectly. By contrast, period data for all age groups become available as soon as vital statistics and population data are tabulated.

Third, cohorts change over time because of deaths and migration. The members of the current cohort of women aged 45–49 who live in a particular area may be quite different from those born there 45–49 years ago, especially in rapidly growing states and local areas. Thus, the foundation of cohort analysis changes over time.

Finally, the theoretical basis of the cohort perspective—that each birth cohort is unique and that this uniqueness persists over time—can be questioned. Some researchers believe that there is little empirical evidence to support this claim (e.g., Bhrolchain, 1992). In particular, all cohorts appear to respond similarly to the factors affecting fertility behavior during any particular period of time. Figure 5.2 illustrates this point. ASBRs for all age groups between 15 and 39 in the United States rose during the 1940s and 1950s and declined during the 1960s and 1970s. A number of researchers have concluded that completed cohort fertility is not significantly different from an average of period fertility rates (e.g., Brass, 1974; Bhrolchain, 1992; Foster, 1990; Lee, 1974).

Period analyses capture year-to-year changes in fertility behavior and use readily available data. Cohort analyses are consistent with the sequential nature of childbearing and reflect cumulative fertility levels. Each perspective illuminates particular aspects of fertility behavior and provides insights that can be useful for projecting births. In the next section, we describe some of the assumptions and techniques that can be used to project fertility rates in a cohort-component projection model.

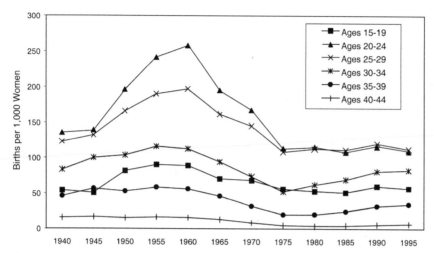

Figure 5.2. Birth rates by age of mother, United States, 1940–1995. (*Sources*: National Center for Health Statistics, *Monthly Vital Statistics Report*, Vol. 46, No. 1, Supplement 2, Rockville, Maryland, 1997; U.S. Bureau of the Census, *Statistical Abstract of the United States: 1961, 1985, and 1997*, Washington, DC, 1961, 1985, 1997)

APPROACHES TO PROJECTING FERTILITY RATES

Births in cohort-component models are typically projected by applying projected age-specific birth rates to projections of the female population by age. Occasionally, CBRs or GFRs are used instead of ASBRs. Most applications follow a period perspective, but a few follow a cohort perspective; some draw on a combination of the two. This section discusses the approaches that are most commonly used for projecting births and describes several models and techniques that can be used with each approach.

Using Period Rates

Constant Rates. One common approach to projecting fertility rates is to hold current ASBRs constant throughout the projection horizon (e.g., Day, 1996a; Treadway, 1997). These rates are often based on the most recent year of data available, but can also be based on an average of several recent years. In many applications, they are calculated separately for different race or ethnic groups.

Holding rates constant can be justified on either of two grounds. One is that future fertility rates are not likely to differ much from current rates. Compared to the first 75 years of the twentieth century, ASBRs did not change a great deal during the last 25 years. Furthermore, the TFR in 1995 was virtually identical to the CFR for women who had largely completed their childbearing by 1995 (i.e., children born from 1950–1954). Period and cohort measures thus provide virtually the same estimate of completed family size. The mean age of fertility was also about the same for both period and cohort measures. Therefore, a reasonable argument can be made that fertility behavior in the United States has stabilized, at least for the time being (Ryder, 1990). If this is true, the continuation of current period fertility rates is a legitimate projection technique.

A second justification for holding birth rates constant is the belief that neither the direction nor the magnitude of future changes can be predicted accurately. The argument here is not so much that current rates *will* remain constant, but rather that scientific theories and past history do not provide a reliable basis for predicting *how* those rates will change. If upward or downward movements are equally likely, current rates provide a reasonable forecast of future rates. This argument is supported by the generally lackluster forecasting performance of past fertility projections. The vast majority of demographers, for example, failed to foresee the timing and magnitude of either the baby boom or baby bust. This does not speak well for our ability to predict the course of future birth rates.

Trend Extrapolation.   Another projection approach is based on the extrapolation of historical trends. This approach is useful when birth rates have been changing steadily and are expected to continue to change steadily in the future. One early application of this approach calculated the rates of decline in ASBRs between the 1920s and early 1930s and extrapolated those rates into the future; adjustments allowed for a gradual slowing in the rates of decline over time (Thompson & Whelpton, 1933).

This approach may be useful for countries in the midst of the demographic transition from high to low fertility rates, but it is risky for countries in which this transition has already been completed. If no long-run trends are clearly discernible, recent changes in fertility rates may simply reflect short-run fluctuations. Extrapolating those changes into the future is likely to create large forecast errors, especially for long-range projections.

Time series techniques have also been used to project births or birth rates. Some time series models have focused directly on births, ignoring age-specific rates, the age structure of the population, and even the total size of the population (e.g., McDonald, 1981). Others have focused on summary fertility indexes which are then converted into schedules of ASBRs (e.g., Carter & Lee, 1986; Lee & Tuljapurkar, 1994). Time series techniques have two advantages compared to simple extrapolation methods: they use more historical information and provide prediction intervals. Projections from time series models, however, are strongly affected by the structure of the models themselves and by the changes in births or birth rates that occurred during the base period. These projections tend to move toward constant levels or converge toward constant oscillations within a fairly short time; they also tend to have very wide prediction intervals. Time series models seem to be more useful for short-range projections (e.g., less than five years) than for long-range projections, especially for countries that have already gone through their demographic transitions from high to low fertility (Land, 1986; Lee, 1993).

Targeting. The targeting approach is based on the assumption that birth rates in the population to be projected will converge over time toward those found in another population (i.e., the target). The target rates can be those currently observed in the target population or those projected for some future time. This approach is similar to the targeting of mortality rates described in Chapter 4.

The targeting approach can be implemented by forming ratios of current birth rates in the areas to be projected to current birth rates in the target population. Those ratios can then be projected to gradually move toward 1.0 over time. For example, one set of state projections produced by the Census Bureau assumed that ratios of state to national birth rates (by age and race) would move linearly to 1.0 by 2020 (U.S. Bureau of the Census, 1979). For each state, the projected ratios were applied to projected national birth rates to provide projections of state birth rates. A similar approach has been used to tie county birth rates to national rates (Nakosteen, 1989).

The notion of convergence has some intuitive appeal, given the homogenizing influences of popular culture, mass communication, and interregional migration. The analysis of historical fertility trends in the United States, however, affords only modest empirical support for the validity of this assumption. State and regional fertility rates have converged during some time periods but diverged during others (Alhburg, 1986; Isserman, 1985). The analyst must judge whether the assumption of convergence is reasonable for the areas to be projected. Recent Census Bureau projections have *not* assumed that state birth rates will converge toward national birth rates over time (e.g., Campbell, 1996; Wetrogan, 1990).

Synthetic Projection. Synthetic projections can be created by forming ratios of birth rates in one area to those in another and applying those ratios to the birth rates projected for the second area (called the *model population*). Although any two areas could be used, ratios are typically based on a smaller area and the larger area in which it is located (e.g., county/state or state/nation). ASBRs and GFRs are the measures most commonly used in constructing these ratios. For example, if (ASBR of County X) / (ASBR of State Y) = 1.1, the projected ASBR for County X would be obtained by multiplying the projected ASBR for State Y by 1.1.

Synthetic projection implicitly assumes that birth rates in the population to be projected will change at the same rate as birth rates in the model population; the analyst must decide whether this is a reasonable assumption. The synthetic approach is similar to targeting but does not assume that birth rates in different areas will converge over time. This approach has been widely used for state and local projections (e.g., Campbell, 1996; Wetrogan, 1990).

   Structural Models.   Fertility is one of the most thoroughly studied topics in demography. Using the tools of economics, sociology, psychology, anthropology, biology, and other disciplines, researchers have tried to determine why fertility rates are higher for some individuals and populations than for others, and why those rates have changed over time. Many theories of fertility behavior have been developed, critiqued, challenged, and revised (e.g., Becker, 1960; Easterlin, 1987; Freedman, 1975; Friedman, Hechter, & Kanazawa, 1994; Lesthaeghe, 1983; Mason, 1997; Schoen, Kim, Nathanson, Fields, & Astone, 1997). Empirical investigations have considered the effects of income, education, religion, wages, female labor force participation, marriage, race/ethnicity, and other variables on fertility. The insights gained through these studies have been incorporated into several structural models for projecting population (e.g., Ahlburg, 1986, 1999; Sanderson, 1999).

   Using structural models for projecting fertility rates has several problems (Isserman, 1985; Land, 1986). First, despite years of study, the determinants of fertility behavior are still not completely understood. Consequently, the theoretical foundations of structural models are somewhat weak. Second, using a structural model requires the availability of projections of the model's independent variables. Such projections are often not available, especially for small areas; when they are, they may not be very accurate. Third, projections from structural models are typically based on the assumption that the regression coefficients estimated from historical data will remain constant throughout the projection horizon; this is not likely to be true. Finally, the data needed to construct structural models for small areas are seldom available.

   Structural models of fertility are useful for many analytical purposes, including simulations and policy analysis. They may also be useful for population projections in countries going through the demographic transition from high to low birth rates. However, they are not particularly useful for state and local projections in countries that have completed this transition. We believe that the resources needed to develop structural models of fertility can be better used elsewhere. Chapters 9 and 10 provide a detailed discussion of structural models but focus on migration and total population rather than fertility.

   Judgment.   Many analysts base projections of fertility rates on their own judgment regarding future fertility trends. Expert judgment is typically informed by evaluating current fertility rates; past trends in those rates; historical and projected rates in other areas; and various economic, social, psychological, and cultural factors that might cause those rates to deviate from current levels. A judgmental approach is often used for state and local projections (e.g., Department of Rural Sociology, 1998; Smith & Nogle, 1999, 2000).

Using Cohort Rates

Birth Expectations Approach. The cohort perspective is more difficult to apply than the period perspective because complete data do not become available until a cohort has completed its childbearing years. Data for cohorts that have not completed—or even begun—their childbearing years literally do not exist. To fill this gap, data on the birth expectations of young women have sometimes been used (e.g., U.S. Bureau of the Census, 1975). The following steps summarize the procedures that may be followed for population projections using cohort fertility rates (Shryock & Siegel, 1973, p. 786):

1. Select the completed fertility levels for each cohort of women that will reach childbearing age during the period covered by the projections. These completed fertility levels will depend on the analyst's judgment regarding historical trends in completed family size and on survey data regarding birth expectations.

2. Assign age patterns to these completed fertility levels to represent the distribution of births by age of mother. These age patterns can be based on historical age patterns associated with different completed fertility levels and/or data on birth expectations.

3. Develop assumptions regarding completed fertility levels for each cohort of women that had reached childbearing age (15–44) before the launch year of the projections. These assumptions must be consistent with the cumulative fertility level already reached by each cohort before the launch year. Completed fertility levels may vary by cohort according to each cohort's historical fertility behavior and their birth expectations.

4. Based on the cumulative fertility rates for each cohort at each age, calculate the implied ASBRs for each future year covered by the projections.

5. Apply these ASBRs to the female population projected for each year to obtain the projected number of births.

As these steps illustrate, the cohort approach presents a number of challenges. Changes in the timing of births may occur even when completed fertility levels remain unchanged. In addition, projections for states and local areas are complicated by the lack of relevant data and by the effects of migration, which may have a significant impact on the composition of an area's population and its fertility behavior. Most important, final results depend heavily on the level of completed fertility projected for each cohort, but these levels are very uncertain. Birth histories of past cohorts do not necessarily provide reliable forecasts of births for current and future cohorts. Furthermore, surveys on birth expectations seldom provide reliable forecasts of a cohort's future fertility behavior (e.g., Ahlburg, 1982; Chen & Morgan, 1991; Schoen et al., 1997; Westoff & Ryder, 1977).

These problems make it difficult to use a cohort fertility model, especially for projections of states and local areas. Statistics Canada experimented with a cohort fertility model in the 1970s but abandoned the effort because of operational difficulties (George, 1998). The Census Bureau used a cohort perspective in several sets of national fertility projections (e.g., Spencer, 1989; U.S. Bureau of the Census, 1975, 1977), but its most recent set used period rates (Day, 1996a). In addition, some analysts have questioned whether the Census Bureau's earlier cohort fertility models were really much different from period models (e.g., Lee, 1974).

Easterlin Approach.   Many cohort fertility models have no clear theoretical basis for projecting fertility rates. Future values are based solely on birth expectations data and the analyst's judgment regarding the direction and pace of fertility trends. Using exactly the same information, different analysts may develop sharply contrasting views regarding future fertility rates. A cohort model developed by Richard Easterlin, however, provides a theoretical basis for predicting the direction (but not the magnitude) of changes in fertility rates (e.g., Easterlin 1961, 1978, 1987).

Easterlin's theory is based on the interplay between cohort size and relative economic status. According to Easterlin, a person's tastes and aspirations are formed largely during adolescence. These tastes and aspirations determine the standard of living adolescents hope to achieve when they become adults and set up their own households. The ability of a cohort to meet or exceed those aspirations (i.e., its relative economic status) affects many aspects of its behavior, including marriage and family size decisions.

Easterlin believes that there are clear advantages to being born into a small cohort. Because of their low numbers, members of small cohorts face less job competition than members of large cohorts. This affords them relatively high wages, low unemployment rates, and rapid advancement up the career ladder; that is, their relative economic status is high. In response, they tend to marry younger and have more children. In contrast, members of large cohorts face more job competition and experience relatively poor labor market conditions. In response, they tend to marry later and have fewer children.

Easterlin's theory provides an explanation of the baby boom and bust. The boom was caused by the favorable economic conditions experienced by young adults during the 1940s and 1950s. Because they were able to meet or exceed their lifestyle aspirations, members of the small cohorts born during the 1920s and 1930s tended to marry early and have relatively large numbers of children. The bust was caused by the relatively poor economic conditions experienced by the large cohorts born during the baby boom. Because they were unable to meet their lifestyle aspirations during the 1960s and 1970s, members of these cohorts tended to marry later and have fewer children.

An important feature of Easterlin's theory is that it provides a basis for predicting future fertility rates. Members of small birth cohorts are expected to have relatively large numbers of children, and members of large birth cohorts are expected to have relatively small numbers of children. Thus, birth rates are subject to long-term boom and bust cycles. Easterlin-type assumptions have been incorporated into several sets of national population projections (e.g., Ahlburg, 1983; Lee, 1976; U.S. Bureau of the Census, 1975), but have rarely (if ever) been used at the state or local level.

Easterlin's theory is intuitively appealing, and the predicted effects of cohort size on wages, unemployment, relative economic status, and other measures of well-being have been confirmed in a number of empirical studies (e.g., Berger, 1985; Smith & Welch, 1981; Wachter, 1975; Welch, 1979). An inverse relationship between the size of a birth cohort and its subsequent fertility has also been noted (e.g., Ahlburg, 1983; Lee, 1976). Of course, not everyone agrees that this theory presents a convincing explanation of the baby boom and bust or that it provides a useful basis for predicting future fertility rates (e.g., Butz & Ward, 1979; MacDonald & Rindfuss, 1978; Ryder, 1990).

Easterlin's theory provides an alternative to birth expectations data as a basis for predicting future fertility rates. However, actually using this theory for population projections is not easy. The theory focuses on *deviations* from long-term trends rather than the trends themselves, and predicts the direction of changes in fertility rates but not the magnitude of those changes. It focuses on completed cohort fertility rates, but not on the timing of births over a woman's reproductive life span. Although it provides a thought-provoking view of the future, Easterlin's theory does not solve the numerous operational problems involved in using a cohort approach for projecting fertility rates.

IMPLEMENTING THE FERTILITY COMPONENT

We believe most practitioners will be better off using the period approach rather than the cohort approach for projecting fertility rates. Data for states and small areas are more readily available for period measures than for cohort measures. More important, recent period data are available for all age groups, whereas complete cohort fertility data become available only after women have passed through their childbearing years. The theoretical basis of the cohort perspective—that each birth cohort is unique and that this uniqueness persists over time—is particularly questionable for small areas in which migration substantially alters the composition of a birth cohort over time.

The analyst should, however, keep the cohort perspective in mind when formulating assumptions about future fertility rates. Does the current TFR provide a reasonable forecast of completed family size over the course of the projection horizon? Do current ASBRs provide a reasonable forecast of the future distribution of those rates by age? If recent changes have occurred, do they reflect a shift in the long-run trend or were they simply short-run deviations from that trend? These questions can be answered only after considering trends in cohort fertility rates.

Sources of Data

Whenever possible, fertility projections should take into account the age and sex structure of the population. To construct ASBRs and TFRs, the analyst must have data on the number of births by age of mother (for the numerator) and counts or estimates of the population by age and sex (for the denominator). Birth data are available annually for states and counties from state offices of vital statistics. Because of fluctuations in fertility behavior over time (especially for small counties), it is common to use a three-year average of births (e.g., 1999–2001) instead of births from a single year (e.g., 2000) in constructing ASBRs.

Population counts by age and sex for states and counties are available every 10 years from the decennial census. For non-census years, population estimates must be used. State and county estimates by age and sex are available from the Census Bureau, state demographic agencies, and private data companies. Because population estimates are less accurate than decennial census data, ASBRs based on estimates are less reliable than those based on decennial census data.

For subcounty areas, the data needed for constructing ASBRs are seldom available. State offices of vital statistics generally do not tabulate fertility data for geographic regions below the county level. In addition, population data by age and sex for subcounty areas are uniformly available only for decennial census years. Consequently, ASBRs for cities, school districts, census tracts, traffic analysis zones, and other subcounty areas are rarely available.

What can the analyst do when ASBRs are not available at the subcounty level? The most common solution is simply to use county- or state-level ASBRs. This approach will often provide reasonable proxies for small-area ASBRs. In some circumstances, however, it is important to account for differences in race/ethnic composition between the small area and the county (or state) because birth rates differ considerably by race and ethnicity. This can be done in several ways. One is to make separate projections for each race/ethnic group, using the appropriate ASBRs for each group. Another is to calculate a weighted average of the ASBRs for race/ethnic groups, with the weights determined by the proportion of the area's female population in each race/ethnic group, by age. For example, if whites account for 75% of the female population of an age group and blacks for 25%, that ASBR could be calculated as

$$ASBR = 0.75 \text{ (white ASBR)} + 0.25 \text{ (black ASBR)}.$$

It is also possible to develop small-area birth rates using indirect estimation techniques. One application of indirect techniques starts with small-area population counts by age and sex from the most recent census (Bogue, 1998). Reverse survival techniques are applied to children aged 0–4 to develop estimates of the number of births that occurred during the five years immediately before the census. Adjustments for migration, census errors, and definitional mismatches are made by comparing births estimated in this manner with registered births in a larger area for which vital statistics data have been collected. Reverse survival techniques are applied to the female population aged 15–44 to provide an estimate of the number of women of childbearing age during the five years before the census. Finally, GFRs are calculated from these estimates of births and females aged 15–44.

Births can be projected using GFRs, but it is more common to use ASBRs. Various techniques can be used to derive estimates of ASBRs and TFRs from the GFRs described earlier. One is to form a ratio of the small-area GFR to the county (or state) GFR and apply that ratio to the set of county (or state) ASBRs. In essence, this technique adopts the age pattern of ASBRs observed for the county (or state) but adjusts those rates to be consistent with the GFR calculated specifically for each small area. Bogue (1998) describes another technique for transforming GFRs into TFRs and ASBRs.

Views of the Future

As was true for mortality projections, the methods and assumptions chosen for projecting fertility rates will be affected by the analyst's views of the future. Uncertainty regarding the course of future rates, however, is considerably greater for fertility than for mortality. Fertility rates fluctuated up and down during the twentieth century. Which way will future rates move? Will they change a lot or only a little? The analyst's answers to these questions will have a large impact on the choice of techniques and assumptions used in projecting fertility rates.

The starting point, then, must be the development of an informed outlook regarding future fertility trends. Demographic transition theories have offered explanations for the decline in fertility rates in more developed countries during the last two centuries, focusing on factors such as rising costs and declining economic benefits of children, declines in infant and child mortality, and changes in female roles in the household and society (e.g., Caldwell, 1982; Mason, 1997; Stolnitz, 1964). Economists, sociologists, anthropologists, psychologists, and others have offered explanations for the reasons that people continue to have children in post-transition societies (e.g., Becker, 1960; Friedman et al., 1994; Lesthaege, 1983; Schoen et al., 1997). The question was framed most starkly (and sardonically) by Joseph Schumpeter: "Why should we stunt our ambitions and impoverish our lives in order to be insulted and looked down upon in our old age?" (cited in Weeks, 1999).

Why indeed? What social, psychological, cultural, economic, and religious factors cause people to continue having children at the outset of the twenty-first century? The answer to this question lies at the heart of any discussion of future fertility rates. Have fertility rates in the United States reached some sort of equilibrium level at which they will remain for a long time to come? Will they fluctuate in long-term cycles, as suggested by Easterlin? Will there be another baby boom? Will fertility rates decline to the low levels currently found in a number of European countries? (See Golini, 1998, for a discussion of the lower bounds of fertility rates). The analyst must answer these questions before choosing projection techniques and assumptions.

Additional factors must be considered when making projections for states and local areas. Income, education, and occupation—as well as race and ethnicity—vary considerably from one area to another. These characteristics often have significant effects on fertility rates. The presence of a college, prison, military base, or other institution may also have a substantial impact on aggregate fertility measures. Differences in fertility rates are much greater among subcounty areas than among states or even counties (Bogue, 1998). Realistic projections will be possible only if these differences are accounted for in the projection process.

We suggest that the analyst thoroughly review the underlying data before developing fertility assumptions. Birth rates by age of mother should be examined to determine whether any unusual patterns are present and, if so, what caused them. Trends in TFRs and individual ASBRs must be considered. If large institutional populations are present, their impact on ASBRs must be accounted for. Changes in race, ethnic, and socioeconomic characteristics must be considered and decisions made regarding their impact on future fertility rates (e.g., will rates for whites and blacks converge over time?). Especially for small areas, the quality of the underlying fertility and population data must be evaluated and adjustments made if errors are found. Even the most prescient set of assumptions will be useless if there are errors in the underlying data.

Examples

The following examples illustrate several different ways to project fertility rates. The first is a set of national projections produced by the Census Bureau in the mid-1970s (U.S. Bureau of the Census, 1975). These projections were based on a cohort fertility model, using historical fertility data and birth expectations data from 1971–1974. Three fertility assumptions were made. The medium assumption projected an ultimate cohort fertility rate of 2.1, and the low and high assumptions projected ultimate rates of 1.7 and 2.7, respectively. Cohort fertility rates for whites and blacks were projected to converge over time, with the ultimate rates achieved first by the 1965 birth cohort for whites and by the 1970 birth cohort for blacks. Period ASBRs for the years before reaching the ultimate rates were

Table 5.4. Projected Age-Specific Birth Rates and Total Fertility Rates, United States, 1975–2015

Assumption–Year	Age of mother								TFR
	10–14	15–19	20–24	25–29	30–34	35–39	40–44	45–49	
Low–1975	1.0	54.3	110.0	99.5	49.1	20.0	5.0	0.4	1.70
Low–1995	0.2	38.8	116.5	114.4	50.4	16.6	4.0	0.3	1.71
Low–2015	0.2	37.3	116.8	117.8	50.1	14.2	3.3	0.2	1.70
Medium–1975	1.1	58.5	120.1	107.4	51.4	20.6	5.2	0.3	1.82
Medium–1995	0.3	47.5	143.8	142.0	63.0	20.7	4.9	0.3	2.11
Medium–2015	0.3	46.1	144.4	145.5	61.9	17.6	4.0	0.3	2.10
High–1975	1.1	65.4	131.1	117.0	54.7	21.7	5.4	0.4	1.98
High–1995	0.4	60.4	184.3	183.0	81.6	26.7	6.1	0.4	2.71
High–2015	0.4	59.2	185.6	198.1	79.6	22.6	5.2	0.4	2.70

Source: U.S. Bureau of the Census, Population Projections of the United States, 1975 to 2050, *Current Population Reports*, P-25 No. 601, Washington, DC, 1975.

calculated by interpolating linearly between the values for 1973 (the last year for which estimates were available) and their ultimate values. They were adjusted so that observed and projected rates would be consistent with completed cohort fertility rates. Projected ASBRs and TFRs for 1975, 1995, and 2015 are shown in Table 5.4.

It is interesting to compare these projections with a set produced more recently (Day, 1996a). The more recent projections were based on a period fertility approach rather than a cohort approach. The reason for this change was that "since the end of the baby boom, completed cohort fertility has remained about the same. Therefore, there appears to be no reason to assume a change from current fertility levels for any race-ethnic group" (Day, 1996a, p. 27). The starting point for these projections was a set of ASBRs calculated for five race-ethnic groups: Hispanic, non-Hispanic white, non-Hispanic black, non-Hispanic American Indian, and non-Hispanic Asian. These rates were based on fertility data from 1990–1992 and population data for July 1, 1991. The population data were adjusted for net census coverage error by using demographic analysis.

These projections used three fertility assumptions. Under the medium assumption, projected ASBRs remain constant for each race-ethnic group; that is, the projected rates did *not* converge over time. Because the five race-ethnic groups were growing at different rates, however, the medium assumption produced overall ASBRs and TFRs that increased gradually over time (see Table 5.5). Under the high fertility assumption, ASBRs for each race-ethnic group were projected to rise by 15% by 2010. In the low series they were projected to decline by 15%.

Table 5.5. Projected Age-Specific Birth Rates and Total Fertility Rates,
United States, 1995–2050

Assumption–Year	Age of mother								TFR
	10–14	15–19	20–24	25–29	30–34	35–39	40–44	45–49	
Low–1995	1.4	59.4	115.4	117.8	78.9	31.6	5.6	0.3	2.06
Low–2020	1.4	55.5	103.0	102.8	69.1	28.2	5.3	0.3	1.82
Low–2050	1.6	59.3	108.9	105.7	70.9	29.8	5.7	0.3	1.91
Medium–1995	1.4	59.4	115.4	117.8	78.9	31.6	5.6	0.3	2.06
Medium–2020	1.6	64.8	121.0	120.9	81.3	33.2	6.2	0.3	2.14
Medium–2050	1.8	69.7	128.2	124.2	83.8	34.9	6.8	0.4	2.24
High–1995	1.4	59.4	115.4	117.8	78.9	31.6	5.6	0.3	2.06
High–2020	1.9	74.1	139.1	139.0	93.5	38.2	7.2	0.3	2.47
High–2050	2.1	80.0	147.5	142.7	95.6	40.1	7.8	0.4	2.58

Source: J. Day (1996a). Population Projections of the United States, 1995 to 2050, U.S. Bureau of the Census, *Current Population Reports*, P25-1130, Washington, DC.

A comparison of Tables 5.4 and 5.5 shows that the TFRs for the medium projections are quite similar: 2.11 (1995) and 2.10 (2015) in Table 5.4 and 2.06 (1995) and 2.14 (2020) in Table 5.5. The age patterns of childbearing are considerably different, however. ASBRs for women aged 30–39 are consistently higher in Table 5.5 than Table 5.4, and ASBRs for women aged 20–29 are consistently lower. These differences reflect the trend toward delayed childbearing seen in recent decades.

The Census Bureau also produced a set of state projections consistent with each of these sets of national projections. The first set started with ASBRs by race for each state, based on data from 1970–1975 (U.S. Bureau of the Census, 1979). Ratios of state/national rates were constructed and were projected to move linearly toward 1.0, reaching that level by 2020. In other words, it was assumed that state ASBRs would converge toward national ASBRs over time. As a final step, state ASBRs were calculated by multiplying the interpolated ratios by the medium set of national ASBRs.

The second set of state projections started with ASBRs calculated for five race-ethnic groups for each state (Campbell, 1996). These rates were based on the annual average number of births in each state between 1989 and 1993. Rates were held constant throughout the projection horizon; that is, it was *not* assumed that state rates would converge toward national rates over time.

In these examples—and most other applications of the cohort-component method—births are projected by multiplying projected ASBRs by the projected female population. Chapter 7 provides illustrations of several ways this process can be carried out. A final example will be given here to show how births can be projected indirectly *without* using ASBRs.

Suppose that projections are to be made for census tracts, but there are no birth data at the tract level. What can be done? One option is to use county-level ASBRs for every census tract. However, if there is a great deal of socioeconomic and demographic diversity among census tracts, this may not be a reasonable approach. Another option is to develop a set of unique tract-specific ASBRs using the techniques suggested by Bogue (1998). Developing these rates, however, requires a great deal of time.

A third option is to construct a set of child–woman ratios and apply them to the projected female population aged 15–44. These ratios can be based on tract-specific data from the most recent census. To take a hypothetical example, suppose that a tract's year 2000 population included 400 children aged 0–4 and 1,100 women aged 15–44, yielding a child–woman ratio of 0.3636. Suppose that the projected female population aged 15–44 in 2005 is 1,200. Then, the population aged 0–4 projected for 2005 could be calculated as

$$1,200 \times 0.3636 = 436$$

In this example the child–woman ratio was held constant over time. This may not be a realistic assumption in some situations. An alternative assumption is that the child–woman ratio in the census tract changes at the same rate as the child–woman ratio in some larger area, such as the county, or converges toward some target level over time. For example, if the child–woman ratio for the county was 0.35 in 2000 and was projected to decline to 0.33 by 2005, the population aged 0–4 in 2005 could be projected as

$$1,200 \times 0.3636 \, (0.33 / 0.35) = 411$$

This adjustment allows the analyst to account for projected changes in fertility rates and age structure and still incorporate the impact of tract-level fertility characteristics. This approach offers a compromise between spending a large amount of resources developing tract-specific fertility rates and completely ignoring the existence of tract-level differences in fertility rates.

CONCLUSIONS

It has often been noted that fertility is the most problematic part of national population projections (e.g., Keyfitz, 1982; Ryder, 1990; Siegel, 1972). For example, Long (1989) found that differences in fertility assumptions accounted for more of the variation in long-range projections of the U.S. population than differences in either mortality or immigration assumptions. During the last 50 years, births have been more volatile and have had a larger impact on population growth than either deaths or foreign immigration. Clearly, fertility assumptions are critical to the preparation of national population projections.

For states and local areas, however, fertility is less important than migration in explaining differences in rates of population growth (e.g., Congdon, 1992; Smith & Ahmed, 1990). At the subnational level, migration rates vary more from place to place and change more dramatically over time than fertility rates. Fertility rates, however, display more variation than mortality rates. As a result, fertility rates can generally be forecasted more accurately than migration rates, but not as accurately as mortality rates. Fertility assumptions typically have more impact on state and local population projections than mortality assumptions, but not as much impact as migration assumptions.

Whereas differences in mortality assumptions have their largest impact at older ages, differences in fertility assumptions have their largest impact at younger ages. Changes in mortality and migration rates affect all age groups, but changes in fertility rates immediately affect only the youngest. Long (1989) found that fertility is the major cause of variability in national projections of the youngest age groups. In a study of county population projections, Isserman (1993) found that differences in fertility assumptions have a much larger impact on the youngest age group than on the population as a whole. Over time, of course, the effects of differences in fertility rates are cumulative, and fertility assumptions have a major impact on both the size and the age structure of the population.

CHAPTER 6

Migration

America is a nation of movers. In a typical year about one-sixth of the U.S. population changes residence, moving to a new location nearby or far away. Table 6.1 provides an overview of mobility and migration in the United States between 1997 and 1998. Of the 42.5 million movers in that year, about 64% moved to a different residence in the same county, 18% moved to a different county in the same state, 15% moved to a different state, and 3% moved to the U.S. from abroad (Faber, 2000).

Although U.S. mobility rates have declined a bit in recent decades, they remain high compared with those in most industrialized countries. Only Canada, Australia, and New Zealand have rates similar to those in the United States. These four countries share long-standing traditions of foreign immigration, cultural values that emphasize personal freedom, and public policies and housing markets that facilitate mobility (Gober, 1993). At current rates Americans will average 11–12 changes in residence during their lifetimes (Kulkarni & Pol, 1994).

Mobility rates and migration patterns vary from place to place. Almost two-thirds of the population of Nevada changed its place of residence between 1985 and 1990, compared with barely one-third of the population of West Virginia. More than 80% of the residents of Pennsylvania and 79% of the residents of Louisiana were living in their state of birth in 1990, compared with only 22% of the residents of Nevada, 30% of the residents of Florida, and 34% of the residents of Alaska and Arizona. Florida had 1,072,000 more interstate in-migrants than out-migrants between 1985 and 1990, whereas New York had 821,000 more interstate out-migrants than in-migrants (Hansen, 1993).

Differences in mobility rates and migration patterns are even greater at the county and subcounty levels than at the state level. In Florida, for example, 53% of the residents of Flagler County in 1990 had lived in a different county in 1985, compared with only 17% of the residents of Bradford County (U.S. Bureau of the Census, 1993b). More than 73% of the residents of the city of Appalachicola

Table 6.1. Geographic Mobility, United States, 1997–1998 (Numbers in Thousands)

| | Population aged 1 and older | Total movers | Same county | Different county | | Movers from abroad |
				Same state	Different state	
Number	265,209	42,507	27,082	7,867	6,335	1,203
Percent	100.0	16.0	10.2	3.0	2.4	0.5

Source: C. Faber (2000). Geographical Mobility, March 1997 to March 1998, U.S. Census Bureau, *Current Population Reports* P20-520, Washington, DC.

lived in the same housing unit in 1990 as in 1985; in Sun Valley, only 5% of the residents lived in the same housing unit in both years (U.S. Bureau of the Census, 1993b). As a result of dramatically different mobility and migration patterns, some places look much as they did a generation ago, whereas others have apartment complexes and gated residential communities springing up seemingly overnight in what once were cornfields and cow pastures.

Differences in mobility rates and migration patterns—combined with their potential for rapid change—make forecasting migration a difficult task. In this chapter, we discuss a variety of concepts, measures, and definitions of mobility and migration (see Box 6.1). We describe data sources that provide historical information on migration trends and characteristics. To set the stage for developing assumptions regarding future migration patterns, we consider the determinants of migration and some of the characteristics of migrants. Then, we describe the data and techniques that can be used to project future migration, focusing on issues with particular importance for states and local areas. We close with an assessment of the impact of migration on population projections.

CONCEPTS, MEASURES, DEFINITIONS

Place of Residence

For many Americans the simple question, "Where do you live?" does not have an equally simple answer. Many retirees spend their summers in New York, Illinois, or Minnesota and their winters in Florida, Texas, or Arizona. Itinerant farm workers follow the harvest from place to place over the year. Dual-career couples may have one spouse who works in New York City and the other in Washington, DC. Children of divorced couples may spend alternating weeks or months with each parent. Some college students spend the school year in Madison, Wisconsin, or Ithaca, New York, and the summer in Milwaukee or Buffalo. Members of rock bands and professional basketball teams spend much of the year moving from city to city. Where do these people live?

Box 6.1

Some Common Migration Definitions

Mover. A person who changes his/her place of usual residence from one address (e.g., house or apartment) to another.

Migrant. A person who changes his or her place of usual residence from one political or administrative area to another. All migrants are movers, but not all movers are migrants.

Gross Migration. The movement of migrants into or out of an area.

Net Migration. The difference between the number of in-migrants and the number of out-migrants.

Internal (Domestic) Migration. Migration from one place to another within the same country.

International (Foreign) Migration. Migration from one country to another.

Immigrant. A citizen or permanent resident of one country who moves into the reference country to establish permanent residence there.

Emigrant. A citizen or permanent resident of the reference country who moves to a different country to establish permanent residence there.

Migration Interval. The period of time over which migration is measured.

The answer to this question is crucial because mobility and migration typically refer to changes in a person's place of usual residence. In the United States, a person's usual residence is defined as "the place where he or she lives and sleeps most of the time or the place where the person considers to be his or her usual home" (U.S. Bureau of the Census, 1992, p. D-1). Under this definition, people who have two or more homes are counted at the one they consider their usual residence. College students are counted at the place they are staying while attending school and members of the armed forces are counted at the location where they are based. If a person has no usual place of residence, he or she is counted at the place he or she was staying on census (or estimation) day.

Because of this focus on changes in usual (or "permanent") residence, traditional measures of geographic mobility and migration miss many types of temporary population movements such as commuting to work, splitting time between weekday and weekend homes, seasonal migration, business trips, vacations, and life on the road in a recreational vehicle. These nonpermanent moves are large in number and can have a substantial impact on both the sending and receiving regions (e.g., Behr & Gober, 1982; McHugh, Hogan, & Happel, 1995;

Smith, 1989). Unfortunately, the data needed to measure and evaluate these moves are sorely lacking, and temporary population movements have received relatively little attention in the scholarly literature.

Population projections typically refer solely to the permanent resident population of a state, county, or subcounty area. Thus, the measures of mobility and migration discussed in this chapter focus solely on changes in one's place of permanent residence. Temporary, cyclical, and seasonal migration are important research topics, but lie outside the scope of this book.

Mobility and Migration

Although *mobility* can refer to changes in social or occupational status, to a demographer it generally means changes in geographic location. The *residential mobility rate* can be defined as the proportion of the population that moves from one housing unit to another during a particular time period (Long, 1991, p. 133). It measures the rate at which people move from one place to another, regardless of whether the move is to an apartment across the street or a house across the country.

Migration, on the other hand, refers to moves across some type of political or administrative boundary (Shryock & Siegel, 1973, p. 617). This distinction is meant to differentiate between local moves within a particular community and moves from one community to another (or, more broadly, to differentiate between short- and long-distance moves). Migration and local mobility cannot be perfectly distinguished from each other, however, because "local community" has no rigid definition. Operationally, moves across state or county lines are almost universally deemed to be migration, although they may cover very short distances for people who live near those lines. Intracounty moves are even more difficult to classify. If a person moves from one town to another within the same county or from one neighborhood to another within the same town, does that move reflect migration or local mobility?

Distinctions between migration and local mobility are critical for some types of analyses but not for the topics addressed in this book. Our focus is on population projections for states, counties, and subcounty areas such as cities, school districts, census tracts, market areas, and traffic analysis zones. We define all moves into or out of the geographic areas that are projected as migration, regardless of the distance moved, the degree of change in the living environment, or the size of the area. Given this focus, there is no need to differentiate between migration and local mobility.

Length of Migration Interval

Migration data are typically derived from censuses, surveys, or administrative records that report the current place of residence and the place of residence at some earlier time. A *migrant* is a person whose current place of residence is different from his/her earlier place of residence. National statistical offices traditionally use either one- or five-year intervals for developing migration statistics (Long & Boertlein, 1990).

What length of interval is best? There is no definitive answer to this question. Migration data that cover different intervals simply reflect different aspects of the migration process. Short intervals pick up practically all moves but are heavily affected by chronic movers and moves that prove temporary. Long intervals cancel out some of the effects of chronic and temporary movers; consequently, they may provide a better measure of long-term population mobility. However, they miss the impact of multiple moves within the time interval and introduce measurement errors for people who are unable to recall accurately the timing or location of earlier moves. Long intervals also miss more people who move but die than do short intervals. For any particular project, the appropriate length of interval will depend on the availability of data and the purposes for which the data will be used. One-year data may provide better estimates of the number of *moves*, whereas five-year data may provide better estimates of the number of *movers*.

Because of the impact of multiple moves and deaths of migrants, migration data based on one length of interval (e.g., five years) are not always directly comparable to migration data based on an interval of a different length (e.g., one year). This lack of comparability has important implications for the production of population projections. Whereas birth and death data can be converted easily into intervals of different lengths, attempting to convert migration data can be a complex and somewhat capricious undertaking. We discuss this issue more fully later in this chapter.

Gross and Net Migration

Migration can be viewed from either of two perspectives. *Gross migration* is the movement of people into or out of an area; *net migration* is the difference between the two. Table 6.2 shows in-, out-, and net migration for every state and the District of Columbia between 1985 and 1990. These numbers refer strictly to internal migrants, or people who moved from one state to another within the United States. Although the decennial census collects data on immigration from abroad, it does not collect data on emigration to foreign countries. This makes it impossible to construct overall net migration estimates for states (or any other regions) using decennial census data.

Florida had 2,130,613 internal in-migrants and 1,058,931 internal out-migrants between 1985 and 1990, yielding a net migration balance of 1,071,682. During the same period, New York had 727,621 internal in-migrants and 1,548,507 internal out-migrants, a net migration balance of −820,886. These were the largest positive and negative internal net migration balances of any state. Twenty-two states had more internal in-migrants than out-migrants between 1985 and 1990, and 29 had more internal out-migrants than in-migrants.

Table 6.2. In-, Out-, and Net Migration for States, 1985–1990[a]

	In-migrants	Out-migrants	Net migration
Northeast			
Connecticut	291,140	342,983	−51,843
Maine	132,006	98,688	33,318
Massachusetts	444,040	540,772	−96,732
New Hampshire	191,130	129,070	62,060
New Jersey	569,590	762,123	−193,533
New York	727,621	1,548,507	−820,886
Pennsylvania	694,020	771,709	−77,689
Rhode Island	105,917	93,649	12,268
Vermont	74,955	57,970	16,985
Midwest			
Illinois	667,778	1,009,922	−342,144
Indiana	433,678	430,550	3,128
Iowa	194,298	288,670	−94,372
Kansas	272,213	295,663	−23,450
Michigan	473,473	606,472	−132,999
Minnesota	320,725	316,363	4,362
Missouri	448,280	420,223	28,057
Nebraska	141,712	181,662	−39,950
North Dakota	56,071	107,018	−50,947
Ohio	622,446	763,625	−141,179
South Dakota	69,036	91,479	−22,443
Wisconsin	307,168	343,022	−35,854
South			
Alabama	328,120	292,251	35,869
Arkansas	240,497	216,250	24,247
Delaware	94,129	68,248	25,881
District of Columbia	109,107	163,518	−54,411
Florida	2,130,613	1,058,931	1,071,682
Georgia	804,566	501,969	302,597
Kentucky	278,273	298,397	−20,124
Louisiana	225,352	476,006	−250,654
Maryland	531,803	430,913	100,890
Mississippi	193,148	220,278	−27,130
North Carolina	748,767	467,885	280,882
Oklahoma	279,889	407,649	−127,760
South Carolina	398,448	289,107	109,341
Tennessee	500,006	368,544	131,462
Texas	1,164,106	1,495,475	−331,369
Virginia	863,567	635,695	227,872
West Virginia	123,978	197,633	−73,655

Table 6.2. (*Continued*)

	In-migrants	Out-migrants	Net migration
West			
Alaska	105,605	154,090	−48,485
Arizona	649,821	433,644	216,177
California	1,974,833	1,801,247	173,586
Colorado	465,714	543,712	−77,998
Hawaii	166,953	187,209	−20,256
Idaho	137,542	157,121	−19,579
Montana	84,523	137,127	−52,604
Nevada	326,919	154,067	172,852
New Mexico	192,761	204,218	−11,457
Oregon	363,447	280,875	82,572
Utah	177,071	213,233	−36,162
Washington	626,156	409,886	216,270
Wyoming	62,286	118,979	−56,693

[a]Interstate migrants only.

Source: U.S. Bureau of the Census, *1990 Census Special Tabulations, County-to-County Migration Flows*, SP 312, Washington, DC, 1993.

Migration data can also be tabulated for specific place-to-place migration flows. For example, 361,295 Florida residents in 1990 had been living in New York in 1985, and 64,214 New York residents in 1990 had been living in Florida in 1985 (Hansen, 1993). Thus, the New York-to-Florida net migration flow from 1985 to 1990 can be calculated as 297,081. Specific state-to-state (or county-to-county) migration flows can be useful for analyzing the determinants and consequences of migration and for developing multiregional projection models.

Using gross rather than net migration data for population projections has several advantages (Smith & Swanson, 1998). First, gross migration is closer to the true migration process than net migration. Some people move into an area, some move out, and others stay put. Therefore, people may be classified as movers or nonmovers and as in-migrants or out-migrants, but there is no such thing as a "net migrant." Net migration is an accounting process rather than a migration process.

Second, focusing on net migration may mask the existence of large gross migration flows. For example, Indiana had net interstate migration of 3,128 between 1985 and 1990 (see Table 6.2). But this does not mean that only a few people were moving into or out of Indiana. Indiana attracted 433,678 in-migrants during this period, even as 430,550 out-migrants opted to leave. Gross migration data illuminate these moves, net migration data obscure them.

Third, gross migration data can be related to the size of the source population from which migrants originate, providing migration rates that approximate the probability of migrating. For example, Indiana's 430,550 out-migrants were drawn from a total state population of 5,459,000 in 1985 and its 433,678 in-migrants were drawn from a total U.S. population of 232.5 million (excluding Indiana). Because net migration simply reflects a residual, it has no identifiable source population and migration rates cannot be constructed that reflect migration probabilities. As we show later, this can have significant implications for population projections.

Fourth, when net migration is calculated from the demographic balancing equation as the difference between total population change and natural increase, it captures all the measurement errors found in birth, death, and total population data. These errors may be substantial (Isserman, Plane, & McMillen, 1982).

Finally, population projections based on net migration data may lead to unrealistic forecasts of future population. When in- and out-migration flows are projected separately, the projection model can account for differences in the demographic structures and rates of growth of origin and destination populations and for the different influences on each flow. When in- and out-migration flows are combined to form net migration, however, the model cannot account for these differences. As Isserman (1993), Plane (1993), and Smith (1986) demonstrate, projections based on net migration sometimes differ considerably from projections based on gross migration, particularly for rapidly growing areas. Although these differences are often small for five- or 10-year horizons, they can be large for longer horizons. Little empirical research has addressed this issue, but the possibility of unrealistic population forecasts seems to be greater for net migration models than for gross migration models.

Gross migration models clearly have a number of advantages over net migration models; however, net migration models also have several advantages. They require much less data and are considerably simpler (and cheaper) to apply than gross migration models. Perhaps more important, they can be used when the data required by gross migration models are difficult or impossible to obtain. Consequently, there are circumstances in which net migration models are more useful than gross migration models. We give examples of both types of projection models in Chapter 7.

Migration Rates

A fundamental methodological problem in constructing cohort-component projections is choosing the appropriate population base (i.e., the denominator) for calculating migration rates. Theoretically, the appropriate base for any rate is the population at risk of the occurrence of the event under consideration. For mortality and fertility, the choice is clear: For purposes of projection, the population at risk of dying or giving birth is the population of the state or local area being projected (adjusted to reflect the total number of person-years lived during the time period).

For migration rates, however, the choice is not so clear. What is the population at risk of migrating?

A number of studies have addressed this question. Most, however, focused primarily on whether the initial, terminal, or midpoint population should be used to calculate migration rates, and what adjustments for births, deaths, and migration during the time period should be made to estimate the total number of person-years lived (e.g., Hamilton, 1965; Thomlinson, 1962). These are important research issues, but they do not cover some of the critical questions regarding the appropriate population at risk.

In fact, most studies have simply used the population of the area under consideration as the denominator in constructing migration rates, regardless of whether those rates referred to in-migration, out-migration, or net migration. Yet the population of the area itself is clearly not the population at risk of in-migration; after all, those people are already living in the area. For net migration, the issue is even more difficult because net migration is a residual rather than an actual event; consequently, it has *no* true population at risk. Only for out-migration does the area under consideration represent the population at risk; therefore, only for out-migration can rates properly be interpreted as migration rates. For in-migration and net migration, rates calculated in this manner are simply migration/population ratios. They provide a measure of the contributions of migration to population size but provide no information on the propensity to migrate.

In-migration rates that approximate the probability of migrating can be developed, however, by basing them on the population of the area of origin rather than the area of destination. We illustrate the construction of such rates using 1985–1990 migration data for Florida.

Table 6.3 shows the number of interstate migrants who entered and left Florida between 1985 and 1990 by age and sex and the 1985 populations of Florida and the rest of the United States. Migration rates for each age-sex group are calculated by dividing the number of migrants by the mid-decade population. For example, there were 119,326 female in-migrants aged 25–29 in 1990, and 10,185,380 females aged 20–24 who lived in the rest of the United States in 1985, yielding an in-migration rate of

$$119,326 / 10,185,380 = 0.01172$$

or 11.7 migrants per 1,000 persons. For the same age-sex group, there were 76,219 out-migrants and a mid-decade population of 356,137 in Florida, yielding an out-migration rate of

$$76,219 / 356,137 = 0.21402$$

or 214.0 migrants per 1,000 persons. Out-migration rates are much larger than in-migration rates because the denominators for out-migration rates are much smaller than the denominators for in-migration rates.

Table 6.3. Interstate Migration Rates by Age and Sex, Florida, 1985–1990

Age in 1985	Age in 1990	In-migrants	Out-migrants	1985 population			In-migration rate[b]	Out-migration rate[c]
				United States	Florida	Adjusted United States[a]		
Male								
0–4	5–9	71,888	44,783	9,127,012	377,854	8,749,158	8.22	118.52
5–9	10–14	63,266	33,288	8,528,135	350,192	8,177,943	7.74	95.06
10–14	15–19	69,952	41,791	8,718,552	319,998	8,398,554	8.33	130.60
15–19	20–24	104,848	65,980	9,553,088	343,077	9,210,011	11.38	192.32
20–24	25–29	127,712	87,519	10,723,099	367,747	10,355,352	12.33	237.99
25–29	30–34	111,658	75,937	10,848,411	439,816	10,408,625	10.73	172.66
30–34	35–39	86,863	52,074	9,944,520	432,744	9,511,776	9.13	120.33
35–39	40–44	73,821	37,615	8,677,614	393,172	8,284,442	8.91	95.67
40–44	45–49	53,626	25,209	6,916,485	353,242	6,563,243	8.17	71.36
45–49	50–54	43,163	16,854	5,672,096	285,624	5,386,472	8.01	59.01
50–54	55–59	46,406	12,597	5,244,376	244,685	4,999,691	9.28	51.48
55–59	60–64	64,701	11,609	5,326,773	238,148	5,088,625	12.71	48.75
60–64	65–69	73,886	11,359	5,057,189	260,993	4,796,196	15.41	43.52
65–69	70–74	42,605	9,543	4,180,381	276,096	3,904,285	10.91	34.56
70–74	75–79	21,069	7,456	3,158,466	252,503	2,905,963	7.25	29.53
75–79	80–84	9,141	5,292	2,128,950	220,121	1,908,829	4.79	24.04
80+	85+	5,394	4,049	1,924,535	312,463	1,612,072	3.35	12.96
Total		1,069,999	542,955	115,729,712	5,468,475	110,261,237	9.70	99.29

Female							
0–4	68,904	42,408	8,714,687	359,841	8,354,846	8.25	117.85
5–9	58,783	32,416	8,136,657	334,213	7,802,444	7.53	96.99
10–14	63,016	36,545	8,308,853	305,701	8,003,152	7.87	119.54
15–19	93,540	55,560	9,174,201	326,439	8,847,762	10.57	170.20
20–24	119,326	76,219	10,541,517	356,137	10,185,380	11.72	214.02
25–29	106,309	65,797	10,822,648	433,120	10,389,528	10.23	151.91
30–34	86,241	45,624	10,080,899	437,206	9,643,693	8.94	104.35
35–39	71,614	32,336	8,926,099	403,625	8,522,474	8.40	80.11
40–44	54,394	21,040	7,171,005	366,699	6,804,306	7.99	57.38
45–49	47,340	15,097	5,933,498	303,177	5,630,321	8.41	49.80
50–54	54,242	12,851	5,609,230	264,631	5,344,599	10.15	48.56
55–59	73,207	13,655	5,901,847	263,416	5,638,431	12.98	51.84
60–64	70,856	15,614	5,849,343	307,475	5,541,868	12.79	50.78
65–69	40,755	14,433	5,162,821	337,482	4,825,339	8.45	42.77
70–74	23,903	13,707	4,356,975	313,915	4,043,060	5.91	43.66
75–79	14,834	11,591	3,381,940	284,786	3,097,154	4.79	40.70
80+	13,350	11,083	4,122,106	456,000	3,666,106	3.64	24.30
Total	1,060,614	515,976	122,194,326	5,853,863	116,340,463	9.12	88.14

[a]Adjusted U.S. population = U.S population − population in Florida.
[b]In-migration rate = (In-migrants/adjusted U.S. population) × 1,000.
[c]Out-migration rate = (Out-migrants/population in Florida) × 1,000.
Sources: Custom tabulation of U.S. Bureau of the Census, *1990 Census Special Tabulations, County-to-County Migration Flows*, SP 312, Washington, DC, 1993; F. Hollman (1993). U.S. Population Estimates by Age, Sex, Race, and Hispanic Origin: 1980 to 1991, *Current Population Reports*, P25-1095, U.S. Bureau of the Census, Washington, DC.

Migration rates calculated in this manner show the proportion of the population that lived in an area at one time and lived in a different area at a later time. They typically refer to one- or five-year intervals, depending on whether the migration data cover a single year or a five-year period. They are not perfect measures of the probability of migrating because they do not account for deaths during the period. In addition, they do not pick up the effects of multiple moves that occurred during the migration interval. The latter issue is not a problem for population projections, however, because the focus is on where people lived at the *end* of the migration interval, not on how many moves they made *during* the interval.

Although the rationale for constructing gross migration rates is clear, the same cannot be said for net migration rates. Because net migration is simply a residual, rates that reflect migration probabilities cannot be constructed. We follow conventional terminology by referring to net migration ratios as "rates," but the reader is reminded that they are not rates in the true sense of the word.

There are two basic methodological questions regarding the construction of net migration rates: (1) Should the denominator reflect the population at the beginning, middle, or end of the migration interval? and (2) Should the denominator reflect the population of the region itself or the population of the rest of the country?

With respect to the first question, we favor using the population at the beginning of the interval as the denominator because it is unaffected by migration during the interval and corresponds to the launch-year population used for making projections. It is also common to use the beginning population "survived" to the end of the migration period using the appropriate survival rates (e.g., Irwin, 1977; Pittenger, 1976). This approach is a bit more complicated to apply but has the advantage of accounting explicitly for deaths of migrants. Both approaches are acceptable and yield similar results. The most important thing to remember is that migration rates must be applied in a manner consistent with the way they were computed; for example, if rates were based on the population at the beginning of the migration interval, they must be applied to the population at the beginning of the projection interval.

The answer to the second question depends on the characteristics of the area to be projected. For regions that are losing population or growing fairly slowly, we favor using the population of the region itself for constructing net migration rates. For regions that are growing very rapidly, however, there are advantages to using the population of the rest of the country. Because there are more in-migrants than out-migrants, the rest of the country rather than the region itself is the base population for the larger number of total migrants; this provides a theoretical justification for this choice. Perhaps more important, using the national population as the denominator reduces the impact of very high migration rates, which may be more realistic for projections of rapidly growing areas. Smith (1986) provides a more detailed discussion of these issues.

We illustrate the calculation of net migration rates using the data in Table 6.3. For females aged 25–29 in 1990, net interstate migration for 1985-1990 can be calculated as

$$119,326 - 76,219 = 43,107$$

This can be expressed as a rate by dividing it by the appropriate population. The most common practice is to use the population of the region under consideration. Using the female population aged 20–24 living in Florida at the beginning of the time interval as the denominator, the net migration rate for females aged 25–29 in 1990 can be calculated as

$$43,107 / 356,137 = 0.12104$$

or 121.0 per 1,000 persons.

Net migration rates can also be calculated using the population of the rest of the country as the denominator. Following this approach the net migration rate for females aged 25–29 in 1990 can be calculated as

$$43,107 / 10,185,380 = 0.00423$$

or 4.2 per 1,000 persons. We believe that migration rates based on the first approach are acceptable for projections in places that have slow or moderate growth rates, but the second is better for projections in rapidly growing places. When the first approach is used, migration is projected by multiplying the migration rates by the population to be projected; when the second approach is used, migration is projected by multiplying the migration rates by the national population (minus the population of the area to be projected).

International and Internal Migration

A final distinction is between international (or foreign) and internal (or domestic) migration. *International migration* refers to moves from one country to another, whereas *internal migration* refers to moves from one place to another within a particular country. The data shown in Tables 6.2 and 6.3 refer solely to internal migrants. Although internal migration has more impact than international migration on population growth and demographic change in most states and local areas, international migration is growing in importance and has a substantial impact in some places. People who move into a country are called *immigrants*, and people who leave a country are called *emigrants*.

It is no exaggeration to describe the United States as a nation of immigrants. More than 65 million people have immigrated to the United States since 1820 (U.S. Immigration and Naturalization Service, 1999). Foreign immigration to the United States rose steadily throughout the nineteenth century, peaked in the early part of the twentieth century, declined through the 1930s, and has since risen to

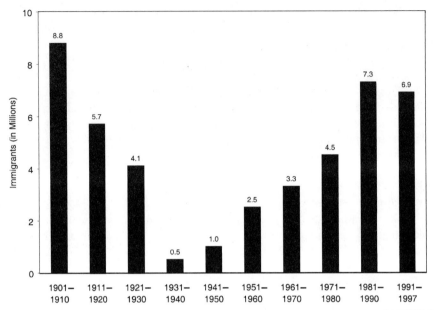

Figure 6.1. Documented immigrants to the United States, 1901–1997. (*Source*: A. Schmidley and C. Gibson [1999]. Profile of the foreign born population in the United States: 1997. U.S. Census Bureau, *Current Population Reports*, P-23, No. 195. Washington, DC)

levels nearly equal to those of 100 years ago. When undocumented entrants are added to legal immigrants, current levels may be the highest ever. Figure 6.1 shows the number of documented, legal immigrants to the United States, by decade, since 1900. The United States receives more foreign immigrants than any other country in the world (Weeks, 1999).

Several categories of international migrants have been defined for the United States (Martin & Midgley, 1999). Technically, *immigrants* are citizens of other countries who have been legally admitted for permanent residence. *Refugees* and *asylees* are persons who have been granted entry because they fear persecution for religious, political, or other reasons in their home countries. Although they are not initially classified as immigrants, many refugees and asylees later become immigrants by attaining permanent resident status. Legal immigration to the United States (including refugees and asylees) has averaged about 800,000 per year in recent years (U.S. Immigration and Naturalization Service, 1999).

In addition, many undocumented or unauthorized aliens enter the country surreptitiously or violate the terms of their temporary visas. Although the number of undocumented entrants is not known exactly, it was estimated to be about

250,000–300,000 per year during the 1990s (e.g., Center for Immigration Studies, 1995; Martin & Midgley, 1999). In addition, each year millions of foreign citizens are granted temporary visas to enter the country for a specific purpose, such as a vacation, a business trip, a temporary job, or to attend school. They may stay for a few days or weeks or remain for many years.

Immigration affects both the size and the racial and ethnic makeup of the U.S. population. Before the late 1800s, most immigrants came from northern and western Europe. During the late 1800s and early 1900s, southern and eastern Europe provided the largest numbers. European countries maintained their dominance until the 1960s, when U.S. policy changes began to alter the immigration flow. During the 1970s, 1980s, and 1990s, the largest numbers of immigrants came from Asia and Latin America. Now, Europe contributes only about 15% of the immigrants to the United States (U.S. Immigration and Naturalization Service, 1999).

Immigrants are not evenly distributed throughout the United States. Nationally, the estimated foreign-born population in 1997 was 25.8 million. Of these, 8.1 million were living in California, 3.6 million in New York, 2.4 million in Florida, 2.2 million in Texas, and just over 1.0 million each in New Jersey and Illinois (Schmidley & Gibson, 1999). Almost three-quarters of all foreign-born residents were living in those six states. The distribution of immigrants is even more variable for cities and counties than for states.

Cohort-component projections often distinguish between international and internal migration (see Chapter 11 for an example). This distinction may be important because international and internal migration are affected by different causal factors and follow different patterns. In addition, international migrants differ from internal migrants in age, sex, race, ethnicity, and other population characteristics. These differences immediately affect the demographic composition of the population; over time, they may also affect fertility and mortality rates (Edmonston & Passel, 1992).

Neither the Immigration and Naturalization Service (INS) nor the Census Bureau collects data on the emigration of U.S. residents to foreign countries. The number of emigrants is currently estimated to be around 200,000 per year (Martin & Midgley, 1999). Separating international from internal migration allows one to make adjustments for this missing link in the underlying base data. Such adjustments are particularly important for states and local areas that have received large numbers of immigrants in the past, because emigration from the United States occurs primarily among the foreign-born population (Edmonston & Passell, 1992).

Assessing the Issues

Migration is a complex process, and the concepts involved are difficult to measure or even to define. Any measure of migration is somewhat arbitrary in its treatment of distances traveled, time intervals covered, geographic boundaries crossed, distinctions between temporary and permanent moves, and definitions of place of usual residence. We have described migration as it is typically defined in the United States, but conventional measures understate the full extent of population mobility and may distort its character due to data inadequacies (Zelinsky, 1980).

Although these shortcomings have serious implications for analyses of the determinants and consequences of migration, they do not necessarily present a problem for constructing cohort-component projections. The objective of population projections is generally *not* to project the total number of moves or to classify them as temporary, permanent, repeat, return, internal, or international. Rather, it is to project the overall impact of migration on the permanent-resident population of a particular geographic area during a particular period. As long as the data accurately reflect this aspect of the migration process, their inadequacies in capturing other aspects of the process are irrelevant. The critical issues are to find data sources that accurately reflect historical migration trends and to develop realistic yet tractable models for projecting those trends into the future. We turn to these issues next.

SOURCES OF DATA

Birth and death data are readily available for states and counties and are generally considered quite accurate, but the same cannot be said for migration data. The ideal migration data set would include at least the following (Wetrogan & Long, 1990):

1. Data on the origins and destinations of migrants.
2. Data disaggregated by age, sex, and race/ethnicity.
3. Data available in one-year age groups.
4. Data available annually for a large number of time periods.
5. Data available in a timely manner.
6. Data consistent with the relevant population base for calculating migration rates.

Ideally, these data would be available for states, counties, and a variety of subcounty areas. Unfortunately, no single data set comes anywhere close to meeting all these criteria. In fact, no central agency in the United States directly tracks population movements, as is the case in a number of European countries that have population registers (e.g., Rees & Kupiszewski, 1999). Rather, migration data must be derived from a variety of sources; each has its own strengths and weaknesses.

Decennial Census

The most commonly used source—and the most comprehensive in terms of demographic and geographic detail—is the decennial census of population and housing. Mobility and migration data have been collected in every decennial census since 1940. They are based on responses to a question asking place of residence five years ago (except for the 1950 Census, which asked place of residence one year ago). The Census Bureau reports mobility data down to the city and census tract level: whether the person lived in a different housing unit five years ago and whether that unit was located in a different place, county, state, or abroad. The Census Bureau also tabulates in- and out-migration data by age, sex, and race for all states and counties.

There are a number of problems with the migration data collected in the decennial census. First, they do not pick up the effects of multiple moves during the five-year period. For example, a person may have lived in an apartment in Chicago in 1995, moved to a house in the suburbs in 1996, been transferred to a job in Atlanta in 1998, and retired to Sarasota in January 2000. Census migration data would show only the move from Chicago to Sarasota, completely missing the other moves. Another person may have moved from Portland to Seattle in 1997 and moved back to Portland in 1999; census data would show this person as a nonmigrant from 1995–2000. Data from the decennial census substantially understate the full extent of mobility and migration during a five-year period (e.g., DaVanzo & Morrison, 1981; Long & Boertlein, 1990).

Second, the question regarding residence five years ago is included only in the long form of the census questionnaire, which is sent to one in six households in the United States. Sample size can create problems of data reliability for small places, especially when data are broken down into age, sex, and race/ethnic groups. This problem was exacerbated in the 1980 Census, when only half the mobility/migration responses were coded and processed because of budgetary considerations (Isserman, Plane, & McMillen, 1982). Reliability problems may also be created by the respondent's lack of knowledge regarding geographic boundaries or an inability to accurately remember his/her place of residence five years earlier (Wetrogan & Long, 1990).

Third, data covering both in- and out-migration are not available below the county level. In-migration data are available directly from the census questionnaire but are tabulated only for states, counties, and census tracts. Out-migration data present an even greater problem because they must be collected from questionnaires filled out by residents throughout the country. For example, the number of out-migrants from Cook County, Illinois, for 1995–2000 can be determined only by counting the number of in-migrants to all other counties in the United States who reported in 2000 that they had been living in Cook County in 1995.

This process requires a major effort. The Census Bureau currently tabulates out-migration data for states and counties but not for subcounty areas.

Fourth, the decennial census does not cover emigration to foreign countries. Consequently, it provides no information on the number of persons who lived in the United States at mid-decade but moved to a foreign country before the following census. Although this is not a major problem in many places, it is significant in some places (e.g., counties near the Mexican border).

Fifth, migration data from the decennial census are subject to the same undercount problems that affect all census data. Because the undercount varies by age, race, ethnicity, and several other characteristics, the number of migrants will be undercounted more for some population subgroups than for others.

Finally, migration data from the decennial census suffer from several timing problems. They are available only once every 10 years, leaving large gaps of time with no new data. In addition, there is a lengthy processing time between the collection of the data and the release of the results (three to five years in recent censuses). Consequently, even the most recent data may be outdated.

Although these problems reduce its usefulness for some purposes, the decennial census is by far the most comprehensive (and commonly used) source of gross migration data in the United States. It provides information on the numbers of migrants and on their origins, destinations, age, sex, race, ethnicity, income, education, and other characteristics. It provides data at the state, county, and (for in-migration) census tract levels. The Public Use Microdata Sample (PUMS) is widely used to analyze the determinants of migration. No other source of migration data in the United States comes close to the decennial census in terms of demographic detail and geographic coverage.

Administrative Records

Administrative records kept by various agencies of the federal government are a second source of migration data. The records most commonly used for migration estimates come from the Internal Revenue Service (IRS). By matching the addresses listed on annual income tax returns and adjusting for the number of exemptions claimed on each return, the Census Bureau is able to create an annual set of state-to-state and county-to-county migration flows. These data have several advantages over decennial census data. They are available every year instead of every 10 years, they cover one-year intervals rather than five-year intervals, and they are available on a more timely basis.

IRS migration data have several limitations, however. Not everyone files an income tax return. In particular, people who have low incomes are not required to file. People who move to or from abroad are also likely to be missed. The address listed on a tax return may be that of a bank, law office, accounting firm, or post office box rather than the home address of the filer. This may lead to an inaccurate distribution of the population at the local level. The methodology assumes that people listed as exemptions on a tax return actually live (and move) with the filer; this may not be true (e.g., college students living away from home). Finally, IRS migration data provide no information on the characteristics of migrants and are not available below the county level.

Despite these limitations, IRS data provide useful information on annual migration flows for states and most counties. They cover more than 90% of the U.S. population, provide information on both the origins and destinations of migrants, and remain relatively stable over time. IRS data have been used by the Census Bureau to produce population estimates since the 1970s and to produce population projections since the 1980s. For further discussion of the strengths and weaknesses of IRS migration data, see Engels and Healy (1981); Isserman, Plane, and McMillen (1982); and Wetrogan and Long (1990).

The Immigration and Naturalization Service (INS), located in the Department of Justice, is the major source of international migration statistics in the United States. The INS began collecting immigration data in 1892. Before that time, immigration statistics were collected by the Department of State and the Treasury Department, beginning in 1820. Incomplete data on emigrants were collected for a number of years, but those collection efforts were discontinued in the late 1950s (Shryock & Siegel, 1973).

The INS produces annual statistics on the number of legal immigrants by type, country of origin, place of intended residence, age, sex, marital status, occupation, and several other characteristics. INS data, however, are based on the year in which a person was granted legal immigrant status, which is not necessarily the same as the year in which that person entered the United States. This distinction had a particularly large impact on immigration statistics for 1989–1992, when many aliens who were residing in the country illegally were granted permanent resident status under the provisions of the Immigration Reform and Control Act of 1986 (U.S. Immigration and Naturalization Service, 1999).

Sample Surveys

Sample surveys conducted by the federal government and other agencies are the final source of migration data. The survey most commonly used for migration data is the Current Population Survey (CPS), a monthly survey conducted by the Census Bureau that covers some 50,000 households. This survey was designed primarily to obtain labor force information, but every March the interviewers ask supplementary questions on migration and other topics. The survey collects information on geographic mobility for the United States as a whole, including the number and characteristics of migrants. However, below the national level, the CPS collects statistically valid data only for census regions, states, and large metropolitan areas. It provides no data for most counties or subcounty areas. Even for states, the relatively small sample size makes it impossible to use the migration data in any but the most highly aggregated form.

Other surveys that have been used to study mobility and migration include the American Housing Survey and the Survey of Income and Program Participation. These surveys provide data that are useful for many types of analyses but do not provide a sufficient basis for projecting state and local migration because of a small sample size, the nature of the sample universe, the definitions of mobility and migration employed, and the levels of demographic and geographic detail provided.

The American Community Survey offers a great deal of potential for future migration research. Begun on a limited basis in 1996, this monthly survey collects data from a large rolling sample of households. Eventually, it is intended to cover every place in the United States during a five-year period. It collects information on many of the economic and demographic characteristics contained in the long form of the decennial census. Migration data, however, are based on place of residence one year ago rather than five years ago, which complicates comparisons with migration data from the decennial census. Although issues regarding definitions and data comparability with the decennial census remain to be worked out, this survey (if it continues to be funded) will eventually provide a valuable source of migration data for population projections.

Residual Estimates

The sources discussed before provide data on gross migration, or unidirectional population movements into and out of an area. Estimates of net migration can be derived from these gross migration data by subtracting the number of out-migrants from the number of in-migrants. For example, Table 6.2 shows net migration calculated for states in the United States from 1985 to 1990.

However, there are many circumstances in which gross migration data are not available. Under these circumstances, indirect estimates of net migration can be made by comparing a region's population at two times, measuring the change due to natural increase, and attributing the residual to net migration. Several methods can be used to calculate net migration in this manner.

One is the *vital statistics method*, in which net migration (NM) is calculated by rearranging the terms of the demographic balancing equation described in Chapter 2:

$$NM = P_l - P_b - B + D$$

where P_l is the population in a given year, P_b is the population in some earlier year, and B and D are the number of births and deaths that occurred between times b and l. For example, Duval County, Florida, had a population of 571,003 in 1980 and 672,971 in 1990. It recorded 114,878 births and 53,618 deaths during the decade. Thus, net migration for 1980–1990 can be estimated as

$$672,971 - 571,003 - 114,878 + 53,618 = 40,708$$

The vital statistics method can be used to calculate net migration for the entire population and for specific subgroups of the population (e.g., age, sex, race, ethnicity). However, this process is very cumbersome and requires collecting a great deal of data. In addition, the birth and death data required by the vital statistics method are frequently unavailable for subcounty areas. To avoid these problems, a second residual method is often used.

This is called the *survival rate method.* Instead of explicitly accounting for births and deaths, this method uses survival rates to estimate the expected population of each age group at the end of a particular period. Estimates of net migration are then calculated as the difference between the expected population and the actual population. The most common form of this method is called the *forward survival rate method,* in which net migration is estimated as

$$NM = {}_nP_{x+y,l} - {}_nS_x\left({}_nP_{x,b}\right)$$

where ${}_nP_{x,b}$ is the population age x to $x+n$ in year b, ${}_nP_{x+y,l}$ is the population age $(x+y)$ to $(x+n+y)$ in some later year l, y is the number of years between b and l, and ${}_nS_x$ is the y-year survival rate for age group x to $x+n$. For example, Florida had 484,538 residents aged 40–44 in 1980, 594,288 residents aged 50–54 in 1990, and a 10-year survival rate of 0.9542, yielding a net migration estimate of

$$594,288 - 0.9542\,(484,538) = 131,942$$

This is the most basic formulation of the survival rate method. It can easily be extended to cover other population subgroups (e.g., sex, race, ethnicity). Other approaches to calculating survival rates and deriving net migration estimates can also be applied. Detailed discussions of the survival rate method and other indirect estimates of net migration can be found in Bogue, Hinze, and White (1982) and Shryock and Siegel (1973).

The major advantage of indirect methods of estimating net migration is that they can be applied when no direct data on in- and out-migration are available (Smith & Swanson, 1998). Consequently, they are particularly useful for projections of small areas. However, the accuracy of these estimates depends heavily on the accuracy of the underlying population estimates (or counts) and the vital statistics (or survival rate) data. Vital statistics and survival rate data in the United States are generally quite accurate, but the accuracy of population estimates and census counts varies over time and from place to place. In particular, because net migration is often estimated for decades (e.g., 1990–2000), changes in the undercount from one census to another may make estimates of net migration too high or too low. Changes in geographic boundaries over time may also affect net migration estimates. This generally will not be a problem for states and counties but may be significant for cities, school districts, ZIP code areas, and other subcounty areas.

Estimates of net migration by age, sex, and race for states were produced for each decade from 1870 to 1950 (Lee, Miller, Brainerd, & Easterlin, 1957). These estimates were extended to counties for the 1950s, 1960s, and 1970s (Bowles & Tarver, 1965; Bowles, Beale, & Lee, 1975; White, Mueser, & Tierney, 1987). Estimates of total net migration for states, regions, and counties for the 1980s and 1990s have been produced by the Census Bureau and are available on the Internet. To our knowledge, however, no breakdowns of these estimates by age, sex, and race have been produced for all states and counties in the United States. Analysts who choose to use net migration data for population projections may have to start by producing the base data themselves.

DETERMINANTS OF MIGRATION

Why do people move? How do they decide when and where to move? Perhaps more important for population projections, why do some areas have more people moving in than out whereas others have more moving out than in? What changes an area's migration patterns over time? It is helpful to consider some possible answers to these questions before attempting to construct projections of future migration flows. Although we cannot provide a complete discussion of the determinants of migration in this chapter, we can point out some of the theoretical perspectives and empirical findings that are particularly relevant to population projections. More complete discussions can be found elsewhere (e.g., Greenwood, 1997; Lee, 1966; Long, 1988; Mohlo, 1986; Zelinsky, 1980).

Theoretical Foundations

Studies of the determinants of migration are often based implicitly or explicitly on the theory of utility maximization (e.g., DaVanzo & Morrison, 1981; Greenwood, 1997; Lee, 1966; Plane, 1993; Rothenberg, 1977; Sjaastad, 1962). The basic idea is that each person (or household) considers all the advantages and disadvantages of living at the current location, the advantages and disadvantages of living at all other possible locations, and the full costs of moving from one location to another (i.e., money, time, and psychological costs). If the person (or household) determines that moving would raise the overall level of utility by more than the cost of the move, the decision is made to move, presumably to the location that will yield the greatest gain in utility. Although terminology and areas of emphasis vary from study to study, this basic theoretical foundation has been used by migration researchers in economics, sociology, anthropology, geography, and other disciplines.

Many factors influence decisions regarding whether, where, and when to move. Some are personal characteristics such as age, education, marital status, health status, occupation, social/psychological ties to the community, and perceptions of risk. Others are characteristics of various locations, including labor market conditions (e.g., wages, unemployment rates, rates of job creation), costs of living (e.g., state and local taxes, housing prices), and amenities (e.g., climate, topography, air and water quality, cultural and recreational opportunities, availability of public services). Moving costs—including opportunity and information costs as well as direct out-of-pocket expenses—are also important. An individual or household weighs all these factors when making migration decisions.

Migration decisions are strongly affected by one's age. Young children typically move with their parents, often with little input into the migration decision. In early adulthood, young people move out of their parents' homes to establish their own households, attend college, enter military service, and so forth. Moves are frequent for young adults as they embark upon their careers, marry, divorce, establish families, and seek better housing. Moves become less frequent as age increases but still occur in response to changes in economic conditions, job status, marital status, family size, and neighborhood characteristics. Retirement from the labor force provides a new opportunity to move, perhaps to an area that has a different climate or mix of amenities. Finally, declining health or the death of a spouse may induce additional moves in the latter years of life.

These life cycle influences are clearly reflected in age-specific mobility and migration rates. As shown in Figure 6.2, annual mobility rates in the United States are high for young children, decline until the late teens, explode upward for people in their 20s, and decline steadily until the oldest ages, when they increase slightly. These are typical patterns found in many countries throughout the world. They persist over time and from place to place even though overall migration levels may vary considerably. They are so pervasive that model migration schedules have been developed to summarize and codify their regularities (e.g., Plane, 1993; Rogers & Castro, 1984; Rogers & Woodward, 1991).

This age profile is consistent with the theory of utility maximization described before. People will migrate if the present value of all future gains in benefits outweighs the full cost of migration. After a person has entered the labor force, further increases in age reduce the remaining number of years over which to reap the benefits of migration; consequently, migration rates would be expected to decline as age increases. That is precisely what the empirical evidence shows. Because this cost–benefit view of migration implies that people will move to the areas that maximize their net benefits, it also provides a basis for projecting migration flows. We return to this idea in our discussion of structural models in Chapter 9.

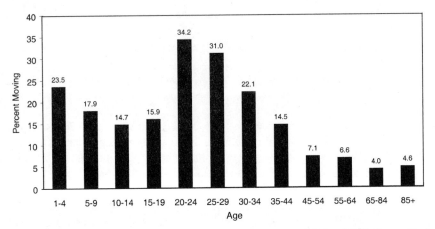

Figure 6.2. Moving rates by age, United States, 1997–1998. (*Source*: C. Faber [2000]. Geographical mobility, March 1997 to March 1998. U.S. Census Bureau, *Current Population Reports*, P20–520. Washington, DC)

Not every state or local area fits this "typical" age pattern, of course. Places that have large universities or military installations may have larger numbers of young adult migrants than is ordinarily the case. Regions that have depressed economies may have large outflows of young adults, but very small inflows. Retirement communities may have unusually large numbers of in-migrants aged 55 and older. The unique characteristics of each state and local area must be considered when developing assumptions for projecting migration rates.

Reasons for Moving

There are two basic approaches to studying reasons for moving (Lichter & DeJong, 1990). One is simply to ask movers about their reasons for moving. As Table 6.4 shows, about 47% of respondents to the American Housing Survey in 1995 cited family reasons, 39% cited housing reasons, and 27% cited employment reasons. (These numbers add up to more than 100% because respondents were allowed to report more than one reason for moving.) The individual reasons most frequently cited were to establish one's own household (15.0%); get a larger house or apartment (13.6%); take a new job or respond to a job transfer (11.9%); and be closer to work, school, or some other location (11.7%). It should be noted that family and housing reasons are often closely related; for example, an increase in family size may lead to a desire for a larger house or apartment. Consequently, distinctions between family and housing reasons for moving are somewhat blurred.

Table 6.4. Reasons for Moving,
United States, 1995

	Number[a]	Percent
Employment	4,726	26.8
New job, transfer	2,097	11.9
Closer to work/school/other	1,775	10.1
Other employment-related	854	4.8
Family	8,295	47.0
Establish own household	2,646	15.0
Larger house or apartment	2,393	13.6
Change in marital status	1,396	7.9
Other family-related	1,860	10.5
Housing	6,812	38.6
Better housing	2,062	11.7
Change in owner/renter status	1,352	7.7
Displacement, disaster	1,200	6.8
Lower rent or maintenance	1,035	5.9
Other housing-related	1,163	6.6
Other	3,196	18.1
Not reported	562	3.2
Number of respondents	17,655	—

[a]Numbers do not add to total because respondents were allowed to
report more than one reason for moving.
Source: U.S. Bureau of the Census (1997). American Housing Survey for the United States, *Current Housing Reports*, H150/95RV,
Washington, DC.

The data shown in Table 6.4 are largely determined by the behavior of local movers because they constitute the majority of movers in the United States. Housing and family factors are by far the most important motives for local moves; employment or job-related factors fall far behind (Lichter & DeJong, 1990). For longer distance moves, however, employment-related factors predominate. Long (1988) reported that a job transfer, taking a new job, looking for work, and other employment-related factors were the primary reasons for moving for more than half the interstate migrants between 1979 and 1981. Marriage, divorce, being closer to relatives, and other family-related factors were the primary reasons for about 15%, and the desire for a change of climate was the primary reason for 6%. Housing and neighborhood factors were the primary reasons for less than 5% of the interstate migrants. These results suggest that analysts should pay special attention to local housing trends when making projections for very small areas (e.g., census tracts, block groups) and should consider the potential impact of changing economic conditions when making projections for large areas (e.g., states, metropolitan areas).

Reasons for moving vary considerably by age. In one study, employment-related factors were the primary reasons for moving for more than half the interstate migrants younger than age 50 (Long, 1988). For persons older than age 65, however, they accounted for only a tiny proportion of interstate moves. Climate, on the other hand, accounted for less than 5% of the interstate moves for persons under age 50, 13–15% for persons aged 50–64, and 30% for persons aged 65–69 (Long, 1988). Health considerations also have a much larger impact on migration decisions for older persons than for younger persons (Rogers, 1992). Therefore, changes in the age distribution of the population may lead to substantial changes in migration patterns, especially for some states and local areas (e.g., retirement areas).

Survey data on reasons for moving provide valuable insights into migration behavior, but have several limitations (Lichter & DeJong, 1990; Long, 1988). Some respondents may not know exactly why they moved or may be unable to articulate their reasons. Some may be unable to disentangle and prioritize within a web of multiple reasons. Some may lie, mislead, or rationalize regarding their true motives. Others may simply forget. Indeed, one study reported that only 54% of migrants gave the same primary reason for moving both before and after they moved (McHugh, 1985). Perhaps most important for projection purposes, survey data add little to our understanding of the reasons that some states and local areas grow rapidly while others grow slowly or decline.

Statistical Analyses

Inferring motives from statistical analyses is a second way to determine reasons for moving. By this approach, analysts seek to uncover systematic relationships between migration and personal characteristics (e.g., age, education, marital status) and/or regional characteristics (e.g., wage rates, unemployment rates, climate). Studies of individual behavior typically find that age is the most important predictor of mobility (e.g., Gober, 1993; Long, 1988). Education is also important: People who have higher levels of education generally migrate more frequently than people who have lower levels, especially for long-distance moves (e.g., Gober, 1993; Greenwood, 1997). Being married and having children living at home tends to reduce the probability of moving (e.g., Greenwood, 1997).

Structural models may also be developed, in which migration is the dependent variable and various characteristics of migrants and the areas of origin and destination are independent variables. The theoretical basis of these models is frequently the concept of utility maximization, as described before. Empirically, they can be tested using data at the individual or household level (e.g., Clark, Knapp, & White, 1996; DaVanzo, 1983; Graves & Linneman, 1979; Morrison, 1971) or using aggregate data for counties, states, or other geographic areas (e.g., Clark & Hunter, 1992; Foot & Milne, 1989; Greenwood & Hunt, 1989; Schachter & Althaus, 1989).

Besides adding to our understanding of the determinants of migration, structural models help explain why some states and local areas grow faster than others and why migration levels increase or decrease over time. Results generated by these models can be incorporated into population projection models to provide projections that are consistent with various theories of migration or with alternative scenarios regarding changing economic conditions. We discuss structural models in Chapters 9 and 10.

MIGRATION MODELS

We have now described many of the concepts, measures, definitions, data sources, and theoretical approaches used to analyze migration. Next, we turn to the construction of migration models that can be used for cohort-component projections. We begin with models using gross migration data and close with models using net migration data.

Gross Migration

There are two basic approaches to projecting gross migration in cohort-component models. The first is based on historical data on in- and out-migration without reference to the places of origin and destination of those migrants; the second is based on data covering specific place-to-place migration flows. Both approaches were developed to provide a consistent set of projections for a large number of places, such as all states in the United States or all counties in a state. Both require a great deal of base migration data and a large number of calculations. We also discuss simplified versions of both approaches that retain a number of their useful features but require less data and fewer calculations.

Migrant Pool Models. The first approach is based on applying out-migration rates and in-migration proportions for each area to be projected. This approach was used by the Census Bureau during the 1960s and 1970s for state population projections (e.g., U.S. Bureau of the Census, 1966, 1972, 1979). We describe this approach using states as the unit of reference and migration data based on five-year intervals. The same approach could be used for other types of geographic areas and different lengths of migration interval, if data were available.

Out-migration rates by age and sex are calculated for each state using out-migration data from the decennial census as numerators and state populations by age and sex (five years earlier) as denominators. Table 6.3 shows these calculations for Florida. These rates form the basis of the projections. They can be used as they are or, as described in the next section, can be adjusted to fit with alternative views of the future.

With or without adjustments, out-migration rates are applied to the launch-year populations of each state, providing projections of out-migrants from all states during the five-year projection horizon. These numbers are then summed, providing a "pool" of potential in-migrants for each state. This pool is allocated to each state by applying the proportion of all interstate migrants that went to each state during the base period. For example, suppose that a state had 100,000 male in-migrants aged 20–24 between 1995 and 2000 and that nationally there were 2 million interstate migrants in this age-sex group. Based on this proportion, it would be projected that the state would receive 5% of the projected pool of male interstate migrants aged 20–24. Adjustments to the in-migrant proportions to account for changing assumptions regarding future migration patterns could also be made, with the constraint that state proportions add up to 100%.

By basing projections of in-migration on the pool of available out-migrants, migrant pool models ensure that the total number of interstate in-migrants is exactly equal to the total number of interstate out-migrants. Thus, state migration projections are consistent with each other and with national projections, in which net internal migration must be zero. This is an important and useful characteristic of migrant pool models.

International migration is generally projected separately from internal migration in migrant pool models. We describe several ways to project foreign immigration and emigration in Chapter 11.

Migrant pool models can also be developed for counties, but the process is extremely data-intensive, time-consuming, and tedious. To project the pool of potential in-migrants, out-migration rates by age and sex (and perhaps by race and ethnicity) would have to be constructed and applied to more than 3,100 counties (or county equivalents) in the United States. A simplified version, however, can be developed.

Suppose that projections are to be made for all of the counties in a particular state. Out-migration rates calculated and applied to each county provide a projection of the pool of out-migrants. This pool can be reduced by the number of migrants who leave the state (using historical proportions), which provides a pool of migrants going to other counties within the state. Intrastate migration into each county can then be based on this pool and historical data that show the shares going to each county. Migration from other states can be based on the national number of interstate migrants (excluding those from the state under consideration) and historical data that show the proportions of those migrants going to each county.

Multiregional Models. The second approach to projecting gross migration uses multiregional models based on specific place-to-place migration flows (e.g., Rogers, 1985, 1995a). In these models, migration is viewed as part of an integrated system of mortality, fertility, and origin-destination-specific population flows by age and sex (and sometimes by other characteristics as well). For example, interstate migration in a multiregional model could be represented by a 51 × 51 matrix that shows the number of people moving from each state to every other state (including the District of Columbia), by age and sex. Migration rates are calculated by dividing destination-specific gross migration flows by the population of each state of origin, giving each state 50 sets of age-sex-specific out-migration rates, one for each other state in the nation. Because they are based on the population at risk of migration, these rates approximate the probabilities of moving from one state to another during a given period. Multiregional models have been used by the Census Bureau in several recent sets of state projections (e.g., Wetrogan, 1988; Campbell, 1996).

Two-Region Models.   The multiregional model used by the Census Bureau is extremely data-intensive and requires many thousands of calculations. A greatly simplified version can be developed by focusing on two regions, one representing the area to be projected and the other representing the rest of the country. The population of the area to be projected provides the base for calculating out-migration rates, and the population of the rest of the country provides the base for calculating in-migration rates.

Isserman (1993) developed a two-region model for counties in West Virginia. Out-migration rates were calculated for each county by dividing the number of out-migrants by age and sex from 1975–1980 by the county's 1975 population for each age-sex cohort. In-migration rates were calculated by dividing the number of in-migrants by the 1975 population of the United States (minus the county's population), by age and sex. These migration rates were calculated in a manner similar to that shown in Table 6.3 for Florida. Projections of out-migration were made by applying out-migration rates to the county's population, and projections of in-migration were made by applying in-migration rates to the U.S. population (minus the county's population). Foreign immigration was lumped in with internal in-migration, but no separate projection was made for foreign emigration.

Two-region models retain many of the benefits of full-blown multiregional models while avoiding much of their cost. We believe that they are easier to apply than the simplified version of the migrant pool model described before. An example of a two-region model is given in Chapter 7.

Net Migration

Top-Down Models.   The first approach to projecting net migration distinguishes between the components of population growth (i.e., natural increase and net migration) but focuses on estimates of total net migration rather than separate estimates for each age-sex cohort. It requires two steps. First, projections of total net migration are made, based on recent levels, historical trends, structural models, or some other procedure. Second, these projections are broken down into age-sex categories, based on distributions observed in the past. We call this a "top-down" approach because projections for individual age-sex groups are derived from projections of total net migration. This was the approach taken in the earliest sets of cohort-component projections made for states and regions in the United States (e.g., Thompson & Whelpton, 1933; U.S. Bureau of the Census, 1957).

The Census Bureau's 1957 state projections illustrate this approach. Three migration assumptions were made, one based on the continuation of the average annual net migration levels observed for each state from 1950 to 1955, one based on the levels observed from 1940 to 1955, and one based on the levels observed from 1930 to 1955. These projections of total net migration were then broken down into age-sex groups for each state according to the distributions observed during the base period. The three migration assumptions provided the basis for developing several alternative sets of population projections (U.S. Bureau of the Census, 1957).

This is the approach currently used for the international migration component of national population projections in the United States. Projections of the level of total net foreign immigration are based on historical data and expectations regarding future levels; they are broken down into age, sex, and race/ethnic categories according to the distributions observed in recent historical data (Day, 1996a). A similar approach has been used for county projections in some states (e.g., Department of Rural Sociology, 1998; Nakosteen, 1989). Many economic models focus on levels or rates of total net migration for analyzing the determinants of migration and for projecting future net migration (e.g., Clark & Hunter, 1992; Greenwood & Hunt, 1989; Lee & Hong, 1974; Murdock, Leistritz, Hamm, Hwang, & Parpia, 1984).

Bottom-Up Models. The second approach to projecting net migration focuses on developing separate net migration rates for each age-sex cohort in the population (cohorts can also be broken down into race or ethnic categories). Projections are based on applying age-sex-specific net migration rates to the base population by age and sex. We call this a "bottom-up" approach because the total volume of net migration projected for an area is the sum of the individual values projected for each age-sex group.

Most applications of this approach use the population of the area to be projected as the denominator for the net migration rates. For example, projections for Ohio would use the population of Ohio as the denominator for calculating net migration rates. We illustrate the bottom-up approach using state projections published by the Census Bureau in 1983 (Wetrogan, 1983).

Demographers at the Census Bureau used a combination of vital statistics and survival rate techniques to estimate 1970–1980 net migration flows for each state, by age and sex. They adjusted these estimates to account for changes in the census undercount between 1970 and 1980, and subtracted changes in the military population to provide net migration estimates for the civilian population. Net migration rates by age and sex were calculated by dividing these civilian net migration estimates by the civilian populations of each state, by age and sex. The denominators used in these rates were the 1970 populations "survived" forward to 1980 using 10-year survival rates by age and sex. Projections for 1990 were made by applying these migration rates to the "survived" 1980 civilian populations of each state. Projections of the military population were added as a final step. The same procedures were repeated to provide projections for 2000.

Net migration models generally combine the effects of international and internal migration. When net migration is calculated as a residual, this is the simplest approach. Separate projections of foreign immigration can be made, however, by subtracting the impact of net foreign immigration from total net migration in the base data and developing separate assumptions regarding future net flows of international and internal migrants.

Earlier in this chapter, we discussed some of the advantages of gross migration models compared to net migration models. An additional drawback of net migration models is that they create inconsistencies in projections for a group of areas. Consider population projections for states, for example. The application of constant net migration rates to states that have rapidly growing populations leads to steadily increasing levels of net in-migration over time, but the application of constant rates to states that have slowly growing (or declining) populations leads to slowly growing (or declining) levels of net out-migration. This creates an inconsistency because net internal migration for states must sum to zero. It can also lead to bias because projections based on net migration rates tend to be too high for rapidly growing places and too low for slowly growing or declining places (e.g., Isserman, 1993; Rogers, 1990; Smith, 1986).

Some of the problems associated with net migration models can be reduced by changing the denominators used in constructing the migration rates. Net migration rates for rapidly growing areas can be based on the population of a larger geographic unit rather than of the area itself. For example, rates for rapidly growing states can be based on the national population rather than the state population. It has been found that this approach greatly reduces projected rates of increase for rapidly growing states (Smith, 1986). Alternatively, projections of net migration (or population) can be constrained or controlled in various ways to prevent unreasonably large increases or declines (e.g., Smith & Shahidullah, 1995). Chapter 11 discusses several ways to control projections for smaller areas to an independent projection of a larger area.

Hamilton–Perry Method. The effects of net migration and mortality can be combined to create a simplified version of the cohort-component method (Hamilton & Perry, 1962). In this method, cohort-change ratios (CCR) that cover the time interval between the two most recent censuses are calculated for each age-sex cohort in the population. These ratios are the same as the census survival rates discussed in Chapter 4, but the notation is slightly different:

$$_nCCR_x = {_nP_{x+y,l}} / {_nP_{x,b}}$$

where $_nP_{x+y,l}$ is the population aged $x+y$ to $x+y+n$ in year l; $_nP_{x,b}$ is the population age x to $x+n$ in year b; x is the youngest age in an age interval; n is the number of years in an age interval; l is the year of the most recent census; b is the year of the second most recent census; and y is the number of years between censuses. For example, there were 284,206 males aged 50–54 living in Florida in 1990, and 234,788 males aged 40–44 in 1980, yielding a cohort-change ratio of

$$_5CCR_{40} = {_5P_{50,1990}} / {_5P_{40,1980}} = 284,206 / 234,788 = 1.2105$$

Cohort-change ratios can be calculated for each age-sex group in the population; they can also be calculated for different race/ethnic groups. Projections can then be made by multiplying these ratios by the launch-year population in each age-sex group:

$$_nP_{x+y,t} = {_nCCR_x} ({_nP_{x,l}})$$

where $_nP_{x+y,t}$ is the population age $x+y$ to $x+y+n$ in target year t. Using the CCR calculated above and the 420,576 males aged 40–44 living in Florida in 1990, we can project the number of males aged 50–54 in 2000 as

$$_5P_{50,2000} = 1.2105 ({_5P_{40,1990}}) = 1.2105 \times 420,576 = 509,107$$

The Hamilton–Perry method is most valuable for census tracts and other small areas for which data on the components of growth are difficult or impossible to obtain. Its drawbacks are the same as those discussed earlier for net migration; an additional limitation is that the effects of mortality and net migration cannot be separated. We will give a numerical example and discuss the strengths and weaknesses of this method in Chapter 7.

IMPLEMENTING THE MIGRATION COMPONENT

Choosing Appropriate Models

The first issue that must be confronted is the choice of the projection model. Should the model be based on gross or net migration data? Should it be a structural model or one based on the extrapolation of past trends? If a structural model is used, what explanatory variables should be included? If an extrapolation model is used, on which migration rates should it be based, and how should those rates be extrapolated into the future? What demographic characteristics should be included?

The answers to these questions will depend primarily on three factors: the expected uses of the projections, the availability of input data, and the amount of time and money available to complete the projections. If the projections will be used to evaluate the demographic effects of different economic scenarios, a structural model is needed. If projections of specific origin-destination migration flows are needed, a multiregional model must be used. If projections by race and ethnicity are needed, the migration data must include race and ethnic characteristics. The expected use of the projections is a major determinant of the choice of projection model and the structure of that model.

The second factor is the availability of input data. At the state level, data are available that can accommodate any approach or migration model. At the county level, however, this is not the case. Detailed gross migration data from recent censuses are available, but annual time series data are much more limited. Data may be available for some socioeconomic characteristics but not for others, or for large counties but not for small counties. Estimates of 10-year net migration by age, sex, and race were produced for all counties in the United States for the 1950s, 1960s, and 1970s, but not for the 1980s or 1990s. Annual net migration estimates for counties are available for some states but not for others.

The availability of data plays an even larger role at the subcounty level. In-migration data from the decennial census are available at the census tract level, but not for smaller areas of geography. Out-migration data are not available for *any* subcounty area. Because of these problems, migration projections for subcounty areas are often based on a net migration model or—closely related—the Hamilton–Perry method.

Time and budget constraints also play a critical role. Time and money costs increase with the complexity of the method and with the level of geographic and demographic detail. Consequently, the amount of time and money available to complete a project will have an impact on the choice of the projection method.

Choosing Data and Assumptions

The migrant pool, multiregional, and net migration models provide an operational framework for calculating and projecting migration rates, but nothing in the models themselves provides any guidance regarding the choice of data or assumptions. Which historical migration rates provide the most realistic foundation upon which to build a set of population projections? Will future rates be higher, lower, or the same as those observed in the recent past? Will migration rates go up for some areas and down for others? Will changes in migration rates be the same for all age, sex, and race/ethnic groups? There are no simple answers to these questions. The analyst will have to develop assumptions based on personal knowledge of historical migration patterns and expectations regarding future trends, including trends in local economic conditions and labor markets.

A number of approaches can be followed in choosing migration rates and projecting them into the future. The simplest (and most commonly used for small-area projections) is to hold migration rates constant at recent levels. For example, one set of state projections published by the Census Bureau assumed that 1970–1980 net migration rates by age and sex would remain constant over the projection horizon (Wetrogan, 1983). In another set, gross out-migration rates and in-migration proportions observed from 1955–1960 were held constant (U. S. Bureau of the Census, 1966).

A closely related approach is to take an average of several recent sets of migration rates. For example, demographers in Florida used an average of age-sex-specific in- and out-migration rates from 1975–1980 and 1985–1990 for a recent set of state projections (Smith & Nogle, 2000). Isserman (1993) followed a similar approach for county projections in West Virginia.

Migration from several different periods can also be used. The Census Bureau's first published set of state projections was based on the average annual levels of civilian net migration observed during three historical periods: 1950–1955, 1940–1955, and 1930–1955. These three migration scenarios were combined with several fertility assumptions to provide four alternative sets of state projections.

Migration rates rarely remain constant over time, of course. Does this imply that holding recent rates constant is a poor assumption? Not necessarily. Will recent rates go up or down? Will they change a lot or only a little? Will the changes be the same or different for various population subgroups? If we cannot answer these questions with some degree of confidence, assuming that a recent set of migration rates will remain unchanged may be the best assumption we can make.

If the analyst chooses to project changes in migration rates over time, what approaches can be followed? One is to assume that a given set of migration rates will gradually converge toward another set over time. For example, the Census Bureau developed a series of state population projections using a migrant pool model in which it was projected that out-migration rates for each state would gradually converge toward the average of all states (U.S. Bureau of the Census, 1966). At the same time, it was assumed that in-migration distributions would converge toward each state's population distribution. Under this approach, state differences in net migration rates decline over time. This outcome is consistent with economic theories in which migration acts as a "equilibrating mechanism" to reduce regional differences in wages and economic opportunities (e.g., Hunt, 1993; Sjaastad, 1962).

Another way to account for changing migration rates is to extrapolate past trends into the future. The Census Bureau followed this approach in several recent sets of state projections (e.g., Campbell, 1996). Using IRS migration data from 1975–1994, the Census Bureau created a series of 19 annual observations on each of 2,550 state-to-state migration streams (i.e., 51 origin states and 50 destination states). They used these data to produce annual state-to-state migration rates and projected those rates into the future using a time series regression model:

$$_{i,j}Y_t = b\left(_{i,j}Y_{t-1}\right)$$

where $_{i,j}Y_t$ and $_{i,j}Y_{t-1}$ represent the first differences of the natural logarithms of the migration rates from state i to state j in periods t and $t-1$, respectively, and b is a coefficient estimated by the regression.

Research conducted at the Census Bureau has shown that projections from time series models become increasingly inaccurate as the projection horizon increases. For horizons of 10 years or longer, extrapolations of average annual values from the base period were found to forecast migration rates more accurately than a time series model (Campbell, 1996). As a result of these findings, the Census Bureau gradually phased out the time series model as the projection horizon increased. For the first five years of the projection horizon, projections were based exclusively on the time series model. For the next 10 years, projections based on the average annual values from the base period were gradually phased in. After 15 years, projections were based exclusively on the average annual values found during the base period. The procedures followed by the Census Bureau illustrate the fact that extrapolating past migration trends can lead to unrealistic projections if carried too far into the future.

Another approach to projecting migration is to develop structural models in which migration is tied to projections of other variables. Economic variables are the most commonly used in these models. We discuss structural models in Chapters 9 and 10.

Accounting for Unique Events and Special Populations

In addition to deciding which data, assumptions, and techniques to use in a set of migration projections, the analyst must decide how to account for unique events and special populations. Unique events are those that have a substantial but short-lived impact on an area's migration patterns. At the state level, these events are generally related to changes in state or national economic conditions. Was the economy unusually strong (weak) during the base period, leading to an unusually large (small) number of in-migrants? If so, the analyst will have to decide if those conditions are likely to continue into the future, and if not, how to make the appropriate adjustments. Changes in foreign immigration policy can also have a major short-term impact on a state's population growth. For example, the Mariel boatlift brought more than 125,000 Cuban immigrants to Florida within a six-month period in 1980, a volume of foreign immigration not seen before or since. Adjustments for these and similar events are somewhat subjective, but must be made to avoid projecting events that are not likely to be repeated.

Unique events often involve changes in special populations. Special populations are defined as groups of people who are in an area because of an administrative or legislative action (Pittenger, 1976, p. 205). Examples include college students, prison inmates, military personnel, and residents of nursing homes. Special populations are affected by a set of causal factors different from those that affect the rest of the population; consequently, changes in special populations are generally unrelated to changes in the rest of the population. If changes in special populations are substantial, it is important to account for them separately. We discuss this issue more fully in Chapter 11.

Accounting for unique events and special populations is especially important for subcounty areas because their impact tends to be highly localized, affecting a few areas dramatically while leaving other areas totally unaffected. Examples include the opening or closing of a military base, prison, or nursing home; the development of a large housing project; the construction of a new road or transportation system; and the growth or decline of a major employer. Events such as these can have a huge one-time impact on migration in very small areas. If no adjustments are made, the analyst in essence will be projecting that these events will be repeated in every future projection interval.

Accounting for Data Problems

Migration data in the decennial census are collected from a sample of households (approximately one in six for the 2000 Census). When these data are subdivided into age, sex, and race/ethnic categories, the number of persons in any given category may be very small, especially for small counties. Sampling variability raises serious questions about the validity of migration data for small counties and can lead to some very strange-looking migration rates. An example from Hardee County, Florida illustrates this problem.

Hardee County is a small, largely rural county in southwest Florida. Its 1990 population was 19,499, with 23.4% Hispanic, 5.3% black, and 15.2% aged 65 and older. Table 6.5 shows the number of in- and out-migrants from 1985 to 1990 for males and females aged 65 and older. The numbers appear quite reasonable for ages 65–69 as the net in-migration stream was 65 for males and 77 for females. They still appear reasonable for ages 70–74, although the net in-migration stream is only five for females, compared to 40 for males. For the three oldest groups, however, some of the numbers are of doubtful reliability. Males aged 75–79 reportedly had four in-migrants and 47 out-migrants, and males aged 80–84 reportedly had 77 in-migrants and no out-migrants. These numbers are wildly inconsistent with each other and with the numbers reported for females. If migration rates based on these data were extrapolated into the future, they would quickly lead to some very bizarre projections.

Discrepancies like these are not unique to Hardee County, Florida. They are found in many other counties and are even more glaring when the data are further

Table 6.5. In- and Out-Migrants Aged 65+,
Hardee County, Florida, 1985–1990

Age	Males			Females		
	In	Out	Net	In	Out	Net
65–69	116	51	65	105	28	77
70–74	67	27	40	50	45	5
75–79	4	47	−43	39	19	20
80–84	77	0	77	12	29	−17
85+	22	11	11	8	31	−23

Source: U.S. Bureau of the Census (1993). *1990 Census Special Tabulations, County-to-County Migration Flows*, SP 312, Washington, DC.

subdivided by race or ethnicity. It is essential that the analyst study the base migration data very carefully, especially for small counties or when unusual events occurred during the base period. When anomalies are found, adjustments to the base migration data must be made or alternative migration rates must be developed. Alternative rates can be based on data from an adjacent age group, a corresponding sex or race group, a similar county, or a model migration schedule. Such adjustments are time consuming and somewhat subjective, but we believe that they generally lead to better projections than could be made by slavishly adhering to an official but dubious migration data set.

Converting Data to Alternate Time Intervals

The decennial census is currently the *only* source of migration data providing a high level of demographic and geographic detail for all areas of the United States. Unfortunately, the time periods covered by decennial census data are limited. Gross migration data refer solely to a five-year interval and are fully available only for states and counties. Net migration estimates for states and counties can be developed for either five- or 10-year intervals, but net migration estimates for subcounty areas can be developed only for 10-year intervals (using two consecutive decennial censuses). Can these data be converted into intervals of different lengths?

For net migration, we believe that reasonable conversions can be made, at least for estimates of the total net migration flow. For example, 10-year net migration estimates can be converted into five-year estimates simply by dividing by two. Because net migration data are measured as residuals rather than actual events, the conversion process does not change the interpretation of the data: regardless of the interval covered, the data reflect the net population change due to migration. Conversions can also be made using net migration rates rather than net migration flows (e.g., Pittenger, 1976; Shryock & Siegel, 1973).

Although the procedures for converting net migration data for the total population are relatively straightforward, they are considerably more complicated for converting age-specific data. Consider net migration from 1990–2000, for example. Persons aged 10–14 in 1990 were 15–19 in 1995 and 20–24 in 2000. Thus, this cohort passed through two different five-year age groups during the decade. Which one should be used for calculating five-year net migration estimates for a five-year age group? A common solution is to calculate the 10-year net migration flow for each age cohort, divide by two, and take an average of two adjacent cohorts (Irwin, 1977). In the example mentioned before, one would use 10-year net migration estimates for persons aged 5–9 and 10–14 in 1990, divide each by two, and take an average to obtain an estimate of five-year net migration for persons aged 10–14 in 1990. Although this "adjacent cohort" procedure works fairly well if net migration rates do not vary much from one age group to another, it can lead to large errors when there are large differences between adjacent cohorts (Irwin, 1977).

For gross migration, the impact of multiple moves and deaths of migrants makes it very difficult to convert data from one length of interval to another. For example, one-year migration numbers cannot be calculated simply by dividing five-year migration numbers by five. Studies of this relationship have found that migration data based on five-year intervals greatly understate the total number of moves actually occurring (e.g., DaVanzo & Morrison, 1981; Long & Boertlein, 1990; Rees, 1977). Consequently, dividing a five-year migration number by five will greatly understate the annual numbers of migrants. Although some research on converting one- and five-year data to different intervals has been done, conversion factors vary from place to place and change over time (Long, 1988; Rees, 1977). In addition, the problems related to converting data for age groups are even greater for gross migration data than for net migration data.

The Census Bureau currently deals with these problems by combining data from several data sources to create a "synthetic" migration data set (e.g., Wetrogan, 1990; Campbell, 1996). Annual state-to-state migration estimates by age, sex, and race are developed using data from the decennial census, the March supplements of the Current Population Survey, and annual state-to-state migration flows based on matched IRS tax returns. A number of smoothing procedures and standardization techniques are used to adjust the data. The Census Bureau has used this synthetic data set in constructing its last several sets of state population projections.

The Census Bureau has the resources, technical expertise, and access to data needed to create a synthetic set of migration rates. For most data users and for substate areas, however, we believe it is risky to convert migration data to an interval of a different length (especially for gross migration data). Instead, we favor using projection intervals that are consistent with the length of the migration base period; that is, we favor using five-year projection intervals when using five-year migration data and 10-year projection intervals when using 10-year migration data. If projections for intervening years or more detailed age groups are needed, they can be constructed using various interpolation procedures (see Chapter 11). If one-year migration data are available, of course, projections can be made in one-year intervals.

CONCLUSIONS

Migration has not received as much attention from population researchers as fertility and mortality, perhaps because of the lack of comprehensive data and the difficulties in developing clear definitions and adequate measures (Greenwood, 1997; Zelinsky, 1971). Yet migration is an extremely important component of change that affects the size, composition, and distribution of the U.S. population. It is by far the most volatile component of population growth for states and local areas, both in changes over time and place-to-place differences for a given period. It is frequently the major determinant of state and local population growth as well (Smith & Ahmed, 1990).

State and (especially) local migration are affected by factors that can change abruptly. Migration is considerably more susceptible than either fertility or mortality to changes in economic conditions, employment opportunities, housing patterns, transportation conditions, and neighborhood characteristics. This volatility makes migration rates more difficult to forecast accurately than either mortality or fertility rates (e.g., Irwin, 1977; Kulkarni & Pol, 1994; Nakosteen, 1989). Because of its potential volatility and its impact on total population growth, migration contributes more to the uncertainty of cohort-component projections for states and local areas than either mortality or fertility. In general, the smaller the geographic region, the greater the difficulty in developing accurate migration forecasts.

Migration affects not only the total population of an area but also its age, sex, race, ethnicity, income, education, and other characteristics. In California, for example, non-Hispanic whites accounted for nearly 80% of the population in 1970 but for only 57% in 1990. This rapid decline was caused primarily by foreign immigration, as the Hispanic population rose from 12% to 26% of the total population and the Asian population rose from 3% to 10% (Gober, 1993). In Florida, the population aged 65 and older rose from 8.6% of the total population in 1950 to 18.2% in 1990. This increase was caused by primarily by the huge number of retirees who moved into the state (Smith, 1995). The impact of migration on the demographic characteristics of states and local areas can scarcely be overstated.

CHAPTER 7

Implementing the Cohort-Component Method

We have now discussed mortality, fertility, and migration—the three components of population change. In this chapter, we provide some examples of combining these components in a complete projection model. These examples illustrate three alternative approaches to projecting migration, the most difficult component of population growth to forecast accurately for states and local areas.

We begin with a discussion of several general issues that must be considered when setting up a cohort-component model. Then, we present three step-by-step examples, each based on readily available data and commonly used computational procedures. We close by assessing the strengths and weaknesses of the cohort-component method. Our strategy in this chapter is to describe the simplest, most straightforward applications of the cohort-component method. In Chapter 11, we discuss several additional factors that must be considered in some circumstances: adjusting for international migration and special populations, accounting for census enumeration errors, controlling to independent population or migration totals, and developing temporal or age-group interpolations.

GENERAL CONSIDERATIONS

To preserve the integrity of age cohorts as they progress through time, it is helpful to follow two basic principles: (1) The number of years in the projection interval should be greater than or equal to the number of years in the cohort, and (2) if the number of years in the projection interval is greater than the number of years in the cohort, it should be exactly divisible by the number of years in the cohort. For example, five-year cohorts are well suited for making projections in five- or 10-year intervals but are not well suited for making projections in one-year intervals. The logic is simple: people aged 10–14 in 2005 will be 15–19 in 2010 (unless they die), but there is no way to know exactly how many will be 11–15 in 2006. Models that stray from these principles can be constructed but are more complicated and less precise.

Cohort-component models are often constructed for five-year age groups, starting with 0–4 and ending with 75+ or 85+. Five-year age groups are common because they satisfy the needs of a wide range of data users and work well with the five-year migration data collected in the decennial census. Single-year age groups are also widely used. Single-year models make it easy to calculate customized age groups (e.g., 5–17) and provide a more detailed picture of population aging over time; by focusing on single-year cohorts, they pick up subtleties that may be missed by five-year models.

However, single-year models are considerably more time-consuming and costly to construct and maintain than five-year models. A single-year model with 100+ as the terminal age group has 202 age-sex categories. In contrast, a five-year model that has 85+ as the terminal age category has only 36 age-sex categories. For a 20-year projection horizon, a single-year model requires applying 202 separate birth, death, and migration rates for each of 20 distinct periods. A five-year model requires only 36 birth, death, and migration rates for four periods. Data management issues become even more imposing when three or four race/ethnic groups are added to the projections.

In some circumstances, migration data are available only in 10-year intervals (e.g., net migration between two decennial censuses). A common practice is to transform 10-year net migration rates into five-year rates by dividing by two and averaging two adjacent age cohorts (Irwin, 1977). Pittenger (1976) suggests a similar approach, using geometric interpolation to create five-year migration rates. Both approaches are acceptable, but are subject to a number of problems, as discussed in Chapter 6. We prefer using 10-year migration rates and 10-year projection intervals in these circumstances.

Models that use one-year intervals and single years of age provide the most detail—annual projections for individual ages—but require the most data and computations. Models that use five- or 10-year intervals and age groups require less data and fewer computations but provide less detail. Because more data and computational requirements imply higher costs, a trade-off must be made: level of detail versus costs of production. The optimal choice for any particular project will depend on the amount of time and money available, the availability of reliable data, and the purposes for which the projections are to be used. These issues are discussed more fully in Chapter 12.

Cohort-component models are almost always stratified by age and sex; they are often stratified by race and ethnicity as well. Race categories can be very basic (e.g., white and nonwhite) or more detailed (e.g., white; black; Asian; Native Hawaiian or Pacific Islander; American Indian or Alaska Native). The most commonly used ethnic categories are Hispanic and non-Hispanic. Recent Census Bureau projections were stratified by race and by Hispanic origin (e.g., Campbell, 1996; Day, 1996a). Less frequently, projections are stratified by household relationship and marital status, often in conjunction with projections of households (e.g., Day, 1996b). Raising the level of stratification allows the analyst to take explicit account of the differences in mortality, fertility, and migration rates found among demographic subgroups. It is obvious—but worth repeating—that additional stratification adds to the costs and complexities of model implementation and maintenance.

Typically, each demographic subgroup in a cohort-component model is projected separately (e.g., white females, white males, nonwhite females, and nonwhite males). Then, these projections are combined to create projections of other population groups. For example, projections of males and females are summed to provide projections of the total population, and projections of white males and white females are summed to provide projections of the white population. The female population is typically projected first because projections of females are needed for projecting births. The procedures for applying mortality, fertility, and migration rates are the same for each demographic subgroup.

A final consideration before implementing the cohort-component method is data reliability. Data problems tend to increase as the level of demographic detail increases and as population size declines. It is important to verify historical population data and, if necessary, to adjust base demographic rates before running the projection model. Techniques for adjusting and smoothing base data are described in Shryock and Siegel (1973).

APPLYING THE COHORT-COMPONENT METHOD

We now describe three applications of the cohort-component method, one based on a gross migration model, the second on a net migration model, and the last on the Hamilton–Perry method. We refer to these as Models I, II, and III, respectively. The mortality and fertility assumptions used in Models I and II are identical; these models differ only in their approaches to projecting migration. Model III combines the mortality and migration components and uses a technique for projecting births that differs from that used in Models I and II.

Applications of the cohort-component method involve four basic modules or steps that are computed in the following sequence: mortality (or survival), migration, fertility, and the final projection. The Hamilton–Perry method combines the first two modules. Figure 7.1 provides an overview of these four modules for Model I, the most complex of the three models.

We apply Models I and II using data for white females in Broward County, Florida and apply Model III using data for all females in a census tract in San

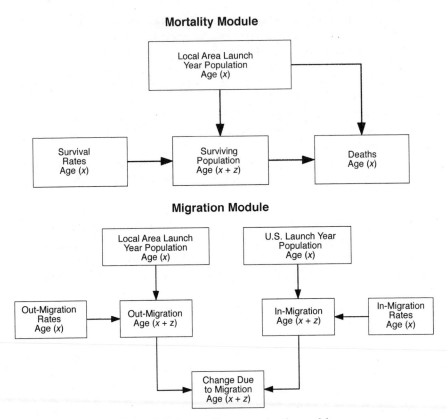

Figure 7.1. A two-region gross migration model.

Diego County, California. All three models use five-year age cohorts; computations using one- or 10-year age cohorts would be similar to those described here. The data were provided by the Bureau of Economic and Business Research at the University of Florida and the San Diego Association of Governments.

Few off-the-shelf software packages are available for applying the cohort-component method. Rather, the analyst will generally have to develop customized computer programs. If small numbers of projections are to be made, they can be implemented fairly easily using an electronic spreadsheet. If large numbers are needed, however, spreadsheets become quite cumbersome. In these instances, it is easier to construct projections using SAS, SPSS, or a similar statistical package, or by using a formal programming language such as FORTRAN or C++. Relational database systems (e.g., Oracle, Microsoft SQL Server) are useful for data docu-

Fertility Module

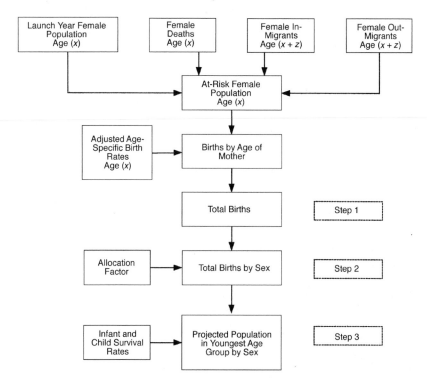

Final Projection Module

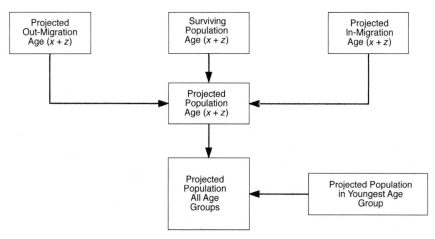

Figure 7.1. (*Continued*)

mentation, storage, retrieval, and management. Software issues are discussed more fully in Chapter 14.

Gross Migration (Model I)

Using 1995 as a launch year and 2000 as a target year, we developed a two-region gross migration model for white females in Broward County, Florida (Model I). The data required for this projection include

1. 1995 population estimates for white females by age (0–4, 5–9, ..., 85+).
2. Age-specific birth rates for white females.
3. Age-specific survival rates for white females.
4. Age-specific in- and out-migration rates for white females.

The 1995 population estimates for white females were produced by the Bureau of Economic and Business Research at the University of Florida. Age-specific birth rates were calculated by using 1990 birth and population data using the procedures described in Chapter 5. Age-specific survival rates were based on Florida life tables for 1990 using the procedures described in Chapter 4. In- and out-migration rates were based on 1985–1990 migration data for Broward County, as reported in the 1990 Census. The procedures for calculating these rates were described in Chapter 6.

Mortality Module.   The first step in the projection process is to calculate the number of people alive in the launch year (1995) who will survive to the target year (2000). This can be done by multiplying the launch-year population by the survival rate for each age group:

$$_n\text{SURVP}_{x+z,t} = {_nP_{x,l}} \times {_nS_x}$$

where x is the youngest age in the age group, n is the number of years in the age group, z is the interval between the launch and target years, t is the target year, l is the launch year, SURVP is the surviving population, P is the population, and S is the probability of surviving for z more years. In this example, l is 1995, t is 2000, and z is 5 years.

Table 7.1 shows the survival rate computations. For example, the surviving population aged 20–24 in 2000 equals the population aged 15–19 in 1995 multiplied by its survival probability (i.e., the probability that a person aged 15–19 lives five more years):

$$22{,}733 \times 0.99740 = 22{,}674$$

Deaths during the projection interval (l to t) in age group x to $x+n$ in the

Table 7.1. Projection of the Surviving Population
for White Females, Broward County, 2000

Age in 1995	Age in 2000	1995 population	Survival rates	2000 surviving population[a]	1995–2000 deaths[b]
0–4	5–9	32,814	0.99566	32,672	142
5–9	10–14	28,867	0.99910	28,841	26
10–14	15–19	26,044	0.99824	25,998	46
15–19	20–24	22,733	0.99740	22.674	59
20–24	25–29	25,987	0.99713	25,912	75
25–29	30–34	34,929	0.99629	34,799	130
30–34	35–39	43,345	0.99508	43,132	213
35–39	40–44	45,137	0.99375	44,855	282
40–44	45–49	42,035	0.99047	41,634	401
45–49	50–54	37,618	0.98501	37,054	564
50–54	55–59	30,660	0.97661	29,943	717
55–59	60–64	24,778	0.96419	23,891	887
60–64	65–69	25,955	0.94788	24,602	1,353
65–69	70–74	32,125	0.92337	29,663	2,462
70–74	75–79	36,920	0.88199	32,563	4,357
75–79	80–84	35,118	0.81223	28,524	6,594
80+	85+	49,077	0.57733	28,334	20,743
Total		574,142		535,091	39,051[c]

[a]Surviving population = 1995 population × survival rate.
[b]Deaths = 1995 population − 2000 surviving population.
[c]Does not include deaths to children born between 1995 and 2000.

launch year are found by subtracting the surviving population z years older (aged $x+z$ to $x+n+z$) from the launch year population aged x to $x+n$:

$$_nD_{x,l \text{ to } t} = {}_nP_{x,l} - {}_n\text{SURVP}_{x+z,t}$$

For example, we can calculate the number of deaths to white females aged 15–19 in 1995 as

$$22,733 - 22,674 = 59$$

It should be noted that this projection does not refer to the number of deaths that actually occurred in Broward County between 1995 and 2000. Rather, it refers to the number of deaths that occurred to people who were living in Broward County in 1995. Although these two numbers will generally be similar, they will not be identical because of the effects of in- and out-migration.

The survival-rate computation for the oldest age group is slightly different from the computations for the younger age groups. The survival rate for the oldest age group in the target year is applied to the sum of the populations in the two oldest age groups in the launch year. For example, if the oldest age group in the target year is 85+, the survival rate is applied to the population 80+ in the launch year (i.e., the sum of the populations 80–84 and 85+).

In a five-year model, the survival routine starts with the launch-year population aged 0–4 to obtain the surviving population aged 5–9 in the target year. The population aged 0–4 in the target year is based on the number of births that occurred between the launch year and target year, as will be described in the fertility module.

Migration Module.   The second step is to project in- and out-migration by applying migration rates to the appropriate at-risk populations. For in-migrants, the at-risk population is the U.S. population (minus the Broward County population), whereas for out-migrants, the at-risk population is the Broward County population. There are two main approaches to developing migration projections in this manner. The first uses migration rates applied to the launch-year population "survived" to the end of the projection interval, and the second uses migration rates applied to the launch-year population. As noted in Chapter 6, either approach is acceptable as long as it is applied consistently. For example, if migration rates are calculated using the population at the beginning of the migration interval, the rates must be applied to the launch-year population. If migration rates are calculated using the surviving population at the end of the migration interval, the rates must be applied to the surviving population in the target year. We use the first approach in our example because it is somewhat simpler to apply and leads to about the same results.

We project gross migration using two equations. For in-migration, we multiply projected in-migration rates by the adjusted U.S. population. The adjusted U.S. population is the U.S. population minus the local area population (i.e., Broward County) measured in the launch year. For out-migration we multiply projected out-migration rates by the local area population, also in the launch year. The equations for projecting gross in- and out-migration are

$$_n\text{AUSP}_{x,l} = {}_n\text{USP}_{x,l} - {}_nP_{x,l}$$
$$_n\text{INMIG}_{x+z,l \text{ to } t} = {}_n\text{AUSP}_{x,l} \times {}_n\text{INMIGRATE}_{x,l \text{ to } t}$$
$$_n\text{OUTMIG}_{x+z,l \text{ to } t} = {}_nP_{x,l} \times {}_n\text{OUTMIGRATE}_{x,l \text{ to } t}$$

where AUSP is the adjusted U.S. population, USP is the U.S. population, P is the population of the area to be projected, INMIG is the projection of in-migrants, INMIGRATE is the z-year in-migration rate, OUTMIG is the projection of out-migrants, and OUTMIGRATE is the z-year out-migration rate. As always, x is the youngest age in the age group, n is the number of years in the age group, t is the

Table 7.2. Projection of In- and Out-Migration for White Females,
Broward County, 1995–2000

Age in 1995	Age in 2000	1995 population		Migration rates		1995–2000 gross migration		
		Broward	Adjusted U.S.[a]	In	Out	In[b]	Out[c]	Difference[d]
0–4	5–9	32,814	7,493,186	0.00100	0.23292	7,493	7,643	−150
5–9	10–14	28,867	7,389,133	0.00094	0.21069	6,946	6,082	864
10–14	15–19	26,044	7,292,956	0.00089	0.29584	6,491	7,705	−1,214
15–19	20–24	22,733	6,953,267	0.00149	0.38934	10,360	8,851	1,509
20–24	25–29	25,987	6,969,013	0.00201	0.38430	14,008	9.987	4,021
25–29	30–34	34,929	7,569,071	0.00169	0.33024	12,792	11,535	1,257
30–34	35–39	43,345	8,880,655	0.00136	0.21651	12,078	9,385	2,693
35–39	40–44	45,137	9,133,863	0.00117	0.16478	10,687	7,438	3,249
40–44	45–49	42,035	8,428,965	0.00098	0.13473	8,260	5,663	2,597
45–49	50–54	37,618	7,451,382	0.00107	0.11496	7,973	4,325	3,648
50–54	55–59	30,660	5,941,340	0.00110	0.15787	6,535	4,840	1,695
55–59	60–64	24,778	4,890,222	0.00137	0.17686	6,700	4,382	2,318
60–64	65–69	25,955	4,545,045	0.00156	0.16010	7,090	4,155	2,935
65–69	70–74	32,125	4,698,875	0.00117	0.09072	5,498	2,914	2,584
70–74	75–79	36,920	4,421,080	0.00101	0.08413	4,465	3,106	1,359
75–79	80–84	35,118	3,535,882	0.00079	0.05891	2,793	2,069	724
80+	85+	49,077	4,944,923	0.00058	0.04513	2,868	2,215	653
Total		574,142				133,037	102,295	30,742

[a]Adjusted U.S. population = 1995 U.S. population − 1995 Broward County population.
[b]In-migration = in-migration rate × adjusted U.S. population.
[c]Out-migration = out-migration rate × Broward County population.
[d]Difference = in-migrants − out-migrants.

target year, l is the launch year, and z is the interval between the launch and target years.

Table 7.2 shows the migration projections. As we did with the survival rate computations, we combine the launch-year populations of the two oldest age groups before applying the migration rates for the oldest group. Also, there is no migration projection for the population aged 0–4 in 2000. Those children were not yet born in 1995; they will be accounted for in the fertility module.

As an illustration, consider the migration of white females aged 40–44 in 1995. We compute the number of in-migrants over the projection interval by multiplying the 1995 adjusted U.S. population aged 40–44 by the in-migration rate for that age group. Similarly, we compute the number of out-migrants by multiplying the 1995 county population by the appropriate out-migration rate. The net change due to migration is simply the difference between the two. From Table 7.2, the specific computations are as follows:

In-migrants: $0.00098 \times 8,428,965 = 8,260$

Out-migrants: $0.13473 \times 42,035 = 5,663$

Net change due to migration: $8,260 - 5,663 = 2,597$

A two-region model is a relatively simple application of a gross migration model. More complex applications might include the calculation of separate in-migration rates for migrants from nearby counties, from other counties within the state, and from places outside the state. Separate rates could also be calculated for in-migrants from each of the four census regions of the United States (Northeast, Midwest, South, West). Recent state projections from the Census Bureau used a 51 × 51 matrix of origin-destination-specific migration rates (Campbell, 1996). The procedures used in more complex models are similar to those described here for a two-region model.

Fertility Module.   The third step is to project the number of births and the net impact of mortality and migration on the youngest age group. This process has three steps. First, we multiply the at-risk female population (by age) by the projected ASBRs and sum the results to obtain a projection of the total number of births (by *at risk*, we mean females of childbearing age). Second, we allocate births between males and females using historical proportions. Finally, we survive the births to the target year to obtain the projection of the youngest age group. The following equations are used in making these calculations:

$$_n\text{ADJASBR}_{x,t} = (_n\text{ASBR}_{x,t} + {}_n\text{ASBR}_{x+5,t})/2$$

$$_n\text{ATRISKFP}_{x,t} = {}_n\text{FP}_{x,l} - (0.5 \times {}_n\text{FD}_{x,l\text{ to }t}) + (_n\text{FINMIG}_{x+z,l\text{ to }t}) - (_n\text{FOUTMIG}_{x+z,l\text{ to }t})$$

$$_n\text{B}_{x,l\text{ to }t} = {}_n\text{ADJASBR}_{x,t} \times {}_n\text{ATRISKFP}_{x,t}$$

$$\text{B}_{l\text{ to }t} = \Sigma \, {}_n\text{B}_{x,l\text{ to }t} \text{ where } \Sigma \text{ is the sum across all age groups}$$

$$\text{MB}_{l\text{ to }t} = \text{B}_{l\text{ to }t} \times \text{PCTM}$$

$$\text{FB}_{l\text{ to }t} = \text{B}_{l\text{ to }t} - \text{MB}_{l\text{ to }t}$$

$$_n\text{M}_{0,t} = \text{MB}_{l\text{ to }t} \times {}_n\text{MS}_0$$

$$_n\text{F}_{0,t} = \text{FB}_{l\text{ to }t} \times {}_n\text{FS}_0$$

where ASBR is the age-specific birth rate, ADJASBR is the adjusted age-specific birth rate, ATRISKFP is the at-risk female population, FP is the female population, FD is female deaths, FINMIG is the projection of female in-migrants, FOUTMIG is the projection of female out-migrants, B is the projection of total births, MB is the projection of male births, PCTM is the proportion of births that are male, FB is the projection of female births, $_n\text{M}_{0,t}$ is the male population projection in the youngest age group, $_nMS_0$ is the male infant and child survival rate, $_nF_{0,t}$ is the female population projection in the youngest age group, and $_nFS_0$ is the female infant and child survival rate. As before, x is the youngest age in an age group, n is the number of years in the age group, l is the launch year, t is the target year, and z is the interval between the launch and target years.

Several of these equations require further elaboration. Females pass from one age group to another during the projection interval. Because they spend half the projection interval in one age group and half in the next higher group (on average), the proper ASBR is the average of the rates for these two groups. We call this the *adjusted* ASBR (ADJASBR). Making this adjustment can have a substantial effect on the projection of births, especially when the population is growing (or declining) rapidly or when some age groups are considerably larger than others.

In addition, some of the original members of each cohort die, others move away, and new members move in. We assume that women who die during the projection interval live through half the interval (e.g., 2.5 years for a five-year interval). Therefore, the population for each age group in the launch year must be reduced by one-half the deaths that occurred during the projection interval. We further adjust the at-risk population by adding in-migrants and subtracting out-migrants. Thus, our final calculation of the at-risk population (ATRISKFP) is the female population in the launch year, minus one-half of the female deaths during the projection interval, plus female in-migrants, minus female out-migrants.

Table 7.3 shows the sequence used for projecting births and the number of whites aged 0–4 in Broward County. Column 3 shows the ASBR for each age group in 1990; these rates are one-year rates multiplied by five to reflect a five-year projection interval. Column 4 shows adjusted ASBRs, calculated as an average of two adjacent sets of ASBRs. For example, the adjusted rate for females aged 10–14 in 1995 is the average of the rates for ages 10–14 and 15–19. The rate for females aged 40–44 in 1995 is one-half the rate for females aged 40–44 because the rate for females older than 44 is assumed to be zero. The at-risk population for each age group (column 9) is calculated by starting with the 1995 population, subtracting one-half the deaths, adding in-migrants, and subtracting out-migrants. For females aged 10–14 in 1995, for example, the at-risk population is calculated as

$$26,044 - (0.5 \times 46) + 6,491 - 7,705 = 24,807$$

Births are computed by multiplying the at-risk population in each age group by the adjusted birth rate for that age group. For example, births for women aged 30–34 in 2000 are calculated as

$$36,121 \times 0.45995 = 16,614$$

The projection of the total number of births is obtained by summing the births projected for all age groups. In the example, this sum is 57,261 for the five-year

Table 7.3. Projection of the White Population Aged 0–4, Broward County, 2000 (Model I)

Age in 1995	Age in 2000	Five-year birth rate Original[a]	Five-year birth rate Adjusted[b]	1995 female population	Deaths	In-migrants	Out-migrants	At-risk population[c]	Births[d]
10–14	15–19	0.00185	0.08040	26,044	46	6,491	7,705	24,807	1,994
15–19	20–24	0.15895	0.28058	22,733	59	10,360	8,851	24,213	6,794
20–24	25–29	0.40220	0.45948	25,987	75	14,008	9,987	29,971	13,771
25–29	30–34	0.51675	0.45995	34,929	130	12,792	11,535	36,121	16,614
30–34	35–39	0.40315	0.28393	43,345	213	12,078	9,385	45,932	13,041
35–39	40–44	0.16470	0.09393	45,137	282	10,687	7,438	48,245	4,532
40–44	45–49	0.02315	0.01158	42,035	401	8,260	5,663	44,432	515
Total				240,210	1,206	74,676	60,564	253,721	57,261

	Share of births	1995–2000 births	Survival rates[g]	2000 population 0–4[h]	1995–2000 child deaths[i]
Males	0.51	29,203[e]	0.99056	28,927	276
Females	0.49	28,058[f]	0.99316	27,866	192
Total	1.00	57,261		56,793	468

[a]5-year birth rate = 1-year birth rate × 5.
[b]Adjusted birth rate = average of birth rates for adjacent age groups.
[c]At-risk population = 1995 population − (0.5 × deaths) + in-migrants − out-migrants.
[d]Births = adjusted birth rate × at-risk population.
[e]Male births = total births × 0.51.
[f]Female births = male births.
[g]Probability of surviving from birth to age 2.5.
[h]Projected population aged 0–4 = births × survival rates.
[i]Child deaths = births − 2000 population aged 0–4.

projection interval. As mentioned in the discussion of the mortality module, this is a projection of the number of births between 1995 and 2000 to women who lived in the county in 2000, *not* of the number of births that actually occurred in the county between 1995 and 2000.

On average, migrants are at risk of giving birth within the county during only half the projection interval. However, our concern is not the number of births that occurred within the county itself, but rather the location at the end of the projection interval of children born during the interval. We assume that babies and young children reside with their mothers at the end of the projection interval, regardless of where they were born. As Isserman (1993) notes, this allows treating births and the migration of young children in a single step. In-migrants and non-migrants are included with the population at risk, but out-migrants are excluded. Thus, births to in-migrants are included in the county population at the end of the projection interval, whereas births to out-migrants are not.

The bottom part of Table 7.3 shows the calculation of the population aged 0–4 in 2000. The first step is to divide the birth projections into males and females, using historical data on the male proportion of total births (0.51). Birth projections for each sex are calculated as

$$\text{Male births: } 57{,}261 \times 0.51 = 29{,}203$$
$$\text{Female births: } 57{,}261 - 29{,}203 = 28{,}058$$

Then, we compute the population aged 0–4 in 2000 by applying survival rates to the birth projections. The survival rates used in Table 7.3 were derived from life tables, but refer to a 2.5-year horizon rather than a five-year horizon because it is assumed that births occur evenly throughout the projection interval. Thus, the average baby is born in the middle of the interval and faces only a 2.5-year horizon to the end of the interval. In Table 7.3, we compute the population aged 0–4 as

$$\text{Male population: } 29{,}203 \times 0.99056 = 28{,}927$$
$$\text{Female population: } 28{,}058 \times 0.99316 = 27{,}866$$

If desired, deaths of children in this age group can be computed by subtracting the 2000 population from the projected births:

$$\text{Male infant and child deaths: } 29{,}203 - 28{,}927 = 276$$
$$\text{Female infant and child deaths: } 28{,}058 - 27{,}866 = 192$$

Final Projection Module. The final calculations combine the results from the mortality, migration, and fertility modules. For all but the youngest age group, the projected population at each age is calculated as the survived population, plus in-migrants minus out-migrants:

$$_{n}P_{x+z,t} = {_{n}}\text{SURVP}_{x+z,t} + {_{n}}\text{INMIG}_{x+z,l \text{ to } t} - {_{n}}\text{OUTMIG}_{x+z,l \text{ to } t}$$

Table 7.4. Projection of White Females, Broward County, 2000 (Model I)

| Age | 1995 population | 2000 surviving population | 1995–2000 | | | 2000 population[b] |
			Deaths	In-migrants	Out-migrants	
0–4	32,814	27,866[a]	192	0	0	27,866
5–9	28,867	32,672	142	7,493	7,643	32,522
10–14	26,044	28,841	26	6,946	6,082	29,705
15–19	22,733	25,998	46	6,491	7,705	24,784
20–24	25,987	22,674	59	10,360	8,851	24,183
25–29	34,929	25,912	75	14,008	9,987	29,933
30–34	43,345	34,799	130	12,792	11,535	36,056
35–39	45,137	43,132	213	12,078	9,385	45,825
40–44	42,035	44,855	282	10,687	7,438	48,104
45–49	37,618	41,634	401	8,260	5,663	44,231
50–54	30,660	37,054	564	7,973	4,325	40,702
55–59	24,778	29,943	717	6,535	4,840	31,638
60–64	25,955	23,891	887	6,700	4,382	26,209
65–69	32,125	24,602	1,353	7,090	4,155	27,537
70–74	36,920	29,663	2,462	5,498	2,914	32,247
75–79	35,118	32,563	4,357	4,465	3,106	33,922
80–84	28,941	28,524	6,594	2,793	2,069	29,248
85+	20,136	28,334	20,743	2,868	2,215	28,987
Total	574,142	562,957	39,243	133,037	102,295	593,699

[a]Surviving births 1995–2000.
[b]Projected population = surviving population + in-migrants − out-migrants.

Although the notation is somewhat different, this equation is similar to the demographic balancing equation discussed in Chapter 2. The only difference is the absence of births, which are accounted for in the fertility module and provide the basis for projecting the youngest age group. The final projection (P_t) is the sum of the projections for all the age groups.

Table 7.4 shows the complete projection for white females, including components of change for each age cohort. Migration for the population aged 0–4 is shown as zero because it is captured in the projection of births. Overall, the number of white females was projected to increase from 574,142 to 593,699 between 1995 and 2000.

It is easy to calculate the overall components of change from Table 7.4. Births can be computed by adding deaths to children less than age 5 (192) to the surviving population aged 0–4 (27,866), yielding a projection of 28,058 white female births. This is considerably less than the projection of white female deaths (39,243), which implies that the projected white female population will experience a natural decrease of 11,185. This is a direct result of Broward County's top-heavy age structure: 27% of the white female population was aged 65 or older in 1995. However, because in-migrants exceeded out-migrants by 30,742, the number of white females in Broward County was still projected to increase by 19,557 between 1995 and 2000.

Again, we emphasize that these projections of births and deaths are approximations for the numbers that actually occurred within the county. For most places, differences between projected births and deaths and those that occurred within the county will be very small. For places that have high rates of in- or out-migration, however, the differences may be substantial.

Net Migration (Model II)

We illustrate Model II using the same data and assumptions as Model I, with one major difference: Instead of using gross migration rates, we use net migration rates. The numerator used for calculating these rates is the difference between the number of in-migrants and out-migrants from 1985–1990. The denominator is the adjusted U.S. white female population in 1985 (i.e., the number of white females in the United States minus the number in Broward County). Alternatively, we could have used the white female population of Broward County as the denominator. Chapter 6 discusses several issues regarding the construction and interpretation of net migration rates.

Mortality Module. The basic steps in the projection process (surviving the population, projecting migration, projecting fertility, and summing the components) are the same for a net migration model as a gross migration model. In fact, the computations in the mortality module are identical and are not repeated here; the reader is referred to Table 7.1 for a review of those computations. In the description that follows, we discuss only the equations in Model II that differ from those in Model I.

Migration Module. Net migration models require only one set of migration rates. For each age group, we project net migration by multiplying net migration rates by the adjusted U.S. population in the launch year:

$$_{n}\text{NETMIG}_{x+z,l \text{ to } t} = {}_{n}\text{AUSP}_{x,l} \times {}_{n}\text{NETMIGRATE}_{x,l \text{ to } t}$$

where NETMIG is the net migration projection, AUSP is the adjusted U.S. population, NETMIGRATE is the z-year net migration rate, x is the youngest age in an age group, n is the number of years in the age group, l is the launch year, t is the target year, and z is the interval between the launch and target years. If the county population had been used as the denominator in constructing the net migration rates, the county population in the launch year (P) would have been used in this equation instead of AUSP.

Table 7.5. Projection of Net Migration for White Females,
Broward County, 1995–2000 (Model II)

Age in 1995	Age in 2000	1995 adjusted U.S. population[a]	Net migration rate	1995–2000 net migration[b]
0–4	5–9	7,493,186	0.00023	1,723
5–9	10–14	7,389,133	0.00028	2,069
10–14	15–19	7,292,956	0.00004	292
15–19	20–24	6,953,267	0.00037	2,573
20–24	25–29	6,969,013	0.00085	5,924
25–29	30–34	7,569,071	0.00049	3,709
30–34	35–39	8,880,655	0.00047	4,174
35–39	40–44	9,133,863	0.00044	4,019
40–44	45–49	8,428,965	0.00028	2,360
45–49	50–54	7,451,382	0.00049	3,651
50–54	55–59	5,941,340	0.00042	2,495
55–59	60–64	4,890,222	0.00065	3,179
60–64	65–69	4,545,045	0.00075	3,409
65–69	70–74	4,698,875	0.00055	2,584
70–74	75–79	4,421,080	0.00025	1,105
75–79	80–84	3,535,882	0.00007	248
80+	85+	4,944,923	−0.00008	−396
Total				43,118

[a]Adjusted U.S. population = 1995 U.S. population − 1995 Broward County population.
[b]Net migration = net migration rate × adjusted U.S. population.

Table 7.5 shows the net migration projections for white females in Broward County. For females aged 45–49 in 2000, for example, we project net migration by multiplying the adjusted U.S. population aged 40–44 in 1995 by the age-specific net migration rate:

$$8,428,965 \times 0.00028 = 2,360$$

As in Model I, we combine the launch-year populations in the two oldest age groups before applying the migration rate for the oldest age group. Also, we do not directly project net migration for the population aged 0–4 in 2000.

A comparison of Tables 7.2 and 7.5 shows how the choice of the migration model can affect population projections. Model I (Table 7.2) generated net migration of 30,742 from 1995–2000, whereas Model II (Table 7.5) generated net migration of 43,118, a difference of about 40%. Although not shown here, net migration rates based on the county population rather than on the adjusted U.S. population would have generated an even larger net migration flow (49,729). The different models also lead to differences in age distribution. These differences illustrate the impact of the choice of the migration model *itself*, independent of the data and assumptions used in applying the model.

Fertility Module. Seven of the eight equations shown in the fertility module for Model I remain unchanged in Model II. The only equation that changes is the one that identifies the at-risk female population, in which projected net migration replaces separate projections of in- and out-migration:

$$_n\text{ATRISKFP}_{x,t} = {_n}\text{FP}_{x,l} - (0.5 \times {_n}\text{FD}_{x,l\text{ to }t}) + {_n}\text{FNETMIG}_{x+z,l\text{ to }t}$$

where ATRISKFP is the at-risk female population, FP is the female population, FD is female deaths, FNETMIG is the projection of female net migration, and n, x, z, l, and t are as defined previously.

Table 7.6 shows the calculation sequence for births and the population aged 0–4 for whites in Broward County. Births for each age group are computed by multiplying the at-risk female population by the ASBR. The total number of births is separated into males and females using historical proportions (0.51 and 0.49). Survival rates are applied, leading to the final projection of the white population aged 0–4 in 2000. Comparing Tables 7.6 and 7.3, we find that Model II leads to 2,912 more births than Model I. This difference is caused primarily by the higher migration projections in Model II, which increased the number of women in their childbearing ages.

Final Projection Module. The final calculations combine the results from the mortality, migration, and fertility modules. The basic equation is about the same as in Model I, except that separate terms for in- and out-migration are replaced by a single term for net migration:

$$_n\text{P}_{x+z,t} = {_n}\text{SURVP}_{x+z,t} + {_n}\text{NETMIG}_{x+z,l\text{ to }t}$$

Table 7.7 presents the complete projection of white females in Broward County in 2000. Comparing Tables 7.7 and 7.4, we find that Model II produced projections that were 13,793 higher than the projections produced in Model I. Although this is only a 2% difference, the differences between projections from the two models would become greater as the horizon extended further into the future. It should also be noted that the impact of model differences is considerably greater when the focus is on population *change* rather than population *size*: Projected population growth from 1995 to 2000 is 71% larger in Model II than Model I. This difference is due almost entirely to the higher migration projection.

Hamilton–Perry (Model III)

Hamilton and Perry (1962) proposed the use of cohort-change ratios as a shortcut for applying the cohort-component method. The Hamilton–Perry method

Table 7.6. Projection of the White Population Aged 0–4, Broward County, 2000 (Model II)

| Age in 1995 | Age in 2000 | Five-year birth rate | | 1995 female population | Deaths | Net migration | At-risk population[c] | Births[d] |
		Original[a]	Adjusted[b]					
10–14	15–19	0.00185	0.08040	26,044	46	292	26,313	2,116
15–19	20–24	0.15895	0.28058	22,733	59	2,573	25,277	7,092
20–24	25–29	0.40220	0.45948	25,987	75	5,924	31,874	14,645
25–29	30–34	0.51675	0.45995	34,929	130	3,709	38,573	17,742
30–34	35–39	0.40315	0.28393	43,345	213	4,174	47,413	13,462
35–39	40–44	0.16470	0.09393	45,137	282	4,109	49,015	4,604
40–44	45–49	0.02315	0.01158	42,035	401	2,360	44,195	512
Total				240,210	1,206	23,051	262,660	60,173

	Share of births	1995–2000 births	Survival rates[g]	2000 population 0–4[h]	1995–2000 child deaths[i]
Males	0.51	30,688[e]	0.99056	30,398	290
Females	0.49	29,485[f]	0.99316	29,283	202
Total	1.00	60,173		58,681	492

[a]5-year birth rate = 1-year birth rate × 5.
[b]Adjusted birth rate = average of birth rates for adjacent age groups.
[c]At-risk population = 1995 population − (0.5 × deaths) + net migration.
[d]Births = adjusted birth rate × at-risk population.
[e]Male births = total births × 0.51.
[f]Female births = total births − male births.
[g]Probability of surviving from birth to age 2.5.
[h]Projected population aged 0–4 = births × survival rates.
[i]Child deaths = births − 2000 population aged 0–4.

Table 7.7. Projection of White Females, Broward County, 2000 (Model II)

Age	1995 population	2000 surviving population	1995–2000 Deaths	1995–2000 Net migration	2000 population[b]
0–4	32,814	29,283[a]	202	0	29,283
5–9	28,867	32,672	142	1,723	34,395
10–14	26,044	28,841	26	2,069	30,910
15–19	22,733	25,998	46	292	26,290
20–24	25,987	22,674	59	2,573	25,247
25–29	34,929	25,912	75	5,924	31,836
30–34	43,345	34,799	130	3,709	38,508
35–39	45,137	43,132	213	4,174	47,306
40–44	42,035	44,855	282	4,019	48,874
45–49	37,618	41,634	401	2,360	43,994
50–54	30,660	37,054	564	3,651	40,705
55–59	24,778	29,943	717	2,495	32,438
60–64	25,955	23,891	887	3,179	27,070
65–69	32,125	24,602	1,353	3,409	28,011
70–74	36,920	29,663	2,462	2,584	32,247
75–79	35,118	32,563	4,357	1,105	33,668
80–84	28,941	28,524	6,594	248	28,772
85+	20,136	28,334	20,743	−396	27,938
Total	574,142	564,374	39,253	43,118	607,492

[a]Surviving births 1995–2000.
[b]Projected population = surviving population + net migration.

is similar to a net migration model in which the denominator for migration rates is the population of the area to be projected. The major difference is that it treats mortality and migration as a single unit rather than separately. In addition, the fertility component is often simplified by using child–woman ratios rather than ASBRs.

The Hamilton–Perry method projects population by age and sex by using cohort-change ratios (CCR) computed from data in the two most recent censuses. These ratios are the same as the census survival rates discussed in Chapter 4, although the notation is somewhat different:

$$_n\text{CCR}_x = {_nP_{x+y,l}} / {_nP_{x,b}}$$

where $_nP_{x+y,l}$ is the population aged $x+y$ to $x+y+n$ in the most recent census (l), $_nP_{x,b}$ is the population aged x to $x+n$ in the second most recent census (b), and y is the number of years between the two most recent censuses ($l-b$). Using the 1990 and 2000 Censuses as an example, the CCR for the population aged 20–24 in 1990 would be

$$_5\text{CCR}_{20} = {_5P_{30,2000}} / {_5P_{20,1990}}$$

In the United States, the Hamilton–Perry method is most commonly used to project five-year age groups in 10-year intervals. In Canada—which has quinquennial censuses—projection intervals of five years could be constructed. The method can easily be adapted to provide projections for additional characteristics such as race or ethnicity.

The following is the basic formula for a Hamilton–Perry projection:

$$_nP_{x+z,t} = {}_n\text{CCR}_x \times {}_nP_{x,l}$$

where z is the number of years between l and t. Using data from the 1990 and 2000 Censuses, for example, the formula for projecting the population aged 30–34 in 2010 is

$$_5P_{30,2010} = ({}_5P_{30,2000} / {}_5P_{20,1990}) \times {}_5P_{20,2000}$$

The quantity in parentheses is the CCR for the population aged 20–24 in 1990. If it is assumed this ratio will remain constant, the projection for the population aged 30–34 in 2010 is the population aged 20–24 in 2000 multiplied by the CCR.

When there are 10 years between censuses, 10–14 is the youngest age group for which projections can be made using CCRs. How can the population aged 0–4 and 5–9 be projected? Hamilton and Perry (1962) used the most recent agespecific birth rates, but this procedure requires data on births by age of mother; these data are seldom available for subcounty areas. We prefer a simpler approach that does not require any data beyond that available in the decennial census (Shryock & Siegel, 1973). This approach uses two child-woman ratios (CWRs) from the most recent census and applies them to the projected female population in the appropriate age groups. To project the population aged 0–4, the CWR is defined as the population aged 0–4 divided by the female population aged 15–44. To project the population aged 5–9, the CWR is defined as the population aged 5–9 divided by the female population aged 20–49. The implementation of these ratios requires four projection equations—two for females and two for males:

$$\text{Females aged 0–4: } {}_5FP_{0,t} = ({}_5FP_{0,l} / {}_{30}FP_{15,l}) \times {}_{30}FP_{15,t}$$
$$\text{Males aged 0–4: } {}_5MP_{0,t} = ({}_5MP_{0,l} / {}_{30}FP_{15,l}) \times {}_{30}FP_{15,t}$$
$$\text{Females aged 5–9: } {}_5FP_{5,t} = ({}_5FP_{5,l} / {}_{30}FP_{20,l}) \times {}_{30}FP_{20,t}$$
$$\text{Males aged 5–9: } {}_5MP_{5,t} = ({}_5MP_{5,l} / {}_{30}FP_{20,l}) \times {}_{30}FP_{20,t}$$

where FP is the female population, MP is the male population, l is the launch year, and t is the target year. For example, the formula for projecting females aged 0–4 in 2010 is

$$_5FP_{0,2010} = ({}_5FP_{0,2000} / {}_{30}FP_{15,2000}) \times {}_{30}FP_{15,2010}$$

Rather than using the CWR from the most recent census, another option is to take an average of the CWRs from the two most recent censuses. A third option is to assume that the trend in the CWRs observed between the two most recent

Table 7.8. Projection of the Female Population,
Census Tract 188.00, San Diego County, 2000 (Model III)

Age	1980 population	1990 population	CCR[a]	2000 population[b]
0–4	159	298	n/a	565
5–9	186	394	n/a	748
10–14	281	408	2.566	765
15–19	288	309	1.661	655
20–24	132	238	0.847	346
25–29	143	286	0.993	307
30–34	190	368	2.788	664
35–39	222	456	3.189	912
40–44	219	448	2.358	868
45–49	197	401	1.806	824
50–54	208	377	1.721	771
55–59	305	346	1.756	704
60–64	275	378	1.817	685
65–69	301	440	1.443	499
70–74	191	321	1.167	441
75–79	93	235	0.781	344
80–84	35	143	0.749	240
85+	38	95	0.572	271
Total	3,463	5,941		10,609

Child–woman ratios (CWR)

	1980	1990	2000[c]
CWR-1: Ages 0–4/15–44	0.13317	0.14157	0.15050
CWR-2: Ages 5–9/20–49	0.16863	0.17934	0.19073

[a]CCR = 1990 population age $(x + 10)$/1980 population age (x).
[b]For ages 0–4, population = CWR-1 × projected population 15–44. For ages 5–9, population = CWR-2 × projected population 20–49. For all other ages, projected population = CCR × 1990 population.
[c]2000 CWR = 1990 CWR × (1990 CWR/1980 CWR).

censuses will continue during the projection interval. Other options could also be developed, such as having CWRs for small areas converge over time with those projected for a larger area.

Table 7.8 illustrates the Hamilton–Perry method (Model III), using the assumption that the historical trends in CWRs will continue. In this example, we prepared a projection for females living in a census tract in San Diego County, California in 2000. As the table shows, the method requires only a limited set of calculations. For example, the female population aged 10–14 in 2000 is calculated as

$$(408/159) \times 298 = 765$$

Projections of the oldest age group differ slightly from projections for the other age groups. The calculations for the CCR require summing the three oldest age groups to get the population aged 75+ in the base year. A ratio of the population aged 85+ in the launch year (1990) to the population aged 75+ in the base year (1980) forms the basis for projecting the population aged 85+ in the target year (2000):

$$CCR_{75+} = P_{85+,l} / P_{75+,b} = 95 / (93 + 35 + 38) = 0.572$$

Population aged 85+ in 2000: $0.572 (235 + 143 + 95) = 271$

We project the two youngest age groups by applying trended CWRs to females aged 15–44 and 20–49:

Females aged 0–4: $0.15050 (655 + 346 + 307 + 664 + 912 + 868) = 565$

Females aged 5–9: $0.19073 (346 + 307 + 664 + 912 + 868 + 824) = 748$

This application of the Hamilton–Perry method holds CCRs constant over the projection interval. It is also possible to average CCRs from several recent censuses or to extrapolate the trends observed between censuses. However, time series approaches can be applied only if geographic boundaries remain constant over time. For many subcounty areas, this is difficult to achieve even for two censuses, much less three or four (Pittenger, 1976). Another approach is to construct ratios of CCRs for small areas (e.g., census tracts) to CCRs for a larger area (e.g., county) and apply those ratios to projections of the larger area's CCRs. This approach is similar to the synthetic methods discussed in Chapters 4 and 5.

COMPARING MODELS I, II, AND III

As a general principle, we believe that gross migration models are preferable to net migration models. Gross migration is closer to the true migration process than net migration. It can be related to identifiable origin and destination populations and provides rates that approximate migration probabilities. In addition, gross migration models may provide more accurate forecasts than net migration models, at least for long projection horizons and rapidly growing areas. Chapter 6 provides a more detailed discussion of these issues.

Gross migration models, however, require numerous computations and a great deal of base data (especially for multiregional models). These data are often unavailable for small areas. Net migration models, on the other hand, require less data and are considerably simpler to apply. Perhaps more important, they can be used when the data required by gross migration models are difficult or impossible to obtain. Although gross migration models are conceptually superior, net migration models are widely used in practice.

The major advantage of the Hamilton–Perry method is that it requires less data than a complete net migration model. Instead of mortality, fertility, migration, and total population data, the Hamilton–Perry method simply requires population data by age and sex from the two most recent censuses. Consequently, it is much quicker, easier, and cheaper to implement than a full-blown cohort-component model.

One caveat should be mentioned, however. The Hamilton–Perry method is essentially a set of cohort growth rates applied to a beginning population. As we show in Chapter 13, the application of constant growth rates often leads to large forecast errors and a strong upward bias for rapidly growing places. Consequently, it is advisable to control Hamilton–Perry projections to an independent set of population projections. When this is done, the Hamilton–Perry method projects only the age-sex composition of the population, not its total size. We provide several examples of controlling techniques in Chapter 11.

CONCLUSIONS

The cohort-component method is a mainstay in the demographer's toolbox and is not likely to relinquish its lofty position any time soon. It provides a theoretically complete model that accounts for the individual components of growth and for the impact of changes in demographic composition over time. It can incorporate many different application techniques, types of data, and assumptions regarding future trends. It can be used at almost any level of geography, from the nation down to states, counties, and subcounty areas. Perhaps most important, it provides projections of the components of growth and changes in demographic composition, as well as projections of total population.

As useful as the cohort-component method is, however, it also has its limitations. Perhaps the most important is that it is very data-intensive and requires a large number of computations. A full cohort-component model requires mortality, fertility, migration, and population data by age and sex (and perhaps other characteristics as well). Collecting, verifying, and cleaning these data is a tedious and time-consuming process, especially when geographic boundaries have changed. The number of computations involved in applying the method is also very large. Consequently, the cohort-component method is relatively expensive to apply. As we show in the next chapter, other projection methods are simpler, less data-intensive, and less costly.

Large data requirements preclude the use of some forms of the cohort-component method at some levels of geography. Although seldom a problem at the state or even the county level, the lack of complete, reliable data presents a formidable problem for constructing cohort-component projections at the subcounty level. Birth and death rates are not routinely available for most subcounty areas. Migration rates are an even greater challenge, as decennial census data are available only for census tracts and cover only in-migration. IRS migration data are not tabulated below the county level. Because of these data problems, the Hamilton–Perry method is often the best cohort-component method to use for subcounty projections.

A final limitation of the cohort-component method is that—although it provides the mathematical framework for making projections for cohorts and components of growth—it provides no guidance regarding the choice of assumptions that will lead to reasonable forecasts. How much will mortality rates decline over the next 20 years? Will fertility rates go up or down? Will migration follow the patterns observed over the last 10 years or revert to the patterns observed during the previous 10 years? What economic, social, psychological, political, or biological factors might cause recent demographic trends to change course? Nothing in the cohort-component method itself provides answers to these questions.

As Chapters 4–6 suggest, we must seek answers to these questions elsewhere. Models based solely on historical population trends are limited in the range of theoretical, policy, and planning questions they can address. However, structural models can be developed that incorporate explanations of the determinants of population growth directly into the projection method (see Chapters 9 and 10). These models can be applied within the framework of the cohort-component method, greatly increasing its usefulness for a variety of purposes. This highlights one of the most important attributes of the cohort-component method; namely, its flexibility. The cohort-component method can accommodate a wide variety of functional forms, application techniques, and data sources. It is not surprising that it continues to be the most widely used of all the population projection methods.

CHAPTER 8

Trend Extrapolation Methods

Early versions of the cohort-component method were developed in the late nineteenth and early twentieth centuries, but the method did not become widely used until the middle of the twentieth century. Before that time projections were typically made by extrapolating historical population trends into the future by using any one of a number of mathematical formulas. Projections based on trend extrapolations were made by such eminent "demographers" as Benjamin Franklin, Thomas Jefferson, and Abraham Lincoln (Dorn, 1950). Despite their simplicity and lack of theoretical content and demographic detail, early applications of this approach often produced reasonably accurate forecasts of total population, even for projection horizons extending well into the future (e.g., Pritchett, 1891; Pearl & Reed, 1920).

Trend extrapolation methods were largely overshadowed by other methods by the middle of the twentieth century, but have made a comeback in recent years as new methods were developed and detailed evaluations of forecast accuracy and utility were conducted. Relatively low costs and small data requirements make these methods particularly useful for small-area projections. We discuss other characteristics of trend extrapolation methods (including their forecast accuracy) in Chapters 12 and 13.

The defining characteristic of trend extrapolation methods is that future values of any variable are determined solely by its historical values. There are many different ways to measure historical values and project them into the future (e.g., Davis, 1995; Irwin, 1977; Isserman, 1977; Pittenger, 1976). In this chapter we describe and illustrate methods that have been commonly used for state and local population projections. Descriptions of trend extrapolation methods used in other fields can be found in Armstrong (2001a), Mahmoud (1984), Makridakis, Wheelwright, and Hyndman (1998), and Schnaars (1986).

We divide trend extrapolation methods into three categories. *Simple extrapo-*

lation methods are those that have simple mathematical structures and require data from only two points in time. We cover three simple methods: linear extrapolation, geometric extrapolation, and exponential extrapolation. *Complex extrapolation methods* are those that have more complex mathematical structures and require data from a number of points in time; they also require an algorithm for estimating the model's parameters (e.g., intercept, slope). We cover four complex methods: linear trend models, polynomial curve fitting, logistic curve fitting, and ARIMA time series models. *Ratio extrapolation methods* are those in which the population of a smaller area is expressed as a proportion of the population of a larger area in which the smaller area is located (e.g., county population as a share of state population). We cover three ratio methods: constant-share, shift-share, and share-of-growth.

We illustrate these methods using data for two planning areas in the unincorporated area of San Diego County, California: Lincoln Acres and San Dieguito. Lincoln Acres is an older, largely built-out neighborhood with little room available for further expansion. San Dieguito is a newer neighborhood that has substantial land available for additional population growth. Population estimates for these two areas from 1975–1995 are shown in Table 8.1 and Figure 8.1. These estimates were adjusted to reflect constant boundaries over the 20-year period. Table 8.1 also shows population estimates for the unincorporated area of San Diego County, which are needed to apply the ratio methods.

For consistency, we use 1975–1995 as the base period for all 10 methods. For the methods that require data for only two time points, we use the population estimates for 1975 and 1995. For the methods that require a large number of time points, we use all of the annual population estimates between 1975 and 1995. Using each of the 10 methods, projections for Lincoln Acres and San Dieguito were made in five-year intervals from 1995 to 2015.

Although there are exceptions, trend extrapolation methods are used much more frequently for projecting total population than for projecting population subgroups (e.g., race or ethnic groups). As discussed in Chapters 4–6, some extrapolation methods can also be used for projecting individual components of growth in the cohort-component method. In this chapter, however, we focus solely on projections of total population.

SIMPLE METHODS

Simple extrapolation methods have small data requirements, simple mathematical structures, and are easy to apply. Given these characteristics, they are particularly useful when few historical data are available, when production times are short, when budgets are tight, or when projections are needed for a large number of areas. In this section, we discuss three commonly used simple methods: linear, geometric, and exponential.

Table 8.1. Population of the Unincorporated Area,
San Dieguito, and Lincoln Acres, 1975–1995[a]

Year	Unincorporated area	Lincoln Acres	San Dieguito
1975	204,922	1,829	3,256
1976	216,899	1,842	3,674
1977	228,247	1,839	4,106
1978	239,416	1,825	4,559
1979	251,408	1,808	5,051
1980	264,270	1,757	5,658
1981	278,445	1,807	5,995
1982	280,524	1,797	6,134
1983	284,147	1,843	6,346
1984	295,367	1,877	6,596
1985	305,483	1,884	6,963
1986	322,607	1,996	7,883
1987	336,876	1,959	8,498
1988	355,522	1,964	9,225
1989	374,788	1,977	9,982
1990	397,783	1,982	10,804
1991	407,081	1,975	10,923
1992	415,096	1,994	11,128
1993	419,669	2,041	11,420
1994	427,078	2,039	11,518
1995	427,564	2,032	11,616
% change 1975–1995	108.6%	11.1%	256.8%

[a]Based on geographic boundaries in 1995.
Source: Unpublished estimates, San Diego Association of Governments, 1997.

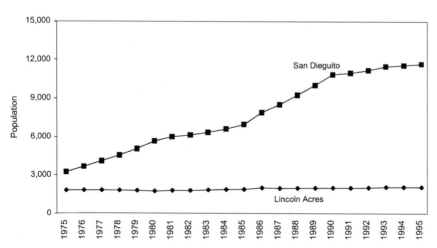

Figure 8.1. Population of Lincoln Acres and San Dieguito, 1975–1995. (*Source*: Unpublished estimates, San Diego Association of Governments, 1997)

Linear

In the linear extrapolation method (LINE), it is assumed that the population will increase (decrease) by the same number of persons in each future year as the average annual increase (decrease) observed over the base period. Average annual absolute change can be computed as

$$\text{AAAC} = (P_l - P_b)/y$$

where P_l is the population in the launch year, P_b is the population in the base year, and y is the number of years in the base period. For example, the AAAC for Lincoln Acres and San Dieguito from 1975–1995 can be calculated as

Lincoln Acres: AAAC = $(2,032 - 1,829)/20 = 10.15$ persons per year

San Dieguito: AAAC = $(11,616 - 3,256)/20 = 418.00$ persons per year

Population projections based on the LINE method can be expressed as

$$P_t = P_l + z\,(\text{AAAC})$$

where P_t is the population in the target year and z is the number of years in the projection horizon. For example, LINE projections for Lincoln Acres and San Dieguito in 2000 can be calculated as

Lincoln Acres: $P_{2000} = 2,032 + 5\,(10.15) = 2,083$

San Dieguito: $P_{2000} = 11,616 + 5\,(418) = 13,706$

Different base periods and projection horizons can be accommodated simply by changing the years used to define the base year (b), launch year (l), and target year (t). For example, projections for 2005 can be calculated as

Lincoln Acres: $P_{2005} = 2,032 + 10\,(10.15) = 2,134$

San Dieguito: $P_{2005} = 11,616 + 10\,(418) = 15,796$

Geometric

In the geometric extrapolation method (GEO), it is assumed that the population will increase (decrease) at the same annual percentage rate during the projection horizon as during the base period. The average annual geometric growth rate r can be computed as

$$r = (P_l/P_b)^{1/y} - 1$$

Given this formula for computing the geometric growth rate, a population projection using the GEO method can be expressed as

$$P_t = P_l(1 + r)^z$$

Average annual geometric growth rates for Lincoln Acres and San Dieguito from 1975–1995 can be computed as follows:

Lincoln Acres: $r = (2,032/1,829)^{1/20} - 1 = 0.00528$, or 0.528% per year

San Dieguito: $r = (11,616/3,256)^{1/20} - 1 = 0.06566$, or 6.566% per year

GEO projections for 2000 can then be calculated as

Lincoln Acres: $P_{2000} = 2,032\,(1 + 0.00528)^5 = 2,086$

San Dieguito: $P_{2000} = 11,616\,(1 + 0.06566)^5 = 15,964$

Exponential

Geometric growth rates are based on compounding at discrete time intervals (in this example, once each year). Another approach to calculating growth rates is based on continuous compounding. This approach more nearly represents the dynamics of population growth because growth generally occurs continuously rather than at discrete intervals. Under this approach, growth rates are computed as

$$r = [\ln(P_l/P_b)]/y$$

where r is the average annual exponential growth rate and ln is the natural logarithm. Average annual exponential growth rates for Lincoln Acres and San Dieguito from 1975–1995 can be computed as follows:

Lincoln Acres: $r = [\ln(2,032/1,829)]/20 = 0.00526$, or 0.526% per year

San Dieguito: $r = [\ln(11,616/3,256)]/20 = 0.06359$, or 6.359% per year

Growth rates based on an exponential model are similar to those based on a geometric model, especially for places that are not growing particularly rapidly. The average annual geometric and exponential growth rates from 1975 to 1995 for slowly growing Lincoln Acres were 0.528% and 0.526%, respectively. For rapidly growing San Dieguito, they were 6.566% and 6.359%, respectively. Exponential growth rates are always smaller than geometric growth rates because they reflect continuous compounding rather than compounding at discrete intervals.

Population projections based on an exponential growth model (EXPO) can be calculated as

$$P_t = P_l e^{rz}$$

where e is the base of the natural logarithm (approximately 2.71828). For example, EXPO projections for Lincoln Acres and San Dieguito for 2000 can be calculated as

Lincoln Acres: $P_{2000} = 2,032\,e^{(.00526)5} = 2,086$

San Dieguito: $P_{2000} = 11,616\,e^{(.06359)5} = 15,964$

For both areas, the EXPO projections are identical to the GEO projections for the year 2000. Even when carried out to 2015, the projections differ by only one for Lincoln Acres and by four for San Dieguito. If applied consistently, the GEO and EXPO methods will provide virtually identical projections. Rather than repeating the same results, we focus primarily on the EXPO method in the remainder of this book.

The EXPO method can lead to very high projections in rapidly growing places. Consider San Dieguito, for example. A continuation of its historical growth rate of 6.4% per year would double its population about every 11 years. In 44 years its population would be about 16 times larger than in the launch year; in 88 years, it would be about 256 times larger. Clearly, these are not reasonable forecasts. The EXPO method must be used cautiously for long-range projections, especially for rapidly growing places.

COMPLEX METHODS

Complex extrapolation methods differ from simple methods in several ways: They incorporate data from a number of points in time, have more complex mathematical structures, and require an algorithm for estimating the method's parameters. Because complex extrapolation methods incorporate more base data, they provide a more complete picture of the historical pattern of population change than simple extrapolation methods. Their more complex mathematical structures provide a wider range of possibilities regarding population trends than simpler methods. In addition, the application of statistical algorithms to estimate the model's parameters provides a basis for constructing prediction intervals. However, complex extrapolation methods are considerably more difficult to implement than simple trend or ratio extrapolation methods. Whether they produce more accurate forecasts than simpler methods will be discussed in Chapter 13.

There are three basic steps for projecting population using a complex extrapolation method. The first is to assemble historical population data at equal time intervals between the base year and launch year. Applications typically use annual time intervals (e.g., Pflaumer, 1992; Saboia, 1974) or intervals between censuses (e.g., Isserman, 1977; Leach, 1981). The data must be based on consistently defined geographic boundaries for each point in time.

The second step is to choose a mathematical model and estimate its parameters through a process called *curve fitting* (Alinghaus, 1994). The choice of the model should reflect the analyst's judgment regarding the nature of population change and the most likely future population trend (Davis, 1995). Analysts use graphs, statistical correlation measures, and the analysis of residuals to evaluate how well the model fits the historical data; however, a close fit does not guarantee an accurate (or even a reasonable) projection.

The essential assumption underlying both simple and complex extrapolation methods is that the functional relationship between population change and time will remain constant during the course of the projection horizon. For complex methods, this implies that the model's coefficients will describe future relationships as well as they described past relationships. If these relationships change, projections are not likely to provide accurate forecasts regardless of how well the model fit the data during the base period.

The final step is to use the mathematical model and estimated parameters to prepare the population projections. In this section, we describe and illustrate four complex trend extrapolation methods: linear trend, polynomial curve fitting, logistic curve fitting, and ARIMA time series models. Except for logistic curve fitting, we estimate each model's parameters using the Statistical Package for the Social Sciences (SPSS). For the logistic curve, we use a lesser known package called the Number Cruncher Statistical System (NCSS). The projections made as illustrations of these methods are based on 21 data points for Lincoln Acres and San Dieguito that represent annual population estimates from 1975 to 1995. For ease of interpretation and calculation, we express time as integers ranging from 1 to 41 (i.e., 1 = 1975, 2 = 1976, ..., 21 = 1995, 22 = 1996, ..., and 41 = 2015).

Linear Trend

The linear trend model is the simplest and most familiar of the complex trend extrapolation methods. It is based on the assumption that the population will increase or decrease by a constant numerical amount, as determined by historical population change. This assumption is identical to the assumption underlying the LINE method discussed earlier, but is operationalized differently. The linear trend model is based on the equation for a straight line:

$$Y = a + bX$$

where Y is the dependent variable (e.g., total population); X is the independent variable (e.g., time); a is the constant term (or intercept); and b is the slope of the line. The terms X and Y are the model's variables. They represent the data used in estimating the model and take on values that vary with each observation. The terms a and b are the model's parameters (or coefficients). They represent the statistical relationships between the model's independent and dependent variables. They take on values that remain constant for any particular application of the model but vary from one application to another.

In a diagram, the intercept reflects the population value where the line estimated by the model crosses the Y-axis (i.e., where $X = 0$). The slope measures the annual change in the population. A positive slope reflects a population that is increasing over time and a negative slope reflects a population that is declining. A slope of zero reflects an unchanging population. Ordinary least squares (OLS) regression techniques are used to estimate the parameters a and b.

For San Dieguito, the estimated intercept and slope are 2,675.976 and 455.14935, respectively. The slope indicates that the population increases by 455 persons per year. The equation has an adjusted r^2 (or coefficient of determination) of 0.980. This is a measure of the "goodness of fit" of an equation, showing the proportion of variation in the dependent variable that can be attributed to variation in the independent variable(s). Values of this measure fall between 0 and 1. The high value in San Dieguito shows that a linear model fits the historical data very well.

For Lincoln Acres, the estimated intercept and slope are 1,761.367 and 13.32597, respectively. The slope shows that the population increases by only 13 persons per year. The adjusted r^2 of 0.798 shows that a linear model does not fit historical population trends as well for Lincoln Acres as for San Dieguito.

We construct population projections by plugging the estimated parameters into the linear trend model as follows:

$$P_t = a + b(\text{TIME}) + \text{CALIB}$$

where P_t is the population in the target year; a and b are the estimated intercept and slope, respectively; TIME is the value corresponding to the target year; and CALIB is an adjustment factor.

The adjustment factor (CALIB) requires explanation. In any curve fitting procedure, it would be an unusual coincidence if the estimated and observed launch-year population values were identical. In San Dieguito, for example, the 1995 population estimated from the linear trend equation is 12,234, or 618 higher than the observed population (11,616). For Lincoln Acres, the two figures are much closer together (2,041 estimated and 2,032 observed). CALIB is an adjustment factor that makes the projected population consistent with the launch-year population. It is computed by subtracting the estimated population from the observed population in the launch year (in this example, 1995):

Lincoln Acres: $2,032 - 2,041 = -9$

San Dieguito: $11,616 - 12,234 = -618$

The CALIB adjustment produces a parallel shift in the trend line that makes it pass directly through the launch-year population. Other adjustment factors can also be used (e.g., Isserman, 1977; Treyz, 1995). We recommend the inclusion of adjustment factors when applying complex extrapolation methods.

The 2000 population projections from the linear trend model for Lincoln Acres and San Dieguito can be calculated as follows:

Lincoln Acres: $P_{2000} = 1,761.367 + 13.32597(26) - 9 = 2,099$

San Dieguito: $P_{2000} = 2,675.976 + 455.14935(26) - 618 = 13,892$

Projections for different target years can be made by changing the TIME value in the equation. For example, a projection for 2007 would use a TIME value of 33 and a projection for 2015 would use a TIME value of 41. In most instances, population projections from the linear trend model are similar to projections from the LINE method.

Polynomial Curve Fitting

Like the EXPO and GEO methods discussed earlier, a polynomial curve can be useful for basing projections on nonlinear patterns (i.e., patterns in which annual population change is not a constant numerical value). The general formula for a polynomial curve is

$$Y = a + b_1X + b_2X^2 + b_3X^3 + \dots + b_nX^n$$

where Y is the dependent variable (e.g., total population) and X is the independent variable (e.g., time). In contrast to the linear trend model, a polynomial curve has more than one term that reflects the independent variable; consequently, there are more parameters to estimate. The coefficients of a polynomial curve (a, b_1, b_2, ..., b_n) can be estimated using OLS regression techniques. These coefficients include both a measure of the linear trend (b_1) and measures of the nonlinear patterns (b_2, b_3, ..., b_n).

Polynomial curves can have any number of terms on the right-hand side of the equation. The highest exponent in the equation is called the *degree* of the polynomial. The linear trend model described earlier may be thought of as a first-degree polynomial; it contains only one term that reflects the independent variable (X). A second-degree polynomial contains X and X^2; a third-degree polynomial contains X, X^2, and X^3; and so forth. Polynomial curves of any degree can be used for population projections.

To illustrate the use of a polynomial curve for population projections, we use a second-degree polynomial (sometimes called a *quadratic function*). This function includes time (the linear term) and time squared (also called the *parabolic term*) on the right-hand side of the equation:

$$Y = a + b_1X + b_2X^2$$

where b_1 is the slope for the linear trend and b_2 is the slope for the nonlinear (parabolic) trend. A quadratic curve can produce a variety of growth scenarios, such as a population growing at an increasing rate, a population growing at a decreasing rate, a population declining at an increasing rate, or a population declining at a decreasing rate. For example, positive values for both b_1 and b_2 reflect a population in which annual increases are growing larger over time. A positive value for b_1 and a negative value for b_2 reflect a population in which annual increases are growing smaller over time; eventually this population will stop growing and start declining. As with EXPO and GEO, projections based on a quadratic curve can lead to very high (or low) projections for places that were growing (or declining) rapidly during the base period.

We use OLS regression techniques to estimate the coefficients in the quadratic equation. Using the resulting coefficients and an adjustment factor similar to CALIB, we develop projections for Lincoln Acres and San Dieguito as follows:

$$P_t = a + b_1 (\text{TIME}) + b_2 (\text{TIME}^2) + \text{CALIB}$$

For example, the 2000 projections for Lincoln Acres and San Dieguito are

Lincoln Acres: $P_{2000} = 1{,}798.681 + 3.59175\,(26) + 0.44247(26^2) - 37 = 2{,}154$

San Dieguito: $P_{2000} = 2{,}746.632 + 436.71751\,(26) + 0.837811(26^2) - 671 = 13{,}997$

For San Dieguito, the quadratic equation has an adjusted r^2 of 0.979, just below that of the linear trend model. This suggests that the parabolic term does not help describe population growth in San Dieguito beyond what can be accounted for by the linear term. Consequently, projections for San Dieguito from the quadratic and linear trend models are similar (Table 8.2).

For Lincoln Acres, the quadratic equation has an adjusted r^2 of 0.816, somewhat higher than the 0.798 for the linear trend model. This suggests that the parabolic term helps describe historical population growth in Lincoln Acres. The impact of the parabolic term is reflected in the projections for Lincoln Acres, which are 15.4% higher in the quadratic model than in the linear trend model in 2015 (Table 8.2).

Although any degree can be used, polynomials higher than third degree are seldom used for population projections. Nonlinearity in the historical data can also be accounted for by using curves based on logarithmic or other transformations of the base data (e.g., Davis, 1995; Isserman, 1977).

Logistic Curve Fitting

The extrapolation methods considered so far are not constrained by any limits to growth. In these methods, population growth (or decline) can go on forever. In many instances, this will not be a reasonable assumption. In particular, the compounding effects of exponential or geometric growth rates and some nonlinear models can lead to very high projections when carried too far into the future.

The logistic curve—one of the best-known growth curves in demography— deals with this problem by including an explicit ceiling (or upper limit) on the size of the population (e.g., Pittenger, 1976; Romaniuc, 1990). It depicts an S-shaped pattern that has an initial period of slow growth, followed by increasing growth rates, followed by declining growth rates that eventually approach zero as population size levels off at its upper limit. The idea of limits to growth is intuitively plausible and is consistent with Malthusian theories of population growth, geographic impediments such as swamps and deserts, growth constraints created by government policies, and the filling up of empty residential sites.

Due in large part to the work of Pearl and Reed (1920) and Yule (1925), the logistic curve was a popular projection method in the early decades of the twentieth century. Although its usefulness for projections has been questioned (e.g., Brass, 1974; Marchetti, Meyer, & Ausubel, 1996), several studies have shown that logistic curves often provide reasonably accurate population forecasts (e.g., Dorn, 1950; Leach, 1981). Other curves that contain asymptotic ceilings on population size include modified exponential and Gompertz models (e.g., Davis, 1995; Pittenger, 1976). In addition, modified exponential and hyperbolic curves may be useful for projecting rapidly declining populations because they set lower limits on population size (Davis, 1995).

Keyfitz (1968) gave a formula for a three-parameter logistic curve:

$$Y = a / [1 + b(e^{-cX})]$$

where Y is population, X is time, a is the upper asymptote or population ceiling, b and c are the other parameters that define the shape of the logistic curve, and e is the base of the natural logarithm. In using a logistic curve for population projections, one must determine the magnitude of the upper asymptote and the time required to reach it. These factors are based on the values of the three parameters (a, b, and c), which can be estimated by using iterative least squares techniques (Keyfitz, 1968). Other computational procedures for estimating these parameters are shown in Pittenger (1976) and Shryock and Siegel (1973).

We used the NCSS statistical package to estimate the parameters of the logistic curve because—unlike SPSS—its algorithm does not require a user-defined value for the upper asymptote. This is a useful feature because it takes some of the guesswork out of the parameter estimation process. For comparison, we ran the logistic model using both statistical packages and obtained almost identical projections when the NCSS estimate of the upper limit was used as input into the SPSS logistic curve algorithm.

For San Dieguito the estimates for the parameters a, b, and c in the logistic curve are 15,697.3, 4.21837, and 0.12639, respectively. The estimate for parameter a implies an upper limit of 15,697 for the population of San Dieguito. This equation has an adjusted r^2 of 0.985, which indicates that the logistic model fits the historical data relatively well.

For Lincoln Acres the parameter estimates are 30,760.1, 16.43878, and 0.00749, respectively. The estimate for parameter a implies an upper limit of 30,760 for the population of Lincoln Acres. This equation has an adjusted r^2 of 0.813, which indicates that the logistic model does not describe population change as well in Lincoln Acres as in San Dieguito.

We use the following formula to project the population for any target year by using the logistic curve model:

$$P_t = a / [1 + b(e^{-cX})] + \text{CALIB}$$

where a, b, and c are the estimated parameters, X is time, and CALIB is an adjustment factor similar to those described earlier. Using this formula, the populations in 2000 for Lincoln Acres and San Dieguito can be projected as follows:

$$
\begin{aligned}
\text{Lincoln Acres: } P_{2000} &= 30{,}760.06 / [1 + 16.43878\,(e^{-0.00749(26)})] - 12 \\
&= 30{,}760.06 / [1 + 16.43878\,(0.82305)] - 12 \\
&= 30{,}760.06 / 14.52994 - 12 = 2{,}105
\end{aligned}
$$

$$
\begin{aligned}
\text{San Dieguito: } P_{2000} &= 15{,}697.3 / [1 + 4.21837\,(e^{-0.12639(26)})] - 489 \\
&= 15{,}697.3 / [1 + 4.21837\,(0.03740)] - 489 \\
&= 15{,}697.3 / 1.15777 - 489 = 13{,}069
\end{aligned}
$$

In San Dieguito, the logistic projections are noticeably lower than the projections from the other trend extrapolation methods, especially for the longer projection horizons (Table 8.2). This occurs because the slowdown in population growth from 1990 to 1995 caused the model to set a relatively low upper asymptote on future population size (see Figure 8.1). In Lincoln Acres, the logistic projections are similar to those produced by the other methods in all target years.

Logistic models are consistent with various theories of population growth and with empirical evidence from many situations in which populations (including nonhuman populations such as yeast cells and fruit flies) move from low to high to low rates of growth. However, they depend on the same basic assumptions as other trend extrapolation methods; namely, that future population changes emerge directly and smoothly from past population changes and that historical relationships remain constant over time. In addition, it is difficult to develop reliable estimates of the model's parameters, and relatively small differences in the parameters (especially the upper limit) can lead to large differences in population projections. Logistic models are no longer widely used for population projections.

ARIMA Model

The last complex extrapolation method is the Autoregressive Integrated Moving Average (ARIMA) time series model. ARIMA models, popularized by Box and Jenkins (1976), have been used extensively for analyzing and projecting demographic attributes measured over time (Land, 1986). They have been applied to individual components of population change (e.g., Carter & Lee, 1986; De Beer, 1993; McKnown & Rogers, 1989) as well as to estimates of total population (e.g., Alho & Spencer, 1997; Pflaumer, 1992; Saboia, 1974). Some analysts claim that ARIMA models are superior to other regression-based time series forecasting methods because they produce more accurate coefficient estimates and smaller forecast errors (e.g., Box & Jenkins, 1976; McCleary & Hay, 1980; McDonald, 1979; Newbold & Granger, 1974). Furthermore, the dynamic, stochastic framework of ARIMA models provides a statistical basis for developing probabilistic prediction intervals to accompany point forecasts (Nelson, 1973).

However, the methods used in ARIMA modeling are quite complicated, which makes them difficult to implement and to explain to data users. These methods are considerably more complex than the other extrapolation methods discussed in this book. We suggest consulting standard texts on ARIMA modeling before attempting to apply this method (e.g., Box & Jenkins, 1976; McCleary & Hay, 1980; Nelson, 1973).

ARIMA models attempt to uncover the stochastic processes that generate a historical data series. These processes are measured using the patterns observed in the data series; the resulting measurements form the basis for developing projections. ARIMA models typically focus on the processes of autoregression, differencing, and moving average.

The autoregressive process has a memory in the sense that it is based on the correlation of each value of a variable with all preceding values. It is assumed that the impact of earlier values diminishes exponentially over time. The number of preceding values explicitly incorporated into the model determines its "order." For example, in a first-order autoregressive process, the current value is explicitly a function only of the immediately preceding value. However, the immediately preceding value is also a function of the one before it, which is a function of the one before it, and so forth. Consequently, all preceding values influence current values, albeit with a declining impact. In a second-order autoregressive process, the current value is explicitly a function of the two immediately preceding values; again, all preceding values have an indirect impact.

The differencing process is used to create a stationary time series (i.e., one with constant differences over time). When a time series is nonstationary, it can often be converted into a stationary time series by calculating differences between values (Nelson, 1973). First differences are usually sufficient, but second differences are occasionally required (i.e., differences between differences). Logarithmic and square root transformations can also be used to convert nonstationary to stationary time series.

The moving average process is, essentially, a "shock" to the system (i.e., an event that has a substantial but short-lived impact on the time series pattern). The order of the moving average process defines the number of time periods affected by the shock.

The most general ARIMA model is usually written as ARIMA (p,d,q), where p is the order of the autoregression, d is the degree of differencing, and q is the order of the moving average. (ARIMA models based on time intervals of less than one year may also require a seasonal component.) The first and most subjective step in developing an ARIMA model is to identify the values of p, d, and q. The d-value must be determined first because a stationary series is required to properly identify the autoregressive and moving average processes. The value of d is the number of times one has to difference the series to achieve stationarity (usually 0 or 1, but occasionally 2). The p- and q-values are also relatively small (0, 1, or—at most—2). The pattern of the autocorrelation function (ACF) and the partial autocorrelation function (PACF) and their standard errors are used to find the correct values for p and q (Box & Jenkins, 1976; Nelson, 1973). For example, a first-order autoregressive model [ARIMA (1,0,0)] is characterized by an ACF that declines exponentially and quickly and a PACF that has a statistically significant spike only at the first lag. Once p, d, and q are determined, maximum likelihood procedures are used to estimate the parameters of the ARIMA model.

The final step in the estimation process is model diagnosis. An adequate ARIMA model will have random residuals, no significant values in the ACF and PACF, and the smallest possible values for p, d, or q. After a successful diagnosis is completed, the ARIMA model is ready to use for population projections.

An ARIMA (1,1,0) model was chosen to project the population of San Dieguito. This model includes a first-order autoregressive term and requires first differences. The choices for the values of p, d, and q were based on our analysis of the historical data series.

It is clear from Figure 8.1 that the time series for San Dieguito is not stationary. A nonstationary series will have an autoregressive parameter estimate close to 1.0, and the autocorrelation function will decline very slowly (Nelson, 1973). Both of these conditions were seen in the population data series for San Dieguito. After first differences were calculated, however, these conditions were no longer apparent. Furthermore, the plot of those differences over time showed no discernible pattern. The patterns of the ACF and PACF indicated that San Dieguito's historical population followed a first-order autoregressive process, whose statistically significant parameter estimate (θ) was 0.52211. Checks of the residuals, the ACF, and the PACF revealed no problems with the adequacy of the model. Population projections from an ARIMA (1,1,0) model are based on θ, population values one and two years before the target year, and the mean of the stochastic process (Nelson, 1973).

The population projections for San Dieguito from the ARIMA (1,1,0) model are similar to those from the linear trend method (Table 8.2). One characteristic of an autoregressive model is that projections will eventually reach and maintain a constant numerical difference similar in value to the mean of the historical series (McCleary & Hay, 1980). Consequently, ARIMA projections are often similar to projections based on linear extrapolation methods (e.g., Pflaumer, 1992; Voss & Kale, 1985). The projected annual population change for San Dieguito shown in Figure 8.2 illustrates the typical pattern. In the early years of the projection horizon (1996–2002), the numerical change in population increases steadily, reaching an annual change of about 400 persons which is then maintained until the end of the projection horizon. This is very similar to San Dieguito's annual population change between 1975 and 1995 (418).

A simpler ARIMA model was chosen to project the population of Lincoln

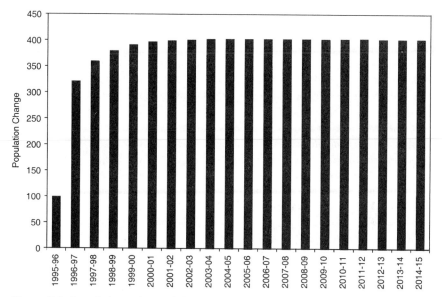

Figure 8.2. Annual change in population, San Dieguito, 1995–2015. (Projections based on an ARIMA (1,1,0) model)

Acres. After analyzing the data we chose an ARIMA (0,1,0) model, also known as a *random walk model* (McCleary & Hay, 1980). Like San Dieguito, first differences were needed to make the data series stationary. Once first differences were taken, both the ACF and PACF revealed a random pattern that had no statistically significant lags. Estimates of the p and q parameters were not statistically significant, which indicated that the autoregressive and moving average processes were not important in explaining the pattern of population change in Lincoln Acres. Checks of the residuals, the ACF, and the PACF revealed no problems with the adequacy of the ARIMA (0,1,0) model.

In a random walk model, the best forecast is the population value immediately preceding the target year, plus a constant term. If the constant term is not statistically significant, it can be ignored; if so, the projected population would be the population value from the launch year, held constant over all target years (McCleary & Hay, 1980). For Lincoln Acres, the estimated constant of 10.15 ($t =$ 1.279) was not considered statistically significant. Consequently, the population projection for Lincoln Acres in any target year is simply the 1995 population estimate (2,032). Had we included the constant term (10.15), the ARIMA projection would have been very similar to the linear trend projection for Lincoln Acres.

The formulas used in computing projections from ARIMA models depend on the specification of the values of p, d, and q. Developing these projections can be tedious, especially if probabilistic prediction intervals are desired. Fortunately, ARIMA modeling software can be used. As is true for other complex extrapolation methods, ARIMA models may require an adjustment factor to bring projections in line with the launch-year population. Nelson (1973) and Box and Jenkins (1976) provide details for computing projections from a wide variety of ARIMA models.

ARIMA models also provide the statistical framework for constructing probabilistic prediction intervals. Using this framework, we developed 95% prediction intervals for the San Dieguito projection (Figure 8.3). The width of the interval increases rapidly over time, from around 1,000 persons in 1996 to more than 11,000 in 2015. This pattern is typical and reflects the growing uncertainty of projections as they move further away from the launch year.

RATIO METHODS

In ratio methods, the population (or population change) of a smaller area is expressed as a proportion of the population (or population change) of a larger area in which the smaller area is located. For example, the population of a census tract may be expressed as a proportion of the population of the county in which the

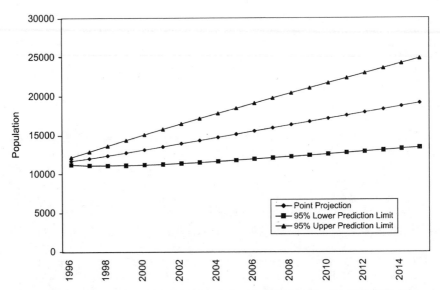

Figure 8.3. Population projection with 95% prediction intervals, San Dieguito, 1996–2015. (Projections based on an ARIMA (1,1,0) model)

census tract is located. Similar to the simple trend extrapolation methods, ratio methods have small data requirements and are easy to apply. In addition, they can be constructed so that the sum of the projections for the smaller areas is equal to the projection for the larger area in which they are located. In this section we discuss three commonly used ratio methods: constant-share, shift-share, and share-of-growth.

All three of these methods require an independent projection of the larger area used in computing the ratios. In the examples shown in this section, the larger area is the unincorporated area of San Diego County. Projections for the unincorporated area were 473,804 for 2000, 519,022 for 2005, 553,621 for 2010, and 600,365 for 2015. These projections were prepared by the San Diego Association of Governments in 1999.

Constant-Share

In the constant-share method (CONSTANT), the smaller area's share of the larger area's population is held constant at some historical level, such as the level observed in the launch year. A projection of the smaller area can then be made by applying this share to the projection of the larger area:

$$P_{it} = (P_{il} / P_{jl}) P_{jt}$$

where P_{it} is the population projection for smaller area (i) in the target year; P_{il} is the population of the smaller area in the launch year; P_{jl} is the population of the larger area (j) in the launch year; and P_{jt} is the projection of the larger area in the target year. For example, the populations for 1995 shown in Table 8.1 were 2,032 for Lincoln Acres, 11,616 for San Dieguito, and 427,564 for the unincorporated area. The population of the unincorporated area of San Diego County in 2000 was projected at 473,804. The projections for the two smaller areas in 2000 can thus be calculated as

Lincoln Acres: $P_{2000} = (2,032 / 427,564) \, 473,804 = 2,252$

San Dieguito: $P_{2000} = (11,616 / 427,564) \, 473,804 = 12,872$

This example held 1995 population shares constant for projections in the target year. It would also be possible to apply the constant-share method using shares from some earlier year (e.g., 1990) or to take an average of the shares observed in several years (e.g., 1990 and 1995).

The CONSTANT method requires historical data from only one point in time; consequently, it is particularly useful for areas where changing geographic boundaries or poor records make it difficult or impossible to construct a reliable historical data series. Another attribute of this method is that projections for all the smaller areas add up exactly to the projection for the larger area (within rounding error). The main drawback of this method, of course, is that it treats all smaller areas exactly the same; that is, it assumes that all the smaller areas will grow at the same rate as the larger area. In many instances, this will not be a reasonable assumption.

Shift-Share

In contrast to the constant-share method, the shift-share (SHIFT) method accounts for changes in population shares over time. Several approaches can be used for extrapolating historical changes in population shares into the future (e.g., Gabbour, 1993). We describe a method in which the projected shares change linearly over time:

$$P_{it} = P_{jt}[P_{il}/P_{jl} + (z/y)(P_{il}/P_{jl} - P_{ib}/P_{jb})]$$

where i denotes the smaller area; j denotes the larger area; z is the number of years in the projection horizon; y is the number of years in the base period; and b, l, and t refer to the base, launch, and target years, respectively.

To apply the SHIFT method, we need the smaller area's share of total population in the launch year (P_{il}/P_{jl}), its share in the base year (P_{ib}/P_{jb}), and a projection of the larger area's population in the target year (P_{jt}). The base-year and launch-year shares for Lincoln Acres and San Dieguito can be computed using the data shown in Table 8.1:

Lincoln Acres, 1975 share: 1,829 / 204,922 = 0.00893

Lincoln Acres, 1995 share: 2,032 / 427,564 = 0.00475

San Dieguito, 1975 share: 3,256 / 204,922 = 0.01589

San Dieguito, 1995 share: 11,616 / 427,564 = 0.02717

Using these data and the projection for the unincorporated area in 2000, SHIFT projections for Lincoln Acres and San Dieguito in 2000 can be calculated as

Lincoln Acres: 473,804 [0.00475 + 5/20 (0.00475 − 0.00893)] = 1,755

San Dieguito: 473,804 [0.02717 + 5/20 (0.02717 − 0.01589)] = 14,209

Lincoln Acres' share of total population declined considerably from 1975–1995, whereas San Dieguito's share grew. Under the SHIFT method these changes continue during the projection horizon. For 2015, the projections would be

Lincoln Acres: 600,365 [0.00475 + 20/20 (0.00475 − 0.00893)] = 342

San Dieguito: 600,365 [0.02717 + 20/20 (0.02717 − 0.01589)] = 23,084

Are these projections reasonable? Perhaps for San Dieguito, but probably not for Lincoln Acres. This points to an important problem inherent in the SHIFT method: it can lead to substantial population losses in areas that grew very slowly (or declined) during the base period, especially when the projections cover long-range horizons (e.g., 25 or 30 years). In fact, SHIFT can even lead to negative numbers. For example, Lincoln Acres' share of the population of the unincorporated area in 2020 would be projected as

$$0.00475 + 25/20\,(0.00475 - 0.00893) = 0.00475 - 0.00523 = -0.00048$$

Because a population can never be less than zero, this obviously is not a reasonable projection. This problem can be dealt with by incorporating constraints on the projected shares or on the projected rates of change in those shares. As mentioned before for EXPO, the SHIFT method must be used cautiously for long-range projections, especially for places whose population shares have been declining rapidly.

Share-of-Growth

The third ratio method focuses on shares of population *growth* rather than population *size*. In this method (SHARE), it is assumed that the smaller area's share of population growth will be the same over the projection horizon as it was during the base period. This method is sometimes called the *apportionment method* (Pittenger, 1976; White, 1954). Using the SHARE method, the population of the smaller area is projected as

$$P_{it} = P_{il} + [(P_{il} - P_{ib})/(P_{jl} - P_{jb})]\,(P_{jt} - P_{jl})$$

For Lincoln Acres and San Dieguito the shares of population growth from 1975–1995 are calculated as

Lincoln Acres: $203/222{,}642 = 0.00091$

San Dieguito: $8{,}360/222{,}642 = 0.03755$

It was projected that the unincorporated area would grow by 46,240 between 1995 and 2000. The populations of Lincoln Acres and San Dieguito in 2000 can thus be projected as

Lincoln Acres: $2{,}032 + (0.00091)\,46{,}240 = 2{,}074$

San Dieguito: $11{,}616 + (0.03755)\,46{,}240 = 13{,}352$

In many instances, the SHARE method seems to provide more reasonable projections than either the CONSTANT or SHIFT methods. However, it runs into problems when growth rates in the smaller and larger areas have opposite signs. For example, suppose that the larger area grew by 1,000 during the base period and the smaller area declined by 100; the smaller area's share of population growth would thus be computed as $-100/1{,}000$, or -0.1. If growth in the larger area were projected as 2,000, the decline in smaller area would be projected at 200. This is probably not a reasonable assumption; if anything, the smaller area is likely to decline *less* rapidly (or even increase a bit) when the larger area is projected to grow more rapidly in the future than in the past.

In situations such as these, the SHARE method should not be used in the manner just described. Rather, it must be adjusted in some way, such as by using a variation of the plus-minus method described in Chapter 11. Some applications of this method have simply projected zero population growth for smaller areas in which growth rates were negative while growth rates for the larger area were positive (e.g., Pittenger, 1976).

Other Applications

Ratio methods are used solely for projecting total population in this chapter, but they have many other applications in demography. Some of these were discussed in Chapters 4 and 5, where we described the use of ratios for relating projections of mortality or fertility rates in one area to rates projected for another area. In Chapter 9, we describe a simple structural model based on ratios.

In addition, ratios are frequently used in combination with population projections to develop projections of other variables or demographic characteristics. For example, projections of households can be made by applying projected "householder" rates by age and sex to population projections by age and sex. Householder rates are simply ratios of the number of householders to the number of persons, calculated for each age-sex group. Similarly, projections of the labor force can be made by applying projected labor force participation rates by age and sex to population projections by age and sex (race and ethnicity can also be used). Labor force participation rates are simply ratios of the number of persons in the labor force to the total number of persons, calculated for each age-sex group. School enrollment, persons with disabilities, and many other variables can be projected by following similar procedures.

ANALYZING PROJECTION RESULTS

Using these 10 trend extrapolation methods and population data from 1975–1995, we projected the total population of Lincoln Acres and San Dieguito for 2000, 2005, 2010, and 2015 (Table 8.2). What can we learn from these projections?

Most of the projections for Lincoln Acres are similar. Except for CONSTANT and SHIFT, the projections for 2015 range only from 2,032 to 2,653. We believe that similar results will be found in most areas that have slow or moderate growth rates: when projections have the same base period and launch year, the results will be about the same for most trend extrapolation methods.

The reasons for the two exceptions are clear. CONSTANT is based on the assumption that Lincoln Acres will grow at the same rate as the unincorporated

Table 8.2. Population Projections Based on Alternative Trend Extrapolation Methods,
San Dieguito and Lincoln Acres, 2000–2015

Extrapolation method	San Dieguito				Lincoln Acres			
	2000	2005	2010	2015	2000	2005	2010	2015
Simple								
LINE	13,706	15,796	17,886	19,976	2,083	2,134	2,184	2,235
EXPO	15,964	21,939	30,151	41,437	2,086	2,142	2,199	2,257
GEO	15,964	21,940	30,153	41,441	2,086	2,142	2,199	2,258
Complex								
Linear	13,892	16,168	18,443	20,719	2,099	2,165	2,232	2,299
Quadratic	13,997	16,419	18,883	21,389	2,154	2,298	2,464	2,653
Logistic	13,069	13,994	14,538	14,845	2,105	2,180	2,258	2,338
ARIMA	13,163	15,162	17,172	19,183	2,032	2,032	2,032	2,032
Ratio								
CONSTANT	12,872	14,101	15,041	16,311	2,252	2,467	2,631	2,853
SHIFT	14,209	17,029	19,726	23,084	1,755	1,381	894	342
SHARE	13,352	15,050	16,349	18,104	2,074	2,115	2,146	2,189
Projection range[a]								
Absolute	3,092	7,946	15,615	26,596	497	1,086	1,737	2,511
Percent	24	57	107	179	28	79	194	734

[a]Projection range is the difference between the highest and lowest projections for each target year.

area; in reality, it grew much more slowly during the base period. SHIFT is based on the assumption that Lincoln Acres' share of the unincorporated area's population will continue to decline rapidly. The steadily declining shares lead to steadily declining projections. These two methods will produce results that are quite different from the other trend extrapolation methods whenever population growth in the smaller area differs substantially from population growth in the larger area. Because the SHARE method is based on shares of population *growth* rather than population per se, it provides projections that are more in line with other trend extrapolation methods.

Projections for San Dieguito show much more variation among the non-ratio methods than projections for Lincoln Acres. The EXPO and GEO methods produce by far the highest projections; in fact, by 2015 those projections are about twice as large as the projections produced by the other methods. This reflects the impact of applying constant growth rates in areas that have been growing very rapidly. In most instances, this will not be a reasonable assumption (especially for projection horizons longer than five or 10 years).

The ratio methods again show a substantial degree of variation, but now SHIFT is the highest of the three, and CONSTANT is the lowest, reversing the pattern seen for Lincoln Acres. This result is not surprising because the population of San Dieguito grew much more rapidly than the rest of the unincorporated area during the base period.

Three of the complex methods produce similar results for San Dieguito, but the logistic projection is the lowest of all the methods. As mentioned earlier, this was caused by the slowdown in population growth at the end of the base period, which led to a relatively low estimate of the upper asymptote.

Several other patterns stand out from Table 8.2. EXPO and GEO projections explode upward in San Dieguito, a rapidly growing area. SHIFT projections explode downward in Lincoln Acres, a slowly growing area. In both areas, projections from linear extrapolation, linear regression, and ARIMA time series models are similar to each other. Are these results unique to this data set or can they be generalized? We provide further evidence and draw some conclusions in Chapter 13.

CONCLUSIONS

Trend extrapolation methods have a long history in demography. In spite of the ascendancy of the cohort-component method and the development of structural models during the last half century, these methods are still commonly used for population projections, especially for small areas. They have a number of useful characteristics but also some serious shortcomings.

Simple trend and ratio extrapolation methods have small data requirements. LINE, GEO, EXPO, SHIFT, and SHARE can be applied by using total population data from only two time points; CONSTANT requires data from only *one* time point. These methods are easy to apply and to explain to data users. They do not require sophisticated modeling or programming skills; in fact, they can be applied rather easily using only a hand calculator. Because of their small data requirements and ease of application, they can be applied in a timely manner and at little cost. Simple methods are particularly useful for small areas, where data availability and reliability create substantial problems for more complex or sophisticated methods.

Complex trend extrapolation methods require data from a number of time points. The lack of sufficient historical data prevents their use in many small areas. Complex methods also require greater modeling skills than simple extrapolation methods, especially for developing logistic and ARIMA time series models. However, compared to cohort-component and structural models, even complex trend extrapolation methods are characterized by low cost, timeliness, and small data requirements. In addition, several of these methods can be used to develop prediction intervals. We return to this point in Chapter 13.

Trend extrapolation methods suffer from several shortcomings. They do not account for differences in demographic composition or the components of growth.

Box 8.1

One Man's View of Extrapolation Methods

In the space of one hundred and seventy-six years the Lower Mississippi has short-ened itself two hundred and forty-two miles. That is an average of a trifle over one mile and a third per year. Therefore, any calm person, who is not blind or idiotic, can see that in the Old Oolitic Silurian Period, just a million years ago next November, the Lower Mississippi River was upward of one million three hundred thousand miles long, and stuck out over the Gulf of Mexico like a fishing-rod. And by the same token any person can see that seven hundred and forty-two years from now the Lower Mississippi will be only a mile and three-quarters long, and Cairo and New Orleans will have joined their streets together, and be plodding comfortably along under a single mayor and a mutual board of aldermen. There is something fascinating about science. One gets such wholesale returns of conjecture out of such a trifling investment of fact.

MARK TWAIN, *Life on the Mississippi* (1874), p. 136

They provide little or no information on the projected characteristics of the population. Because they have no theoretical content beyond the structure of the model itself, they cannot be related to behavioral or socioeconomic theories of population growth (the logistic model is an exception). Consequently, they are not useful for analyzing the determinants of population growth or for simulating the effects of changes in particular variables or assumptions. In addition, they can lead to unrealistic or even absurd results if carried too far into the future (Box 8.1).

The basic assumption underlying trend extrapolation methods is that—in terms of the population change specified by a particular method—the future will be just like the past. Given the changes that have occurred in many places, this assumption may be unrealistic. Just how useful are trend extrapolation methods? How accurate are their projections when used as forecasts? How does their forecast accuracy compare with the accuracy of other methods? We provide some answers to these questions in Chapters 12 and 13.

CHAPTER 9

Structural Models I
Economic–Demographic

OVERVIEW OF STRUCTURAL MODELS

Suppose that a new freeway extending into the sparsely populated outskirts of Denver were built. What impact would that freeway have on the population and housing growth of these outlying areas? Suppose that a large meatpacking plant were built in a small town in Iowa. What impact would the new plant have on the town's population growth and demographic characteristics? Suppose that a large military base in South Carolina were closed as part of a cutback in the federal defense budget. What impact would this base closure have on the population and economy of the county in which the base was located?

Planners, decision makers, and applied demographers frequently face questions such as these, but projection methods based solely on demographic factors or the extrapolation of historical trends cannot provide any answers. This is where structural models—the third approach to population projection—come into play. Structural models relate population change to changes in one or more independent (i.e., explanatory) variables. They are invaluable for many planning and policy-making purposes because they explicitly account for the influence of factors such as employment growth, wage rates, land use, housing, and the transportation system. In some circumstances, in fact, federal transportation and clean air legislation *mandate* the use of structural models.

Structural models can be expressed in many ways. Some focus on total population. For example, population growth for census tracts might be based on the spatial distribution of employment opportunities within a county. Others focus on specific components of population growth, such as migration. For example, county migration projections might be based on wage rates and changes in

185

employment. When structural models focus on a particular component of growth, they are generally linked to a cohort-component model to complete the projection.

Some structural models are relatively simple, containing only a few variables and equations (e.g., Mills & Lubuele, 1995). Others are very complex, having huge systems of simultaneous equations that involve many variables and parameters (e.g., Data Resources Incorporated, 1998). Likewise, the procedures for projecting the model's independent variables range from simple extrapolation and shift-share techniques to formal multi-equation statistical models. We do not discuss the procedures for projecting the model's independent variables in this book; such descriptions can be found elsewhere (e.g., Bolton, 1985; Greenberg, Krueckeberg, & Michaelson, 1978; Putman; 1983; Schroeder, 1987; Shao & Treyz, 1993; Treyz, 1993).

We cover two general categories of structural models in this book. This chapter considers economic–demographic models that are used primarily to project population and economic activities for counties, labor market areas, metropolitan areas, and states. Chapter 10 covers urban systems models that are used primarily to project population, housing, land use, economic activities, and transportation patterns for census tracts, block groups, blocks, and other small geographic areas. Although economic–demographic and urban systems models are distinguished largely by differences in geographic scale, they typically provide different explanations of the causes and consequences of population change.

During the last several decades, structural models have been used more and more frequently for population projections at geographic levels that range from states to counties, cities, census tracts, and other small areas (e.g., Campbell, 1996; Isserman, Plane, Rogerson, & Beaumont, 1985; Murdock et al., 1984; San Diego Association of Governments, 1998, 1999; Treyz, 1993; U.S. Bureau of Economic Analysis, 1995). In this and the following chapter, we describe the variables and relationships specified in several commonly used structural models and discuss some of the issues involved in their implementation. Our objective is to provide a general introduction to the use of structural models in population projections; we do not attempt to provide detailed blueprints for applying any specific model. Descriptions of the processes, techniques, and strategies used to formulate, build, calibrate, test, and implement structural models can be found in Batty (1976), Conway (1990), Foot (1981), Kmenta (1971), Putman (1991), San Diego Association of Governments (1998, 1999), and Treyz (1993).

FOCUS ON MIGRATION

A properly constructed structural model is not simply a collection of disparate variables and equations but represents a distinct theoretical framework that postulates a variety of interrelationships among demographic, economic, and other variables. Such models have been widely used for many years to analyze the determinants and consequences of fertility, mortality, and migration. For population projection, however, structural models generally focus solely on migration. We know of only a few instances in which structural models have been used for projecting fertility or mortality rates (e.g., Ahlburg, 1986, 1999; Sanderson, 1999). These studies dealt with projections of large areas such as nations or regions of the world. Mortality and fertility models for small areas have been proposed (e.g., Isserman, 1985), but have rarely been implemented.

We believe that there are several reasons for focusing on migration, especially when developing projection models for small areas. First, migration rates are more volatile than mortality and fertility rates and have the potential to change much more rapidly within a short time period. Second, in areas that are growing or declining rapidly, migration generally has a greater impact on population change than natural increase. Third—and perhaps most important for small-area projections—economic fluctuations have a greater impact on migration patterns than on mortality and fertility rates. Consequently, we confine our discussion to models that focus on migration or changes in total population.

We begin this chapter by describing several factors that influence state and local migration patterns. Then, we discuss and illustrate two classes of models in which these factors are used to project migration or total population: recursive and nonrecursive. Recursive models account for the impact of one or more independent variables on population change but do not consider the impact of population change on those variables. Nonrecursive models, on the other hand, allow for two-way interactions; that is, they consider both the determinants *and* the consequences of population change. We close with a discussion of the advantages and disadvantages of using structural models for population projections.

FACTORS AFFECTING MIGRATION

Human capital theory has guided many studies of the determinants of migration (e.g., Greenwood, 1997; DaVanzo & Morrison, 1981; Sjaastad, 1962). According to this theory, migration is an investment in human capital that involves both costs and benefits. People migrate if the present value of all future gains in benefits are expected to outweigh the full costs of migration. This theory echoes Ravenstein's words from more than a century ago, that people migrate primarily "to 'better' themselves in material respects" (Ravenstein, 1889, p. 286). Narrowly interpreted, this theory implies that people will move to the area in which their economic opportunities are expected to be the greatest.

Economic opportunities are typically measured by using variables such as job growth, unemployment rates, wages, and income. In addition to their theoretical relevance, data on these variables are readily available for many geographic areas and points in time. The U.S. Bureau of Economic Analysis (BEA), for example, provides time series data for wages, income, and employment for states, counties, and metropolitan areas. The U.S. Bureau of Labor Statistics (BLS) also provides historical labor force and unemployment data for states, metropolitan areas, and many counties.

Economic factors are not the only ones that affect migration decisions, of course. Amenities such as climate, crime rates, and a coastal location play a major role (e.g., Clark & Murphy, 1996; Graves & Linneman, 1979; Schachter & Althaus, 1989). Life-cycle changes such as marriage, divorce, childbearing, and retirement are also important, as are personal characteristics such as age, education, income, family ties, and residential preferences (e.g., Astone & McLanahan, 1994; Bartel, 1979; DaVanzo & Morrison, 1978; Fuguitt & Brown, 1990; Mincer, 1978). In addition, social networks are known to have an important impact on migration flows to some areas (e.g., Massey, Alarcon, Durand, & Gonzalez, 1987).

From a theoretical perspective, a complete migration model should include a variety of economic and noneconomic variables. Using such a model for migration projections, however, would be very difficult. In addition to the problem of finding reliable data for small geographic areas, all of the independent variables in the model must themselves be projected. Such projections are seldom if ever available. Because projections of economic variables are often available from state and regional economic models, narrowly economic models have a distinct advantage over theoretically richer formulations.

In this section we discuss several economic factors that influence state and local migration patterns: employment, the unemployment rate, and wages (or income). Although this list is not exhaustive, these variables can have a substantial impact on migration and are often included in economic–demographic models. We also discuss the role played by amenities, focusing on two theories of migration that offer opposing views regarding the relative importance of economic factors and amenities as determinants of migration. Understanding the determinants of migration is essential before a valid migration model can be constructed.

Employment

Almost 25 years ago the American Statistical Association suggested that studies of internal migration should focus on the relationship between the growth in jobs and the movement of people (American Statistical Association, 1977). Within the human capital framework, it is assumed that potential migrants perceive that areas of increasing employment opportunities are more attractive than areas that have stagnant or declining employment opportunities. It would be expected, then, that areas that have relatively high (low) rates of job growth would have relatively high (low) rates of population growth due to migration. This expectation has been strongly supported by the literature (e.g., Blanco, 1963;

Clark & Hunter, 1992; Greenwood, 1975; Greenwood, Hunt, & McDowell, 1986; Krieg & Bohara, 1999; San Diego Association of Governments, 1999; Treyz, Rickman, & Shao, 1992). Simply put, jobs attract people, and people create jobs. This empirical finding forms the foundation of virtually every economic model of migration in use today.

The exact nature of the employment-migration relationship, however, varies over time and among geographic areas (e.g., Greenwood, 1981; Greenwood & Hunt, 1984, 1991; Plane, 1989). New jobs can be filled by migrants or by the local population (Congdon, 1992). At the extremes, migrants may take all of the new jobs in an area or they may not take any; the reality usually lies somewhere in between. Factors such as changes in commuting patterns and local unemployment rates—plus the time and cost of a job search—help explain the lag often seen between employment changes and subsequent population movements (Greenwood, 1985).

To clarify the roles of commuting patterns and migration as responses to changes in local employment opportunities, it is useful to view an area in terms of its labor market. A *labor market* can be defined as the range of employment opportunities available to workers without changing their places of residence (Fischer & Nijkamp, 1987, p. 3). Labor markets often span a number of counties, extend across state lines, and reflect commuting distances of one, two, or even more hours. Workers can respond to new job opportunities by migrating (i.e., changing residence) *or* by adjusting their daily commuting patterns. For example, some residents of Temecula, a city in southern California, commute more than 100 miles each day (round trip) to jobs in San Diego. Similar examples can be found in labor markets throughout the nation.

Human capital theory implies that in-migration should be positively related to changes in job opportunities. This expectation has been strongly supported in the empirical literature. It might also be expected that out-migration would be inversely related to changes in job opportunities; that is, when the economy is strong and employment is growing, fewer persons would leave the area in search of employment elsewhere. Conversely, when the economy is weak and employment is stagnant or declining, more people would be expected to leave the area to seek jobs elsewhere.

The empirical evidence for out-migration, however, is not nearly as clear as it is for in-migration. Some studies (e.g., Greenwood, 1975; Schachter & Althaus, 1989) have found the expected negative relationship, but others have found a positive relationship (e.g., Plane, 1989). Furthermore, the relationship between employment and out-migration is often found to be nonlinear and weaker than the relationship between employment and in-migration (e.g., Kriesberg & Vining, 1978; Plane, Rogerson, & Rosen, 1984).

Another complicating factor is that a number of studies have found a positive relationship between in- and out-migration; that is, places that have high (low) levels of in-migration also have high (low) levels of out-migration (e.g., Miller, 1967; Morrison, 1971; Mueser & White, 1989; Rogers & Raymer, 1999; Stone, 1971; Tabuchi, 1985). Why might this be true? One explanation is that areas that have large numbers of in-migrants have populations that are relatively migration-prone, thereby raising the probability of out-migration. Another explanation is that in-migration creates its own counterstream, as in-migrants return to their previous places of residence. Whatever the explanation, the empirical evidence provides a stronger basis for projecting in-migration than out-migration.

The overall strength of the employment–migration relationship, however, is quite impressive. This relationship forms the basis of most of the economic–demographic projection models used today.

Unemployment Rate

The unemployment rate is a widely watched measure of the economy, as evidenced by the emphasis attached to the unemployment figures released each month. In general, a high unemployment rate is a sign that the economy is not creating enough jobs for those who want to work. Conversely, a low unemployment rate is a sign of a "healthy" economy that is creating jobs for most of those who want to work. In the human capital framework, migrants perceive that areas that have low unemployment rates are attractive and areas that have high unemployment rates are unattractive. Therefore, areas that have rising unemployment rates would be expected to have increasing levels of out-migration and declining levels of in-migration, whereas areas that have falling unemployment rates would be expected to have increasing levels of in-migration and declining levels of out-migration.

Despite the theoretical rationale and the findings of a few studies (e.g., Foot & Milne, 1989; Haurin & Haurin, 1988), much of the empirical research on this topic has found unexpected signs or insignificant coefficients (e.g., Clark & Hunter, 1992; Greenwood, 1975, 1985; Schachter & Althaus, 1989). What might explain these poor empirical results? Gordon (1985) believes that the rate of employment growth is more important to potential migrants than the unemployment rate, and swamps its observed effect. Haurin and Haurin (1988) believe that employment and wage variables have been improperly included in migration equations, minimizing the true impact of the unemployment rate. Another explanation is based on the statistical issue of simultaneity: When the unemployment rate is measured at the end of the migration period, it not only *influences* migration but is also *influenced by* migration (Greenwood, 1981).

The relatively small size of the unemployed population may also be a factor, masking the true relationship between the unemployment rate and migration. Because high unemployment rates are likely to be of more concern to the unemployed than to the employed, the effects of unemployment on migration may not be apparent in studies using aggregate data (Greenwood, 1985). Consequently, unemployment effects may be more evident in studies using micro data than in studies using aggregate data. For example, DaVanzo (1978) analyzed survey data for individuals and found that the unemployed are more likely to move than jobholders. She concluded that higher unemployment rates did indeed encourage out-migration. Several recent studies have supported DaVanzo's findings, but others have failed to confirm them (Greenwood, 1997).

The empirical evidence for the unemployment–migration relationship is somewhat murky. Some studies have found significant effects that have the expected signs, but others have found no significant effects or even effects that have the wrong signs. Based on this evidence, we believe that unemployment rates will generally not perform as well as other economic variables in economic–demographic projection models.

Wages and Income

The human capital model suggests that wages (or income) should be positively associated with in-migration and negatively associated with out-migration. Therefore, areas that have relatively high wages or incomes would be expected to attract a relatively large number of in-migrants and lose a relatively small number of out-migrants, whereas areas that have relatively low wages or incomes would be expected to attract relatively few in-migrants and lose a relatively large number of out-migrants.

General support for these expectations is found in the literature (e.g., Clark & Cosgrove, 1991; Foot & Milne, 1989; Greenwood et al., 1986; Schachter & Althaus, 1989; Treyz et al., 1992). However, the strength of this effect—although greater than for the unemployment rate—is not as great as it is for measures of job growth (e.g., Greenwood, 1981; Isserman et al., 1985).

Several explanations have been offered for the relatively weak influence of wages (or income) on migration. Household data have shown that wages are not always the most prominent factor in a person's motivation for moving, especially for people in older age groups (Gibbs, 1994; Long & Hansen, 1979). Isserman et al. (1985) postulated that capital may be attracted to low-wage regions, thereby raising employment opportunities and attracting in-migrants; this "employment effect" might obscure the relationship between wages and migration. Vijverberg (1993) suggested that higher wages might actually *discourage* migration because of diminishing marginal returns to income. Krieg and Bohara (1999) found that using aggregate earnings data obscures the effect of wages on migration because unmeasured personal characteristics such as ambition, drive, and the quality of schooling—which have a positive impact on migration—are not picked up in the data. Simultaneity bias has also been suggested as a possible explanation for the lack of a clear relationship between income and migration (Sjaastad, 1960).

The empirical evidence for the relationship between migration and wages (or income) is generally consistent with the predictions of human capital theory but is not especially strong. Wage (or income) data have been used successfully in a number of economic–demographic projection models but do not play as large a role as employment as a determinant of migration.

Amenities

The discussion thus far has focused on the economic determinants of migration, but people undoubtedly consider other factors as well. We use the term *amenities* to describe noneconomic characteristics such as climate, topographical features, cultural attractions, recreational opportunities, air quality, and crime rates. Because amenities have a substantial impact on the quality of life in an area, it would be expected that they would also have an impact on migration patterns. How important are amenities compared to economic factors as determinants of migration? This question has given rise to two competing theories of migration.

Before the 1970s, most migration research focused on regional differences in economic variables such as wages, employment, and income. This research was based on the assumption that the economic system was in a constant state of disequilibrium, as reflected by the existence of persistent geographic differences in economic opportunities. Thus, migration acts as an "equilibrating mechanism" that shifts people from one area to another, thereby reducing these differences over time. Geographic differences in economic opportunities tend to persist, however, because the labor market is slow to adjust to changes. Although disequilibrium theorists acknowledge that geographic differences in amenities may affect migration, they believe that differences in economic opportunities are the main explanatory factors (e.g., Hunt, 1993; Sjaastad, 1962).

An alternative theory has gained adherents during the last few decades, in part because of the failure of economic variables to provide consistent explanations of migration in empirical studies (Greenwood, 1997). Equilibrium theory postulates that differences in amenities—rather than differences in economic opportunities—are the main determinants of migration (e.g., Graves, 1979, 1983; Graves & Linneman, 1979). Equilibrium theorists assume that labor markets, land markets, and the migration process itself are efficient. Consequently, migration quickly eliminates any significant geographic differences in economic opportunities and promptly restores equilibrium (e.g., Graves & Knapp, 1988; Schachter & Althaus, 1989). Regional differences in wages or income are simply compensating for regional differences in amenities. Equilibrium theorists believe that failing to account for amenities in migration equations leads to model misspecification and biased parameter estimates (e.g., Graves, 1980).

The empirical evidence related to this debate is mixed. Some studies have found support for the predictions of equilibrium theory (e.g., Clark & Murphy,

1996; Graves, 1979; Graves & Mueser, 1993; Schachter & Althaus, 1989). Others have found support for the predictions of disequilibrium theory (e.g., Carlino & Mills, 1987; Greenwood et al., 1986; Greenwood & Hunt, 1989) or cast doubt on the equilibrium perspective (e.g., Evans, 1990; Harrigan & McGregor, 1993; Henderson, 1982; Hunt, 1993; Treyz, Rickman, Hunt, & Greenwood, 1993).

A substantial amount of empirical evidence shows that both economic opportunities and amenities influence migration and that including both types of variables generally improves migration modeling (e.g., Clark & Cosgrove, 1991; Cushing, 1987; Greenwood, Hunt, Rickman, & Treyz, 1991). The life-cycle literature further suggests that economic variables are more important to working age people and that amenity variables become more important as people become older (e.g., Clark & Hunter, 1992). We believe that both types of variables have important effects on migration.

Despite the importance of amenities and other noneconomic factors, most structural models used for population projections focus solely on the economic determinants of migration. Consequently, our discussion focuses primarily on economic factors. Later in this chapter, we describe a model that accounts for the impact of amenities as well as economic factors.

RECURSIVE MODELS

Recursive models are based on one-way interactions: Independent variables influence dependent variables, but dependent variables do not influence independent variables. In recursive models, the basic assumption is that the economic factors that affect migration are themselves unaffected by migration. Is this a reasonable assumption? Probably not. Over time, migration has a direct impact on job growth, wages, unemployment rates, and a host of other economic and noneconomic variables. From a theoretical perspective, then, recursive models do not reflect the full range of interactions between migration and the economy.

Recursive models have several attractive features, however. They pick up a number of important effects and are simpler to develop and easier to apply than models that account for two-way interactions. Consequently, they are well represented in the literature and in practice. Recursive models have been developed for explaining and projecting net migration flows (e.g., Clark & Hunter, 1992; Greenwood et al., 1986, 1991; Greenwood & Hunt, 1991; Haurin & Haurin, 1988; San Diego Association of Governments, 1999) and gross migration flows (e.g., Greenwood, 1975; Schachter & Althaus, 1989; Tabuchi, 1985). Recursive relationships have also been incorporated into several multiregional migration models (e.g., Campbell, 1996; Cushing, 1987; Foot & Milne, 1989; Isserman et al., 1985; Rogers & Williams, 1986).

We discuss three general approaches to designing and implementing recursive models and provide examples of their use in projecting migration or population. First, we look at models that use regression analysis to project migration as a direct statistical function of a set of economic variables; we call these *econometric models*. Second, we examine an approach that treats migration as a balancing factor by accounting for differences between the projected supply and demand for labor. Finally, we discuss a method that uses population/employment ratios to derive population projections from employment projections.

Econometric Models

Brief Overview.   In the econometric approach, equations are developed in which migration is determined by one or more independent variables. Using historical data and regression techniques, parameters are estimated for each independent variable. Migration projections are made by applying the parameter estimates to projections of the independent variables. The migration equations are typically integrated into a larger structural model that provides projections of the entire economy.

When migration models are used for projections, parameter estimates are typically based on time series data measured at annual intervals. Because the equations are recursive, they can usually be estimated by using ordinary least squares (OLS) regression techniques. However, the presence of autocorrelation—a likely possibility with time series data—may require more complicated techniques such as those described by Cochrane and Orcutt (1949) or Bates and Watts (1988).

Another common practice is to use nonlinear transformations of the variables (e.g., the natural logarithm). Nonlinear transformations help correct statistical problems such as unequal variances in the regression residuals and may provide a better description of the relationship between the independent variables and migration. A crucial assumption in most econometric projection models is that parameters and functional forms do not change over the projection horizon. In other words, it is assumed that the historical relationships between migration and the independent variables remain constant over time.

Net migration models require estimating a single migration equation (unless net migration is divided into several categories to account for differences in demographic characteristics). Gross migration models require estimating two equations, one for in-migrants and one for out-migrants. In either approach, the model builder attempts to construct equations that accurately portray the influence of the independent variables on migration. Variables such as the change in employment and the average wage rate are typically used to measure area-specific economic conditions.

Another strategy is to define economic conditions for one area in relation to those in another area, most often the nation (e.g., San Diego Association of

Governments, 1999; Treyz et al., 1993). This strategy provides a mechanism for capturing the effects of national economic trends on the local economy (Greenwood, 1981). Furthermore, by focusing on relative rather than absolute changes in economic conditions, this strategy is consistent with human capital theory's emphasis on economic conditions in one area compared with those in another.

Multiregional models require the most data because they incorporate specific place-to-place migration flows. For example, a state-to-state model (including 50 states and the District of Columbia) involves 2,550 gross migration flows (51 × 50). In comparison, a net migration model for states requires only 51 migration flows and a two-region gross migration model requires only 102. Large data requirements present a formidable challenge in constructing multiregional models.

Several analytical approaches have been used to construct structural models of multiregional migration flows. The Census Bureau (Campbell, 1996) projected state migration flows as

$$_{i,j}\text{MIG}_t = (b_1 \times {}_i\text{EMP}_t) + (b_2 \times {}_j\text{EMP}_t)$$

where, i and j are the states of origin and destination, t is the target year, MIG is the migration flow, EMP is civilian employment, and b_1 and b_2 are the estimated parameters. In this model, migration flows and employment are measured by first differences in the natural logarithms of the variables. The state employment projections used as independent variables are those produced by the BEA.

Isserman et al. (1985) developed a Markov transition model using annual IRS data to project state migration flows. They constructed a transition matrix reflecting the probability that a person will migrate from one state to another (or remain in the same state). This model incorporated the size and characteristics of the origin population, changes in economic conditions at all potential destinations, and base year migration probabilities by origin and destination. The authors used an economic attractiveness index based on the ratio of employment growth to unemployed workers as a measure of economic conditions, and used this index to estimate the impact of changes in economic conditions on migration.

Gravity models represent another approach to modeling multiregional migration flows. In a gravity model, a migration flow is directly related to the size of the origin and destination populations and inversely related to the distance between the two areas. These models can be adjusted to include other determinants of migration. Rogers (1967) and Foot and Milne (1989) provide examples of gravity models that incorporate economic factors. We discuss gravity models in more detail in Chapter 10.

Demographic and Economic Forecasting Model. Now, we turn to a description of the Demographic and Economic Forecasting Model (DEFM), a projection system developed in the late 1970s for San Diego County. It has been used successfully for more than 20 years and had a major update of the modeling framework, associated software, and database structure in 1997 (San Diego Association of Governments, 1999). The demographic sector of this model contains an econometric model of net migration.

DEFM uses a combination of two methods to produce annual projections for more than 1,000 demographic and economic variables. The demographic sector relies heavily on a cohort-component model that uses single-year age data for males and females, stratified by four race/ethnic groups. In addition to population, the demographic sector includes housing demand and labor force status. On the economic side, time series econometric models project employment, income, output, and other economic variables. These projections are based on assumptions about national, state, and local growth patterns and on local interindustry relationships derived from an input-output table. The economic projections also take into account trends in the national, state, and regional economies. DEFM links all sectors of the model directly to each other through a series of econometric equations.

DEFM produces projections of three mutually exclusive categories of net migration: (1) migration of uniformed military personnel and their dependents, (2) international migration, and (3) domestic (nonmilitary) migration. These distinctions are important because economic models do not apply equally to all categories of migrants. Movements of military personnel and their dependents, for example, are determined by changes in U.S. defense policy and military needs, not by changes in the local economy. Similarly, the local economy typically has little impact on international migration. Domestic migration, however, is strongly affected by changes in local economic conditions.

As shown in Figure 9.1, domestic migration in San Diego County fluctuated considerably between 1980 and 1998, generally following cycles in employment opportunities. International migration, on the other hand, remained relatively stable during this period. Consequently, domestic migration is the only category linked directly to economic conditions in the DEFM model. Unlike some structural models of migration (e.g., Smith & Fishkind, 1985; Treyz, 1993), DEFM does not treat retirement migration separately because it constitutes only a small component of domestic net migration in San Diego County (5% between 1985 and 1990).

The military population in San Diego County has remained fairly constant for many years. Consequently, DEFM holds the military population constant over the projection horizon.

DEFM projects net international migration (including both documented and undocumented immigrants) by assuming that San Diego's share of U.S. net international migration will remain constant. This share is based on a regression analysis of annual data from 1980 to 1997. During this period, the share fluctuated very little and fell between 1.6 and 1.8% each year. The national projections of international migration used in this projection were those produced by the Census Bureau (Day, 1996a).

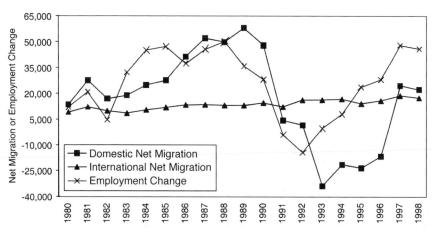

Figure 9.1. Domestic net migration, international net migration, and employment change, San Diego County, 1980–1998. (*Sources*: State of California, Employment Development Department, *San Diego MSA Annual Average Labor Force and Industry Employment, March 1998 Benchmark,* Sacramento, California, March 1999; State of California, Department of Finance, *Historical County Population Estimates and Components of Change, July 1, 1970–1990,* Report E-6, Sacramento, California, December 1998; State of California, Department of Finance, *Historical County Population Estimates and Components of Change, July 1, 1990–1998,* Report E-6, Sacramento, California, January 1999)

Domestic net migration (DMIG) is viewed as a function of employment change (EMPCHG), the labor force (LABOR), and the real wage (RWAGE). The latter two variables are combined to form a composite variable (COMPOSITE). These variables are measured as follows:

$$\text{EMPCHG} = \text{EMP}_{t-1} - \text{EMP}_{t-2}$$
$$\text{LABOR} = (\text{EMP}/\text{P})/(\text{USEMP}/\text{USP})$$
$$\text{RWAGE} = \text{WAGE}/\text{USWAGE}$$
$$\text{COMPOSITE} = \text{LABOR} \times \text{RWAGE}$$

where EMP is civilian nonagricultural employment in San Diego County, USEMP is civilian nonagricultural employment in the United States, P is the population of San Diego County, USP is the U.S. population, WAGE is the real wage for San Diego County, and USWAGE is the real wage for the United States.

Figure 9.1 shows a lag between employment change and domestic migration: Changes in domestic migration tend to occur a year or two after changes in employment. This lag is built into DEFM by subtracting employment lagged two years from employment lagged one year. The net migration equation also includes a first-order autoregressive (AR) term. DEFM uses the regression residuals and the size of the AR term to project the pattern of residuals. Annual data from 1970 to 1997 form the basis of the migration equation. Incorporating the autoregressive term leads to an equation that is nonlinear in its parameters; it is estimated by using the Levenberg–Marquardt algorithm for nonlinear least squares (Bates & Watts, 1988). The equation and regression statistics are

$$\text{DMIG} = -80.7 + 0.627\,(\text{EMPCHG}) + 91.586\,(\text{COMPOSITE}) + 0.320\,(\text{AR})$$

$$\text{Adjusted } r^2 = 0.573 \qquad \text{Durbin–Watson Statistic} = 1.955$$

where DMIG is domestic net migration, EMPCHG is the lagged change in civilian nonagricultural employment, COMPOSITE is the composite variable described before, and AR is the autoregressive term. The standard errors were 37.8, 0.226, 37.02, and 0.219 for the intercept and independent variables. The equation has a Durbin–Watson statistic of almost 2.0, which indicates that there are no autocorrelation problems. The adjusted r^2 is moderately high and is similar to previous calibrations of DEFM's net migration equation. Both economic variables have significant coefficients and show the expected relationship to net domestic migration. These results indicate that better economic conditions in San Diego County compared to the United States create job opportunities and attract migrants to the region. The employment change coefficient (EMPCHG) implies that every new job attracts an average of about 0.6 of a migrant (net), when controlling for other factors.

We have now discussed the procedures for projecting the local economy and military, international, and domestic net migration. The projection of total population is obtained by adding the projection of net migration to the projection of the surviving population derived from the cohort-component model. Figure 9.2 provides an overview of DEFM's major components for projecting migration and population.

The migration projections are made by age, sex, and race/ethnicity using a top-down approach. For military and international migrants, the most recent demographic characteristics are held constant over time. For domestic migrants, demographic characteristics are allowed to change over the projection horizon by applying net migration rates by age, sex, and race/ethnicity to the surviving population. The resulting numbers are adjusted to sum to total net migration projected by the model. These net migration rates were developed using in- and out-migration data from the PUMS file in the decennial census. Although some researchers have argued that using net migration rates in this fashion can distort or misrepresent the compositional structure of the migration stream (e.g., Isserman, 1993), this approach works well for San Diego County.

Balancing Labor Supply and Demand

Another widely used recursive model matches independently derived projections of labor supply and demand to determine migration. If labor supply exceeds

Figure 9.2. The Demographic and Economic Forecasting Model (DEFM).

labor demand, it is projected that workers will move out of the area; if labor demand exceeds labor supply, it is projected that workers will move in. In both cases, migration tends to restore the balance between labor supply and demand. We refer to this as a *balancing* model.

A balancing model is a two-part model in which labor supply is determined by using a traditional cohort-component model (with one important difference discussed later) and labor demand is determined by economic factors. Such a model is typically used only to project those migrants who are most affected by changes in employment opportunities; other migrants are projected using alternative techniques. Unlike an econometric forecasting model, a balancing model does not require formal statistical equations or time series data to project future levels of migration. In addition, it does not require implementing a large-scale model of the economy. Consequently, balancing models are less costly to implement and easier to use than econometric models, and are more accessible to a wider range of practitioners.

We present a brief overview of the steps involved in developing and implementing a balancing model (Figure 9.3). Actually applying the model, however, requires detailed computations and specifying a number of assumptions. Murdock and Ellis (1991) provide a simple numerical example illustrating a balancing model.

The first step is to project the demand for labor in the area to be projected. Labor demand is usually represented by some measure of employment opportunities (e.g., the total number of jobs). It is typically projected using export-base models, input-output models, or shift-share techniques (e.g., Center for the Continuing Study of the California Economy, 1997; Greenberg, Kreukeberg, & Mi-

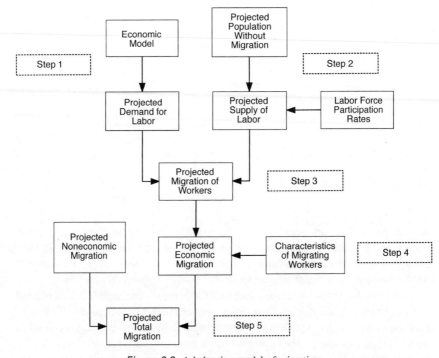

Figure 9.3. A balancing model of migration.

chaelson, 1978; Murdock et al., 1984). In some instances, projections of labor demand are based on large-scale econometric models (e.g., Reeve & Perlich, 1995).

The second step is to project the labor supply. This step usually involves a cohort-component model in which mortality and fertility rates are applied in the usual manner, but migration rates are assumed to be zero; in other words, the population is "closed" to migration. *Labor force participation (LFP) rates*—defined as the proportion of the population working or actively looking for work—are projected by assuming that current rates will remain constant, that local rates will follow national trends, or that some other trend will prevail. The BLS provides national projections of LFP rates by age, sex, race, and Hispanic origin (Fullerton, 1997). Then, labor supply is projected by multiplying the projected population by the appropriate LFP rates. Although labor supply can be projected by using LFP rates based on the total population, it is much more common to use rates broken down by age, sex, race, or other demographic characteristics.

The third step is to derive the migration projection by matching labor supply and labor demand. This can be done by using a variety of procedures that range from relatively simple to relatively complex. For example, Murdock et al. (1984) developed a model specifying four types of labor demand and labor supply broken into age-sex groups. Murdock, Jones, Hamm, and Leistritz (1987) and Murdock and Leistritz (1980) describe this model in detail. In the simplest model, the volume of net migration is equal to the gap between labor supply and demand. Net in-migration occurs when labor demand is greater than labor supply, and net out-migration occurs when labor supply is greater than labor demand. This assumption can be relaxed by setting thresholds that trigger the migration response; that is, it can be assumed that migration will not occur until labor supply and labor demand are out of balance by more than some predetermined amount (Murdock et al., 1984).

The matching procedure in the third step determines the net number of workers that leave or enter the area. The fourth step converts these migrating workers into a projection of all "economic" migrants, including other family members, by applying various characteristics to the migrating workers (e.g., marital status and family size). The assumptions made for these characteristics can markedly influence the size of the migrant population and require careful attention (Murdock & Ellis, 1991).

The economic migrants projected in the fourth step do not represent all migrants; in particular, they exclude groups such as retirees and military personnel whose moves are largely unaffected by changes in economic conditions. The fifth step projects these groups by using procedures such as those described in the previous section and the upcoming section on nonrecursive models. Adding these migrants to those projected in Step 4 completes the migration projection.

A final step (not shown in Fig. 9.3) ascribes demographic characteristics to migrants. This can be done in a number of ways. One commonly used procedure is to give migrants the same characteristics as the U.S. population when net migration for an area is positive, and to give them the same characteristics as the local population when net migration is negative (e.g., Center for Continuing Study of the California Economy, 1997).

Population/Employment Ratios

Our final recursive model does not single out migration or any other individual component of population change. Rather, it develops projections of total population from employment projections and the projected ratio of population to employment (P/E). In its simplest form this model uses projections of total population and total employment and holds the P/E ratio constant at its current value. However, because P/E ratios vary by demographic characteristic and change over time, this simple approach is no longer frequently used (Murdock & Ellis, 1991). More refined approaches can be followed, such as projecting trends in P/E ratios or restricting projections to persons less than age 65 (U.S. Bureau of Economic Analysis, 1995). Despite some drawbacks, this approach offers an easy and inexpensive way to derive population projections from economic projections.

In the mid-1960s, the BEA implemented a P/E ratio model known as "OBERS," an acronym based on the names of two federal agencies: Office of Business Economics—now the Bureau of Economic Analysis—and Economic Research Service in the Department of Agriculture. For more than 30 years, the BEA used this model to project population, employment, and earnings for states, economic areas, and metropolitan areas. Budget cutbacks forced the BEA to stop producing population projections in the mid-1990s, but its model has been widely copied, refined, and applied. We describe the most recent version of the BEA model to illustrate the P/E ratio approach to population projection. We refer to this model as "BEA" rather than "OBERS" because the latter acronym is no longer used (Figure 9.4).

BEA projections divide the population into three age groups: less than 18 (prelabor pool), 18–64 (labor pool), and 65 and older (postlabor pool). The BEA model is based on the assumption that changes in the state or local labor pool are mainly a function of economic opportunities. Although the BEA did not explicitly state any assumptions concerning the prelabor pool population, projections of that age group are indirectly related to the same economic factors that influence the labor pool population. The BEA assumed that changes in the population aged 65+ are independent of state or local economic changes (U.S. Bureau of Economic Analysis, 1995).

The BEA projects an area's labor pool population by following five steps. First, they compute the P/E ratio of the population aged 18–64 to total employment (including self-employed and uniformed military) for the area and for the

Labor Pool Population (18–64)

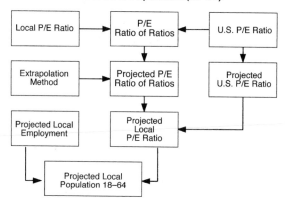

Prelabor Pool Population (0–17)

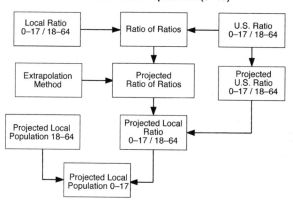

Postlabor Pool Population (65+)

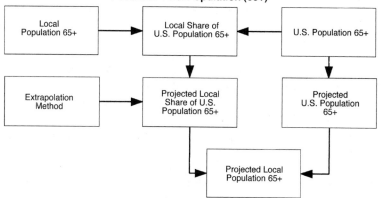

Figure 9.4. The Bureau of Economic Analysis (BEA) model.

nation. Second, they form a ratio of the area's P/E ratio to the national P/E ratio. Third, they use historical trends to project this "ratio of ratios" into the future (they did not publish details regarding their trending method). Fourth, they calculate a projected P/E ratio for the area by multiplying the projected "ratio of ratios" by the projected P/E ratio for the nation. Fifth, they multiply the area's projected P/E ratio by its projected employment figure to obtain a projection of the population aged 18–64.

The independent employment projections upon which the population projections are based are produced by using a combination of export-base and shift-share methods. Clearly, the validity of the projections of the labor pool population depends heavily on the validity of the employment projections.

The area's prelabor pool population (aged 0–17) is projected similarly but uses a different ratio. Instead of being based on a population/employment ratio, the projection for this age group is based on a ratio of the population aged 0–17 to the population aged 18–64. This ratio is multiplied by the area's projected labor pool population and yields a projection of the population aged 0–17.

The area's postlabor pool population (aged 65+) is projected in three steps. First is to calculate each area's share of the national population aged 65+. Second is to project these shares into the future based on their historical trends (again, the BEA did not publish details regarding the trending method). Third is to project the population aged 65+ for each area by multiplying its projected share by the projected national population aged 65+.

To demonstrate the BEA model we prepared a population projection for 2005 for Massachusetts. Our launch year was 1997, corresponding to recent BEA estimates of total employment (U.S. Bureau of Economic Analysis, 1999). Our projection was based on 1997 population estimates for the United States and Massachusetts (U.S. Bureau of the Census, 1998), the middle series of U.S. population projections for 2005 (Day, 1996a), and 2005 total employment projections for the nation and Massachusetts (U.S. Bureau of Economic Analysis, 1995). We used data from 1990–1997 to determine the projected trend for the various ratios in the BEA model.

A problem can arise when the employment projections used in this model are prepared before the current launch year and do not track with recent trends. For example, the unexpectedly strong performance of the economy between 1993 and 1997 caused employment growth during this period to exceed the projections based on a 1993 launch year. Therefore, we created new employment projections for 2005 by using 1997 employment as a base and assuming that the percentage change from 1997–2005 would match the percentage change from 1998–2005 shown in the BEA projections. Practitioners must be aware of this type of problem when using this model and make adjustments when necessary.

Table 9.1 shows the computations for producing a 2005 population projection for Massachusetts. The top panel provides data for the population aged 18–64.

First, we constructed P/E ratios for Massachusetts and the United States in 1990 and 1997:

Massachusetts P/E (1990): 3,848,656 / 3,646,074 = 1.05556

Massachusetts P/E (1997): 3,807,356 / 3,921,971 = 0.97078

U.S. P/E (1990): 154,031,000 / 139,184,600 = 1.10667

U.S. P/E (1997): 163,958,000 / 156,400,400 = 1.04832

Next, we computed the ratio of ratios for both years by dividing the Massachusetts P/E ratio by the U.S. P/E ratio:

Ratio of ratios (1990): 1.05556 / 1.10667 = 0.95382

Ratio of ratios (1997): 0.97078 / 1.04832 = 0.92603

To project the ratio of ratios to 2005, we extrapolated the numerical change observed between 1990 and 1997. To simplify the computations in this example, we did not adjust for the one-year difference between the length of the base period (1990–1997) and the length of the projection horizon (1997–2005):

2005 ratio of ratios: 0.92603 + (0.92603 − 0.95382) = 0.89824

Next, we projected the Massachusetts P/E ratio to 2005 by multiplying the projected ratio of ratios by the 2005 U.S. P/E ratio projected by the BEA:

2005 Massachusetts P/E ratio: 0.89824 × 1.03755 = 0.93197

Finally, we projected the Massachusetts population aged 18–64 in 2005 by multiplying its 2005 P/E ratio by its 2005 employment projection:

2005 Massachusetts population aged 18–64: 0.93197 × 4,259,261 = 3,969,503

The middle panel of Table 9.1 shows the computations for projecting the population aged 0–17. The computational sequence is the same as that described for the population aged 18–64, with two differences. First, instead of a population/employment ratio, we used a ratio of the population aged 0–17 to the population aged 18–64. Second, we applied the projected ratio to the projected population aged 18–64 for Massachusetts, as determined in the top panel of the table. Following these steps yields a projection of 1,520,121 persons aged 0–17 in 2005.

The bottom panel shows the computations for projections of the population aged 65+. This is the simplest part of the BEA model. The first step is to compute the share of the U.S. population aged 65+ that resided in Massachusetts in 1990 and 1997:

1990 share 65+: 817,286 / 31,239,000 = 0.02616

1997 share 65+: 860,853 / 34,198,000 = 0.02517

Table 9.1. The BEA Model:
Population Projection for Massachusetts, 2005

	Labor pool population (18–64)		
	1990	1997	2005
Massachusetts			
Population 18–64	3,848,656	3,807,356	3,969,503[g]
Total employment[a]	3,646,074	3,921,971	4,259,261[e]
P/E ratio[b]	1.05556	0.97078	0.93197[f]
United States			
Population 18–64	154,031,000	163,958,000	177,851,000
Total employment[a]	139,184,600	156,400,400	171,414,857[e]
P/E ratio[b]	1.10667	1.04832	1.03755
Ratio of ratios[c]	0.95382	0.92603	0.89824[d]

	Prelabor pool population (0–17)		
	1990	1997	2005
Massachusetts			
Population 0–17	1,352,452	1,446,231	1,520,121[j]
Population 18–64	3,848,656	3,807,356	3,969,503
Ratio[h]	0.35141	0.37985	0.38295[i]
United States			
Population 0–17	64,169,000	69,588,000	71,964,000
Population 18–64	154,031,000	163,958,000	177,851,000
Ratio[h]	0.4166	0.42443	0.40463
Ratio of ratios[c]	0.84352	0.89497	0.94642[d]

To project the Massachusetts share of the population aged 65+ in 2005, we linearly extrapolated the change in the share that occurred from 1990–1997:

2005 share aged 65+: 0.02517 + (0.02517 − 0.02616) = 0.02418

Finally, the population aged 65+ in Massachusetts in 2005 was calculated by multiplying Massachusetts' projected share by the U.S. population aged 65+ in 2005:

2005 population aged 65+: 0.02418 × 36,166,000 = 874,494

The total population projected for Massachusetts in 2005 is simply the sum of the projections for these three age groups:

2005 total population: 3,969,503 + 1,520,121 + 874,494 = 6,364,118

Table 9.1. (*Continued*)

	Postlabor pool population (65+)		
	1990	1997	2005
Massachusetts	817,286	860,853	874,494[l]
United States	31,239,000	34,198,000	36,166,000
Share of U.S.[k]	0.02616	0.02517	0.02418[d]

[a]Includes civilian wage and salary employment, proprietors (self employed), and uniformed military employment.

[b]P/E ratio = population aged 18–64 / total employment.

[c]Ratio of ratios = Massachusetts ratio / U.S. ratio.

[d]Assumes same numeric change between 1997–2005 as occurred between 1990–1997.

[e]Assumes that the percent change from 1997–2005 is the same as the percent change from 1998–2005 shown in the BEA projection.

[f]2005 Massachusetts P/E ratio = 2005 ratio of ratios × 2005 U.S. P/E ratio.

[g]2005 Massachusetts population aged 18–64 = 2005 Massachusetts P/E ratio × 2005 Massachusetts total employment.

[h]Population ratio = population aged 0–17 / population aged 18–64.

[i]2005 Massachusetts population ratio = 2005 ratio of ratios × 2005 U.S. ratio.

[j]2005 Massachusetts population aged 0–17 = 2005 Massachusetts population ratio × 2005 Massachusetts population aged 18–64.

[k]Share of U.S. population aged 65+ = Massachusetts population aged 65+ / U.S. population aged 65+.

[l]2005 Masssachusetts population aged 65+ = 2005 share of U.S. population aged 65+ × 2005 U.S. population aged 65+.

Sources: U.S. Bureau of Economic Analysis, *BEA Regional Projections to 2045: Volume I, States*. Washington, DC, July 1995; U.S. Bureau of Economic Analysis, Regional Economic Information (REIS) 1969–97 CD. Washington, DC, August 1999; J. Day (1996a). Population Projections of the United States by Age, Sex, Race, and Hispanic Origin: 1995–2050. U.S. Bureau of the Census, *Current Population Reports* P25-1130, Washington, DC; U.S. Census Bureau, Population Estimates by Age, Sex, Race, and Ethnicity for States, Internet release date, December 1998.

How reasonable is this projection? It is very close to the most recent 2005 projection produced by the Census Bureau using the cohort-component method (Campbell, 1996). The projections of total population differ by only 1%; for individual age groups, the projections differ by 2%, 1%, and 5%, respectively, for ages 0–17, 18–64, and 65+. This projection is also similar (about 2% lower) to a projection recently produced by the Massachusetts Institute for Social and Economic Research (http://www.umass.edu/miser).

One additional caveat should be mentioned regarding the use of employment data: Measures of employment vary from one government agency to another. Two federal agencies—the BEA and the BLS—develop estimates of employment by place of work. BEA estimates cover wage and salary workers, the self-employed,

---

Box 9.1

A Ratio-Correlation Method for Short-Range Population Projections

A simple structural model can be developed using the ratio-correlation method. Ratio-correlation is a regression method in which changes in population are based on changes in symptomatic indicators of population change (e.g., births, school enrollment, and employment). It has been used primarily for constructing population estimates (e.g., Namboodiri, 1972; O'Hare, 1976; Swanson, 1980) but can also be used for short-range population projections.

In the ratio-correlation method, an equation is estimated in which variables are measured as ratios of proportions. These ratios typically focus on changes between the two most recent censuses (e.g., county share of state population in 2000 divided by the county share of state population in 1990). The population variable is regressed on the symptomatic indicators, and the resulting regression coefficients are used to form a new equation. For postcensal population estimates, current values for the symptomatic indicators are plugged into this equation to provide an estimate of the current population.

Swanson and Beck (1994) developed a ratio-correlation method for short-range population projections and tested it for counties in Washington State. Using employment, school enrollment, and registered voters as symptomatic indicators, they constructed a model in which population was regressed on values of the symptomatic indicators lagged by two years (e.g., population values for 1970–1980 were regressed on symptomatic indicator values for 1968–1978). They used the resulting regression coefficients and current values for the symptomatic indicators to project the population two years into the future. They found that this simple structural model produced reasonably accurate forecasts but noted that it is best suited for projection horizons of less than five years.

---

and the uniformed military. BLS estimates cover only wage and salary workers but are updated more frequently (monthly as opposed to annually). State government agencies often develop their own adjustments to basic employment data. When constructing projections based on P/E ratios, it is important to make sure that the data used in constructing historical P/E ratios are consistent with the data used in developing the independent employment projections.

Where can employment projections be obtained? For a number of years, the BEA produced employment projections for the nation, states, economic areas, and metropolitan areas. These projections typically spanned a 50-year projection horizon, provided breakdowns by Standard Industrial Code (SIC) groups, and were easily accessible in a variety of formats. However, the last set of BEA projections was produced in 1995. It remains to be seen whether future sets will be produced.

The BLS makes 10-year employment projections for the United States and for states but not for substate areas. Many state governments develop employment projections for their state and for metropolitan areas and counties, but the employment detail, geographic specificity, length of projection horizon, and accessibility vary considerably from state to state. Some private vendors provide long-range employment projections for states, counties, and economic areas. Finally, some of the larger local governments develop long-range employment projections, but they typically cover only a limited number of geographic areas, time horizons, and employment categories.

NONRECURSIVE MODELS

Recursive projection models have been criticized on both statistical and theoretical grounds because they view economic-demographic relationships as one-sided. They account for the influence of economic factors on population growth, but do not account for the influence of population growth on economic conditions. However, a large body of evidence shows that demographic variables not only are affected by economic variables but influence those variables as well (e.g., Borts & Stein, 1964; Greenwood, 1981; Muth, 1971; Plane, 1993).

Nonrecursive models address this problem by incorporating relationships that simultaneously depict both the economic determinants and consequences of demographic change. Although they have more complicated mathematical structures and larger resource requirements, a number of nonrecursive projection models have been developed and implemented (e.g., Conway, 1990; Joun & Conway, 1983; Mills & Lubuele, 1995; Treyz, 1993). In this section, we discuss some of the ways that population growth affects economic conditions and describe a projection model that accounts for both the economic determinants and consequences of population growth. We do not attempt to describe all of the details of a complete nonrecursive modeling system; rather, we focus solely on the migration component of that system.

Economic and Demographic Relationships

Earlier, we explained how economic factors—especially changes in employment—affect migration and population growth. Almost 40 years ago, Borts and Stein (1964) argued that the opposite is also true: Migration is influenced by employment and employment is influenced by migration. They based their argument on the premise that an area's labor demand curve is perfectly elastic; therefore, employment will increase by the same amount as the shift in the labor supply curve. Because labor supply is affected by migration, employment must be affected by migration as well.

A fundamental question has been posed: Do people follow jobs or do jobs follow people? In a groundbreaking study, Muth (1971) found support for both views but concluded that the evidence more strongly favored the Borts and Stein hypothesis. Specifically, he estimated that every 10 new jobs attract between six and seven new migrants, but every 10 new migrants create 10 additional jobs. Steinnes (1982) concluded that causality runs only one way: from a change in residence to a change in employment. His study added further support to the Borts and Stein hypothesis and rekindled the causality debate.

Greenwood et al. (1986) found that migrants fill about five out of every 10 new jobs, a bit lower than Muth's estimate. Turning to migration's impact on employment, they estimated an effect about 36% higher than that found by Muth (1971). Similarly, Clark and Murphy (1996) and Mills and Lubuele (1995) found stronger support for the hypothesis that jobs follow people than for the hypothesis that people follow jobs. Carlino and Mills (1987), however, found that the impact of employment growth on population growth is greater than the impact of population growth on employment growth.

Although these studies reached differing conclusions regarding the magnitude and significance of various economic–demographic relationships, they provide sufficient empirical evidence to conclude that causal relationships run in both directions: Economic conditions affect migration flows and migration flows affect economic conditions. How can both effects be accounted for in a projection model?

The simplest approach is to develop a two-equation simultaneous model of migration and employment:

$$\text{MIG} = \alpha_0 + (\alpha_1 \text{EMP}) + (\alpha_2 \text{WAGE}) + \mu_1$$

$$\text{EMP} = \beta_0 + (\beta_1 \text{MIG}) + (\beta_2 \text{WAGE}) + \mu_2$$

where MIG is migration, EMP is employment, WAGE is wages, α and β are the estimated parameters, and μ is the residual. Migration is the dependent variable in the first equation and an independent variable in the second, whereas employment is the dependent variable in the second equation and an independent variable in the first.

MIG and EMP are referred to as endogenous variables because they are determined within this system of equations. WAGE is an exogenous variable because it is determined outside the system; it influences both migration and employment but is not influenced by them. This model allows for reciprocal causality: employment affects migration, and migration affects employment. The parameters α_1 and β_1 provide estimates of the impact of employment on migration and of migration on employment, respectively.

A full explanation of parameter estimation in nonrecursive models is beyond the scope of this book. In addition to issues such as identification and the use of instrumental variables, statistical problems occur when an explanatory variable in one equation is a dependent variable in another equation. This violates a principal assumption of OLS regression analysis; namely, that explanatory variables are not correlated with residuals. Consequently, OLS regression coefficients are biased and inconsistent. This problem is often handled using special estimation algorithms such as two- and three-stage least squares. However, Joun and Conway (1983) found that OLS techniques often produce accurate forecasts and reasonable simulation results, even for nonrecursive models. Detailed discussions of these issues can be found in Berry (1984) and Kmenta (1971).

Regional Economic Models, Incorporated (REMI)

Now, we turn to a nonrecursive model in which economic conditions affect migration and migration affects economic conditions. REMI is an integrated economic–demographic projection model used since the early 1980s by government agencies, consulting firms, universities, and public utilities throughout the United States (Treyz, 1993). The model is frequently updated in response to new research and data, client requirements, and internal and external evaluations (Treyz et al., 1992; Zhao, Carlson, & Swanson, 1994). Treyz (1993) and Treyz et al. (1992) describe the philosophy, theory, and structure of the REMI model. Most of the material in this section comes from these sources and two other articles (Greenwood et al., 1991; Treyz et al., 1993).

The REMI model is based on two basic assumptions from economic theory: Households attempt to maximize utility and producers attempt to maximize profits. The model integrates demographic and economic processes by using a cohort-component model and a system of simultaneous equations. It includes economic–demographic relationships within local areas, incorporates the impact of amenities, and accounts for economic interdependencies by using input–output techniques.

Figure 9.5 outlines the primary economic–demographic relationships in the REMI model. Output and wage rates determine employment opportunities. New jobs are filled by migrants and by members of the current population. Employment opportunities influence migration and also have a feedback effect on wages. Wages—adjusted for production and housing costs—also influence migration. Migration is combined with the effects of fertility and mortality within the framework of a cohort-component model to complete the population projection. Population growth influences employment opportunities directly and indirectly through its effect on state and local government spending and total output.

Like most economic models of migration, REMI does not assume that all migrants respond equally to changes in economic conditions. International migrants, retirees, and military personnel and their dependents are assumed to be unaffected by the model's economic variables. International migration is projected

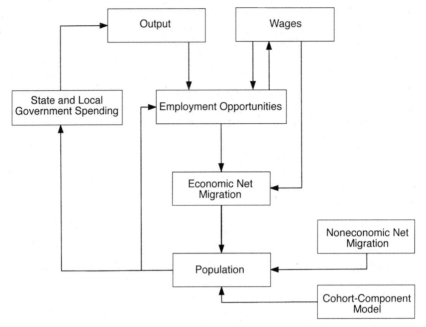

Figure 9.5. The REMI model: Relationships between population and the economy.

by using a fixed local share of U.S. international migration, as in DEFM. The military population and their dependents (including overseas personnel) are treated as a special population. The model specifies a link between changes in the local population and changes in the military population.

Net migration rates for retirees are based on 1980 and 1990 Census data. These rates are stratified by age, sex, and race and are held constant over the projection horizon. If an area's retiree net migration was positive during the base period, projections of retiree net migration are tied to changes in the U.S. elderly population. If it was negative, projections of retiree net migration are tied to changes in the area's elderly population.

Net economic migration of persons under the age of 65 responds to location-specific amenities and to economic opportunities. For example, net in-migration occurs if positive amenity differentials exceed negative economic opportunity differentials or if positive economic differentials exceed negative amenity differentials. The variables in the net economic migration equation assume a semilog functional form, and lags of up to two years are included.

Net economic migration is estimated using two-stage least squares and annual data for each state. Migration projections for substate areas are based on coefficients estimated at the state level rather than for individual substate areas. The equation is calibrated for substate areas through the intercept term, which represents the local amenity value. Calibration of the intercept term is based on the coefficient estimates from the state model and local values for economic net migration and the economic variables in the net migration equation (Treyz, 1993; Treyz et al., 1992).

CONCLUSIONS

Structural models require more resources and are more difficult to implement than many of the projection methods discussed in this book. They often require a large amount of base data, sophisticated model-building skills, complex statistical procedures, and intricate computer programs. These requirements make many structural models very costly to develop and apply. For example, developing an economic–demographic model for the state of Washington cost $250,000 in the late 1980s (Conway, 1990). Costs generally come down once a model is in place, but updating and running an existing model can still be quite expensive. An overhaul and updating of DEFM cost $100,000 in 1997 (San Diego Association of Governments, 1999). Even off-the-shelf systems like REMI can cost tens of thousands of dollars. The costs of developing and implementing urban systems models (discussed in the next chapter) are even greater than the costs of economic–demographic models.

Are structural models worth it? The answer to this question depends on the purposes for which the projections will be used. If the only objective is to create an accurate population forecast, structural models are not worth their cost and complexity. The accuracy of population forecasts from structural models depends heavily on the accuracy of the forecasts of the model's independent variables and the assumption that economic–demographic relationships remain constant over the projection horizon. As we show in Chapter 13, there is no evidence that structural models provide more accurate population forecasts than can be obtained from simpler, less expensive methods.

Structural models can be very useful for other purposes, however. Perhaps their greatest strength is their ability to address a wide range of theoretical, policy, and planning questions (e.g., Klosterman, 1994; Tayman, 1996b; Treyz, 1995). Decision making and planning at all levels require detailed information on a broad array of interrelated variables and geographic areas. Structural models are well suited to meet these needs. In addition, they can provide population projections that are consistent with a variety of employment, transportation, land use, and other types of projections. This attribute is especially important for the urban systems models discussed in the next chapter.

Structural models are more useful than other projection methods for simulation and scenario analysis. Although other methods (particularly the cohort-component method) can be used for developing simulations and analyzing alternative scenarios, structural models permit investigating a wider range of variables and interrelationships. They can also be used to evaluate the implications of particular decisions and to make policy changes when these decisions lead to unintended or undesirable consequences (e.g., Schmidt, Barr, & Swanson, 1997; Tayman, 1996b). In some circumstances, developing simulations and hypothetical scenarios is more important than developing a specific projection or forecast.

For example, what impact would increased labor productivity and higher wages have on migration into an area? How might changes in the age structure affect the demand for housing? What effect would a more restrictive U.S. immigration policy have on the local economy? What effect would a 15% cut in the defense budget have on an area's population growth? How would a policy restricting residential development affect land prices and housing affordability? What impact would increased housing density have on traffic patterns and air quality? What impact would a new baseball stadium have on the geographic distribution of residential and nonresidential development? These and many other questions can be investigated most thoroughly within the framework of a structural model. This may explain why structural models are more widely used today than ever before.

Finally, data users and decision makers may view projections from structural models as more realistic and authoritative than projections from other methods. It is widely understood that wages and employment opportunities affect migration patterns and that land use and transportation characteristics affect decisions regarding where to live. Therefore, it is understandable why some might conclude that models that incorporate these factors are more credible than models that exclude them. Although we believe this perception is ill-founded, it may give structural models an advantage over other projection methods, especially when projections must be defended in a public (and perhaps highly politicized) forum. Appearance is sometimes more important than reality.

CHAPTER 10

Structural Models II
Urban Systems

We began our discussion of structural models in Chapter 9 by focusing on economic–demographic models for projecting migration and total population. In this chapter, we present another important class of structural models known as urban systems models. These models are used for projecting the distribution of residential and nonresidential activities within large urban areas. Like economic–demographic models, urban systems models incorporate economic factors such as jobs, unemployment rates, and income and use historical data to develop statistical parameter estimates. However, they differ from economic–demographic models in several important ways.

First is geographic scale. Economic–demographic models typically focus on relatively large areas such as nations, regions, states, counties, and metropolitan areas. Urban systems models typically focus on much smaller areas such as census tracts, block groups, blocks, and traffic analysis zones (TAZs). (Traffic analysis zones are user-designed areas that are typically composed of one or more blocks.) Second, the variables used in urban systems models differ from those used in economic–demographic models. Along with jobs, unemployment rates, and income, urban systems models typically incorporate land use (e.g., land costs and development potential) and transportation characteristics (e.g., travel costs, times, and distances). Third, urban systems models use a set of statistical tools somewhat different from economic–demographic models. In particular, geographic information systems (GIS) play an important—perhaps essential—role (e.g., Batty, 1992; Landis, 1995; Prastacos, 1991; San Diego Association of Governments, 1998).

Finally, urban systems models generally require more time and resources than economic–demographic models, making them substantially more costly. One agency, for example, estimated that it would cost about $600,000 to implement an urban systems model in the Delaware Valley region near Philadelphia (Delaware Valley Regional Planning Commission, 1996).

Urban systems models have both temporal and spatial dimensions. The temporal dimension divides time into discrete intervals. These are usually five-year intervals such as 2000–2005, 2005–2010, and so forth. The spatial dimension starts with a large region (e.g., a metropolitan area) and divides it into a series of smaller geographic areas we refer to as *zones*. The number, geographic location, and size of zones depends on several factors, including the larger region's geographic size, population size, and employment base, and the economic, demographic, and land use characteristics of the zones themselves. The most effective way to define zonal boundaries is to minimize the variability of these characteristics within zones and maximize their variability across zones. For example, one might identify zones within commercial areas, industrial areas, older neighborhoods, newer suburbs, undeveloped areas of high development potential, and undeveloped areas of low development potential. Applications typically use between 150 and 300 zones, but some have used as few as 15 or as many as 800.

Urban systems models can be used to examine a variety of issues—for example, air quality, traffic congestion, loss of open space, fiscal implications of land use decisions—that cannot be considered in most economic–demographic models. Projections from urban systems models cover population, housing, employment, income, land use, and transportation characteristics. In this chapter we discuss models that focus on small geographic areas, incorporate both residential and nonresidential land uses, and develop links with transportation factors. We do not consider other types of small-area structural models, such as those relying primarily on an econometric approach (e.g., Greenwood, 1981; Levernier & Cushing, 1994).

Urban systems models vary considerably in their theoretical approaches, mathematical algorithms, data requirements, and ease of implementation. Presenting a detailed description of these models is beyond the scope of this book. We simply provide a general overview of urban systems models, highlighting some of the questions they address and some of the answers they provide.

We start with a brief history of urban systems models. Then, we describe the major components of these models, investigate potential data sources, and discuss several issues related to model implementation. Because many readers may be unfamiliar with these models, we provide a numerical example of one particular model. Finally, we present brief summaries of several urban systems models used today: gravity models, optimization models, land pricing models, microgeographic area models, and the California Urban Futures model.

A BRIEF HISTORY OF URBAN SYSTEMS MODELS

The 1950s saw the emergence of computer models that linked land use, residential and nonresidential activities, and the transportation system. Computers made urban systems models possible, but sociopolitical conditions provided the impetus for developing them. For example, the desire to use scientific methods to assess the impact of new highways and analyze urban problems spurred the development of new models (Putman, 1983). Though many early modeling efforts did not succeed, the following 20 years produced a wealth of information about spatial relationships within urban areas. This period saw the publication of several groundbreaking works that revolutionized urban systems models and urban planning practices (e.g., Harris, 1965; Lowry, 1964).

Lee (1973) wrote a scathing critique of urban systems models and predicted their demise. His main criticisms centered on their overly ambitious but mostly unachieved goals, the lack of sufficient data and computing power, and an inadequate understanding of the urban development process. This influential paper—along with factors such as the lack of technical skills among planners and institutional resistance to new methods—slowed the development and implementation of urban systems models in the United States (e.g., Batty, 1994; Harris, 1994). Although some work continued (e.g., Putman, 1979, 1983), the most important theoretical and practical advances occurred in other countries (e.g., Anas, 1982; Batty, 1976; Echenique, 1983; Foot, 1981; Wilson, 1974).

Almost 30 years have passed since Lee predicted the demise of urban systems models, but they are more widely used today than ever before (e.g., Boyce, 1988; Klosterman, 1999; Tayman, 1996a). At least 20 centers on four continents are actively involved in urban modeling research (Wegener, 1994). This resurgence of interest was evidenced in a 1994 symposium on urban systems models published by *The Journal of the American Planning Association*. This was the journal's first collection of articles on this topic in more than two decades. One year later, the U.S. Department of Transportation and the U.S. Environmental Protection Agency cosponsored a conference on urban systems models in Dallas. Several other conferences have been held since that time.

What accounts for this renewed interest? In part, it stems from two pieces of federal legislation—the Intermodal Surface Transportation Efficiency Act of 1991 (ISTEA), which was reauthorized as the Transportation Equity Act for the 21st Century (TEA21) in 1998, and the Clean Air Act Amendments of 1990. ISTEA and TEA21 mandated that transportation plans consider the long-range effects of interactions among land use patterns, residential and nonresidential activities, and the transportation system. The Clean Air Act Amendments specified that the analysis of air quality must take into account interrelationships among travel patterns and the location of homes, businesses, shopping, and recreational activities. Deakin (1995) noted that even without federal mandates, policy makers are under increasing pressure from environmentalists, developers, and the public to address issues associated with urban form, land use, and the transportation system. Urban systems models provide a systematic way to analyze these issues and evaluate policy options.

Equally important are recent increases in computing power, the development of desktop GIS systems, and the greater availability of computer-readable data. These dramatic changes provided the infrastructure needed to develop and implement urban systems models and overcame some of data and technological limitations noted by Lee (1973). Significant advances have also been made in understanding the processes and patterns of urban development (e.g., Anas, 1992; de la Barra, 1989; Echenique 1994; Hunt, 1994a,b; Hunt & Simmons, 1993; Kim, 1989; Putman, 1991; Wegener, Mackett, & Simmons, 1991).

COMPONENTS OF URBAN SYSTEMS MODELS

Urban systems models typically consist of three major components—regional projections, zonal activity and land use, and transportation. These components are themselves models that represent specific parts of the larger system. Figure 10.1 shows these three components and their primary interrelationships. The dashed lines represent issues that involve the transportation model interface, which we discuss later in this section.

Regional Projections

The first component of an urban systems model is a set of economic and demographic projections for the region covered by the model (e.g., a metropolitan area). These projections provide the control totals that are distributed among zones through the zonal activity and land use component of the model. Regional projections are almost always produced by using economic-demographic models similar to those described in Chapter 9, typically in conjunction with a cohort-component model.

Zonal Land Use and Activity Model

The second component is a zonal land use and activity model. This component is composed of a complex set of procedures for distributing changes in "activities" (e.g., population, housing, employment, and income) for a region (e.g., a metropolitan area) to smaller geographic areas within that region. As Figure 10.1 shows, the zonal land use and activity model deals with spatial relationships among residential activities, nonresidential activities, and land uses.

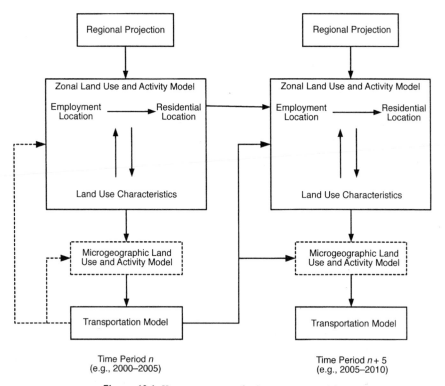

Figure 10.1. Key components of urban systems models.

Land use is a general term covering a variety of characteristics such as housing densities; residential, commercial and industrial designations; zoning restrictions; the supply of vacant land; and local growth policies.

A crucial premise of virtually all zonal land use and activity models is that employment location is a primary determinant of residential location. The link between home and work place is captured through zonal travel probabilities (e.g., time, cost, or distance) and commuting patterns furnished by the transportation model (e.g., Putman, 1983; Zax, 1994). The spatial distribution of residents also influences the location and demand for employment, especially for population-related jobs such as retail trade and services. Many zonal land use and activity models implement this relationship by assuming a lag between population or housing development and the subsequent location of new jobs. The arrow connecting residential location at time *n* to employment location at time *n* + 5 in Figure 10.1 reflects this lag. Transportation characteristics, such as home-to-shopping travel times, also play a role in determining the zonal location of employment.

The two-way arrows between land use characteristics and employment and residential location in Figure 10.1 reflect the iterative nature of these relationships. Land use characteristics influence residential and employment activities, and residential and employment activities influence land use characteristics. For example, local land use policies may spur population and employment growth, and population and employment growth may lead to a decline in the supply of vacant land.

Zonal land use and activity models reconcile the demand for residential and nonresidential activities with the available supply of vacant land. One approach is to hold the supply of vacant land constant (e.g., San Diego Association of Governments, 1998). If demand exceeds supply, growth shifts to an alternate zone that has a sufficient supply of vacant land to support that particular activity. Another approach is to use land pricing mechanisms to reconcile gaps between supply and demand (e.g., de la Barra, 1989; Hunt, 1994b; Waddell, 2000). Excess demand can also be satisfied by allowing housing, population, and employment densities to rise (e.g., Putman, 1995).

Transportation Model

The third component in an urban systems model is transportation, which is often modeled by using a four-step approach: trip generation, trip distribution, mode choice, and traffic assignment (e.g., Southworth, 1995; Yu, 1982). The first step determines the number of trips that occur in each zone. These trips are classified by type (e.g., home to work, home to shopping) for zones of origin and destination. The second step matches origins and destinations, creating an origin–destination matrix for all types of trips and all possible pairs of zones. The third step separates trips by mode of travel, such as one-person auto, carpool, mass transit, biking, or walking. The final step determines the number of trips that occur on particular streets, highways, and transit networks. It takes into account street and highway capacities to prevent more trips from occurring than could realistically be expected; it can also simulate the effects of drivers choosing alternative routes because of traffic congestion.

Linking the Components

Transportation factors affect land use and the location of activities, but activity and land use patterns also affect the transportation system. The opening of a new housing development or shopping center, for example, typically leads to increased traffic flow and greater congestion. These transportation characteristics in turn influence the zonal land use and activity allocation in the following time period. This sequential process is the most widely used method for linking the transportation component of urban systems models with the zonal land use and activity component (Wegener, 1995).

This sequential process has been questioned on the grounds that it may not capture the true extent of interactions between the transportation component and the land use and activity component (Weatherby, 1995). One way to provide closer integration is to have the two components influence each other within the same period (e.g., Putman, 1991, 1994). Figure 10.1 depicts this iterative feedback mechanism with dashed lines that lead from the transportation model back to the zonal activity model within time period *n*. Zonal land use and activity projections feed into the transportation projections in the first iteration. The transportation projections feed back into the zonal land use and activity projections, which produce another set of transportation system projections to complete a second iteration. This process continues until the system (i.e., transportation characteristics and land use and activity patterns) converges or reaches an equilibrium.

The value of the iterative approach remains to be seen. We do not know, for example, whether an iterative process produces better projections than an easier to implement sequential process. There is also a question as to whether complete convergence should or should not occur within a given period, especially for the land use and activity variables (Putman, 1994; Wegener, 1986). We believe that the planning profession is moving toward a closer integration of land use and activity models and transportation models, but further research is required before we can form clear conclusions.

Another critical issue is the nature of the spatial relationship between land use and activity models and transportation models. Urban systems models such as METROPILUS (Putman, 1995), MEPLAN (Echenique, 1994; Hunt, 1994a,b), TRANUS (de la Barra, 1989), and UrbanSim (Waddell, 2000) include both a zonal land use and activity model and a transportation model. Consequently, these models use one zonal system for all of the components. In many models, however, the zonal land use and activity model must interface with a transportation model that is already in place.

Most transportation modeling—at least in the United States—operates at a much smaller zonal level than can be supported by a land use and activity model. For example, the land use and activity model for the San Diego region uses 208 zones, whereas the transportation model uses 4,500 TAZs (San Diego Association of Governments, 1998). Discrepancies among zonal structures are usually not as extreme as in San Diego, but they are prevalent in the majority of urban systems models. These situations require another allocation model to obtain land use and activities for the geographic areas used by the transportation model. We refer to this allocation model as a *microgeographic land use and activity model*, or simply a "micro" model. This is shown as the box with the dotted lines in Figure 10.1.

Micro models take the results from zonal land use and activity models and distribute them to smaller geographic areas such as census tracts, TAZs, or blocks (e.g., Maricopa Association of Governments, 1996; San Diego Association of Governments, 1998). To date, there has been little published research on these models (e.g., Tayman, 1996a,b). Consequently, there are no widely accepted best practices or standard procedures. Most micro models are simply collections of ad hoc procedures that are not well integrated either with zonal land use and activity models or with their underlying databases. Agencies often spend a significant amount of time and resources implementing zonal land use and activity models, only to find themselves with an incomplete solution. This is clearly an area that requires further research.

DATA REQUIREMENTS AND SOURCES

Urban systems models have extensive data requirements and rely on information from a variety of sources. Regional projections—the first component in an urban systems model—are typically made using cohort-component and/or economic-demographic models. We discussed the data requirements and data sources for these methods in previous chapters and do not provide additional information here.

Zonal land use and activity models and transportation models—the second and third components of urban systems models—rely on many of the same types of data: population, housing, income, employment, and land use. Transportation models also require detailed information on commuting patterns, transit networks, transit stops, street capacities, posted speed limits, and the characteristics of intersections. These data come from government agencies and from household surveys that cover modes of transportation, destinations, travel times, and similar types of information.

In this section, we focus on the data requirements for zonal land use and activity models because these models are the ones most directly related to small-area population projections. Moreover, many of the data sources used in zonal land use and activity models are the same as those used in transportation models.

Zonal land use and activity models require huge amounts of input data. Cheaper and more powerful computers, the development of GIS, and improved access to small-area databases have greatly facilitated collecting and manipulating these data. Although data availability, quality, and consistency are clearly less of an issue than they were a decade or two ago, Lee's (1973) characterization of land use and activity models as "data hungry" is still valid.

Collecting and updating data for zonal land use and activity models is a major undertaking and represents a significant impediment to the wider application of these models (Weatherby, 1995). Many models require data for census tracts, block groups, and even individual blocks. Moreover, the trend is toward applications for even smaller areas such as parcels, street faces, and ZIP + 4 areas (e.g., Batty, Cote, Howes, Pelligrini, & Zheng, 1995; Wegener, 1995). Developing the data needed for these increasingly detailed models is very costly, often requiring periods of six to 18 months and a number of person-years of effort.

Although small-area databases have generally become more accurate and reliable, the data come from many disparate sources and frequently must be adjusted to meet the needs imposed by the model. Building these databases requires a judicious blend of interdisciplinary estimation methods, GIS, and mathematical algorithms for handling iterative fitting, rounding errors, and small or zero values. Agencies responsible for creating spatially intensive databases are doing some of the most innovative work in small-area estimation, but their innovations are not well known to most demographers, planners, and other practitioners because they rarely publish articles or present papers at professional conferences.

Batty et al. (1995) provide a useful classification of the data items used in land use and activity models. Their scheme draws distinctions among socioeconomic data, physical data, and cartographic data. Socioeconomic data include items such as population, housing, employment, and income. Physical data include characteristics such as land use, environmental features, and zoning and building conditions. Cartographic data refer to computerized street and line files, such as the Topologically Integrated Geographic Encoding and Referencing (TIGER) system developed by the Census Bureau. TIGER files are available for the entire nation from the Census Bureau and from private companies. Digital street and address files permit identifying activities on virtually any spatial scale.

Population, Housing, Income

The most comprehensive source of population, housing, and income data is the decennial census, which provides many types of data down to the block level. Census data can be updated using a variety of data sources and estimation methods. Population and housing estimates for states, counties, cities, and smaller areas can be made by using the housing unit method (e.g., Smith & Cody, 1994). These estimates are typically based on building permits, electric customers, and similar indicators of population and household changes. Building permit data for cities and counties are available from the U.S. Department of Commerce, but building permit data for smaller areas must usually be obtained from individual municipal governments. Address-level tax assessor data are available for many areas from state and local government agencies and private companies. These computerized records can be used to pinpoint the location of new homes and to produce reliable small-area housing estimates (Rynerson & Tayman, 1998). Tax returns and other symptomatic indicators can be used to generate postcensal estimates of income for relatively small places (e.g., Fonseca & Tayman, 1989;

Rynerson & Tayman, 1997). Private companies also provide census data and current population, housing, and income estimates for small geographic areas (see Batty et al., 1995, for a list of third-party data suppliers).

Employment

County-level data on employment by place of work are readily available from the U.S. Bureau of Labor Statistics (BLS) and the U.S. Bureau of Economic Analysis (BEA). Although they are easily accessible, these data have several shortcomings. Temporary workers are an increasing proportion of the labor force but are often not assigned to a specific industry. In addition, temporary workers are often counted where their employer is located rather than where the employees actually work. There are problems in assigning geographic work locations to employees of multilocation firms. Another challenge is converting employment from the widely used Standard Industrial Classification (SIC) coding system to the new North American Industrial Classification System (NAICS), especially for the time series data required by many econometric models.

Subcounty employment data pose by far the most difficult problem for land use and activity models (Batty et al., 1995; Weatherby, 1995). The *Census Transportation Planning Package* is one source of these data, but coding problems and response errors limit the completeness and usefulness of these data. *County Business Patterns*—another Census Bureau product—provides subcounty employment information annually. However, these data do not include government or railroad workers and are not available below the ZIP code level; most applications require data for smaller geographic areas. Some state government agencies and private companies also maintain address-level employment records. Obtaining these records from state governments, however, is often difficult or impossible due to confidentiality regulations.

Regardless of the source, address-level employment files have many shortcomings (e.g., duplications, missing employment sites, and incorrect address or employee information). Attempting to rectify these shortcomings can add significantly to the costs of developing employment data for small areas (Tayman, 1994).

Land Use

Advances in data collection technology have greatly increased the availability and accuracy of land use data. Through the manipulation of polygons (i.e., a series of connected computerized points representing specific spatial areas), it is now possible to produce land use data at virtually any geographic level. Acquisition costs for these data vary tremendously, from zero to many thousands of dollars, depending on the source. Analysis time, processing costs, and storage requirements, however, are substantial. Public–private partnerships and data sharing agreements are becoming increasingly common to reduce time requirements and defray costs.

Land use and activity models use two types of physical data. Natural characteristics such as elevation contours and floodplains have a substantial impact on population and economic growth and can act as constraints on future development. A great deal of data on these characteristics is now in the public domain and is available digitally from agencies such as the U.S. Geological Survey (USGS) and Federal Emergency Management Authority (FEMA).

Man-made characteristics are important because they track changes in the physical development of an area and help identify areas that have potential for future development. Land use and activity models vary greatly in their use of this information. Some models simply distinguish between developed and undeveloped areas, whereas others differentiate among different types of development (e.g., houses, commercial establishments, factories, schools, office buildings, parks, open spaces, roads, and freeways). Aerial photography is often used to identify land use, but it can determine only the predominant types of use in an area, making it difficult to classify areas that have mixed uses.

Satellite imagery and remote sensing are other technologies for classifying urban land use. Although these technologies cannot yet identify land use in sufficient detail for most land use and activity models, they can be used for detecting changes over time. Some agencies now use change-detection methods to identify only those areas requiring an update to their land uses. Using new technologies in this manner saves time and money and creates more temporally consistent information (e.g., Coulter, Stow, Kiracofe, Langeuin, Chen, Daeschner, Service, & Kaiser, 1999).

ILLUSTRATION OF A RESIDENTIAL LOCATION MODEL

We use a gravity model to illustrate the residential component of the zonal land use and activity model depicted in Figure 10.1. We focus on residential location because it is directly related to population projections. We use a gravity model because such models are widely used for urban forecasting and policy analysis (Putman, 1994; Wegener, 1994).

Gravity models of residential activity typically involve three components: (1) employment by zone, (2) accessibility between zones, and (3) attractiveness of zones for residential purposes. Employment by zone is exogenous to the residential model and is determined in a different component of the model. Accessibility reflects the ease of commuting from one zone to another; it is typically measured by using distances, travel times, or money costs. In our illustration, we use a probability function in which it is assumed that the greater the time and money costs of a commute, the lower the probability of commuting. Attractiveness can be represented by a number of different factors that influence residential decisions, such as neighborhood characteristics and the supply of vacant land.

Gravity models of residential activity can be based on several formulations (e.g., Fotheringham, 1984). In our illustration, projections of where workers live (*residential location*) are based on projections of employment by zone and the accessibility and attractiveness of potential residential zones. This production-constained location model is represented as follows:

$$RES_{ij} = EMP_j \times ATTR_i^\gamma \times ACCESS_{ij} \times BAL_j$$
$$ACCESS_{ij} = TRIP_{ij}^\alpha \times e^{\beta TRIP_{ij}}$$
$$BAL_j = [\Sigma (ATTR_i^\gamma \times ACCESS_{ij})]^{-1}$$

where RES_{ij} is the number of workers in area j that live in area i; EMP_j is the employment projection for area j; $ATTR_i$ is the attractiveness measure for area i; $ACCESS_{ij}$ is the probability of commuting from area i to area j; BAL_j is a constraint or balancing factor; $TRIP_{ij}$ is the cost of commuting from area i to area j; Σ is the sum across all residential zones (i); and α, β, and γ are empirically derived parameters.

In our illustration, commuting costs (TRIP) are transformed into a probability (ACCESS) by using a modified gamma function (e.g., Putman, 1983). Other functions have also been used, such as a declining power function (e.g., Lowry, 1964), a modified lognormal function (e.g., San Diego Association of Governments, 1998), or a negative exponential function (e.g., Watterson, 1990). A balancing factor is incorporated into the model to ensure consistency between the employment and residential projections (e.g., Fotheringham, 1984; Putman, 1983):

$$EMP_j = \Sigma RES_{ij}$$

where Σ is the sum across all residential zones (i). This condition holds for each employment zone (j).

Finally, we derive a projection of the total number of workers who live in each residential zone by summing the allocation of workers from all of the employment zones:

$$RES_i = \Sigma RES_{ij}$$

where Σ is the sum across all employment zones.

We use an example from Putman (1983) to illustrate the computations of a residential gravity model. This example considers a hypothetical region that contains five zones and 600 projected workers. Commuting probabilities ($ACCESS_{ij}$) are based on the following modified gamma distribution:

$$ACCESS_{ij} = TRIP_{ij}^{-1.5} \times e^{-2TRIP_{ij}}$$

Residential attractiveness ($ATTR_i$) is measured by the number of acres of vacant land in each residential zone. For simplicity, we assumed that the value of the γ parameter is 1.0. The objective of the model is to determine the projected zones of residence for the 600 workers projected for the region.

The top panel of Table 10.1 shows commuting probabilities from one zone to another, the attractiveness index for each residential zone, and the total employment projection for each employment zone. Small cell values in the 5 × 5 matrix reflect low probabilities of commuting from one zone to another and large values reflect high probabilities. For example, a person who works in Zone 1 has the greatest probability of commuting from Zone 1 (0.14) and the lowest probability of commuting from Zone 5 (0.03). Attractiveness values were based on land use information. Projected employment for each employment zone was determined at an earlier stage of the zonal land use and activity model (Figure 10.1).

The second panel shows the computation of the balancing factors. The values for each cell in the matrix are obtained by multiplying the attractiveness index by the commuting probabilities, using the values shown in the top panel of the table. For example, the cell value for Employment Zone 4 and Residential Zone 5 is computed by multiplying 0.09 by 5, yielding a value of 0.45. Balancing factors for each employment zone are computed by adding up the cell values for each column. For Employment Zone 3, for example, the balancing factor is 0.68 (0.10 + 0.14 + 0.14 + 0.15 + 0.15).

The third panel combines the effects of commuting probabilities and attractiveness to create a normalized probability distribution within each employment zone; we call these *allocation probabilities*. These probabilities are obtained from Panel 2 by dividing the values for each cell in an employment zone by that zone's balancing factor. For example, the probability that a person who works in Zone 2 lives in Zone 1 is 0.205 (0.18 / 0.88). The allocation probabilities within each employment zone sum to 1.0. This simply means that all of the workers in a particular employment zone have been assigned to a residential zone.

The bottom panel shows the distribution of the 600 employees by employment and residential zones. These numbers are derived by multiplying the employment projections shown in Panel 1 by the normalized allocation probabilities shown in Panel 3. For example, 60 workers from Zone 3 live in Zone 2 (300 × 0.20) and 17 workers from Zone 1 live in Zone 5 (100 × 0.169). Because Zones 4 and 5 have no employment, the residential allocation from these zones is zero.

The final step is to sum the allocations for all employment zones to obtain the total number of employees allocated to each residential zone. For example, 117 employees live in Zone 1 (32 + 41 + 44). It should be noted that adding up the numbers in each column yields employment numbers identical to those shown in the top panel of the table. This indicates that the model successfully preserved consistency between the residential allocation and the zonal employment projections.

Two examples—not shown in the table—illustrate how this gravity model might be used for simulations. The first example represents the effects of a proposed freeway that connects Zones 2 and 5 (without changing employment projections or the amount of vacant land). The freeway would reduce travel costs

Table 10.1. Residential Location Based on a Gravity Model

Commuting Probabilities

	Employment zone					
Residential zone	1	2	3	4	5	Attractiveness
1	0.14	0.09	0.05	0.09	0.03	2.0
2	0.09	0.14	0.09	0.05	0.05	1.5
3	0.05	0.09	0.14	0.05	0.03	1.0
4	0.09	0.05	0.05	0.14	0.09	3.0
5	0.03	0.05	0.03	0.09	0.14	5.0
Employees	100	200	300	0	0	600

Balancing Factors

	Employment zone				
Residential zone	1	2	3	4	5
1	0.28	0.18	0.10	0.18	0.06
2	0.14	0.21	0.14	0.08	0.08
3	0.05	0.09	0.14	0.05	0.03
4	0.27	0.15	0.15	0.42	0.27
5	0.15	0.25	0.15	0.45	0.70
Balancing factor	0.89	0.88	0.68	1.18	1.14

Allocation Probabilities

	Employment zone				
Residential zone	1	2	3	4	5
1	0.316	0.205	0.148	0.153	0.053
2	0.153	0.239	0.200	0.064	0.066
3	0.056	0.102	0.207	0.043	0.026
4	0.305	0.170	0.222	0.357	0.238
5	0.169	0.284	0.222	0.383	0.617
	1.000	1.000	1.000	1.000	1.000

Allocation of Employees to Residential Locations

	Employment zone					Residential
Residential zone	1	2	3	4	5	allocation
1	32	41	44	0	0	117
2	15	48	60	0	0	123
3	6	20	62	0	0	88
4	30	34	67	0	0	131
5	17	57	67	0	0	141
Employees	100	200	300	0	0	600

and thus raise the probability of commuting between these two zones. The probability of workers commuting from Zone 5 to Zone 2 would increase from 0.05 to 0.20. The residential allocation to Zone 5 would jump from 141 to 207, a 47% increase. Residential allocations to all other zones would be reduced proportionately.

The second example investigates the impact of a new regulation restricting floodplain development in Zone 1 (without changing employment projections or commuting costs). This regulation is expected to reduce vacant land from 2.0 acres to 1.0 acre. In this case, allocation probabilities to Zone 1 from all employment zones would decline by 50–55% and the residential allocation to Zone 5 would fall from 117 to 66, a decline of 44%. Under this scenario, residential allocations to all other zones would be increased proportionately.

Our discussion so far has illustrated how the model allocates projected workers from employment zones to residential zones. The next step is to transform these projected workers into a population projection. One widely used technique uses the following formula:

$$P = (ER / ERHH) \times PPH + GQ$$

where ER is the number of employed residents, ERHH is the average number of employed residents per household, PPH is the average number of persons per household, and GQ is the group quarters population. ERHH is based on characteristics such as household composition, the unemployment rate, and the labor force participation rate. Dividing ER by ERHH yields a projection of the number of households. Multiplying the projected number of households by the projected PPH yields a projection of the household population. Adding a projection of the population residing in group quarters facilities provides a projection of the total population.

Table 10.2 shows the calculation sequence for our hypothetical region. Consider Zone 1, for example. The number of employed residents (ER) shown in the bottom panel of Table 10.1 is 117. The projected average number of employed residents per household (ERHH) is 0.8. Dividing ER by ERHH yields a projection of 146 households. Multiplying projected households by projected PPH yields a projected household population of 219. Finally, adding an independent projection of the group quarters population yields a projection of

$$(117 / 0.8) \times 1.5 + 35 = 254$$

ERHH, PPH, and GQ values vary tremendously from one place to another. Whenever possible, applications of this procedure should be based on values projected specifically for each zone. The San Diego Association of Governments (1998) described a number of techniques that can be used for projecting zone-specific ERHH, PPH, and GQ values.

Table 10.2. Deriving a Total Population Projection
from the Employed Resident Allocation

Zone	Employed[a] residents	ERHH[b]	Households[c]	PPH[d]	Household population[e]	Group quarters population	Total population[f]
1	117	0.8	146	1.5	219	35	254
2	123	1.2	103	2.2	227	50	277
3	88	1.5	59	2.0	118	72	190
4	131	1.0	131	3.2	419	1,500	1,919
5	141	1.9	74	2.8	207	72	279
Region	600	1.2	513	2.3	1,190	1,729	2,919

[a]From Table 10.1.
[b]Employed residents per household.
[c]Households = employed residents / employed residents per household.
[d]Persons per household.
[e]Household population = households × persons per household.
[f]Total population = household population + group quarters population.

LAND USE AND ACTIVITY MODELS USED TODAY

At least 20 different land use and activity models have been used in metro-politan areas throughout the world. Although these models share some common features, they represent a wide range of theoretical and empirical perspectives and vary considerably in comprehensiveness, reliability, and implementation (e.g., Klosterman, 1999; Southworth, 1995; Wegener, 1994, 1995). We do not attempt to provide a complete review of all these models or detailed examples of how they are applied; rather, we provide a brief overview of three of the most widely used models—gravity models, optimization models, and land pricing models. Then, we look at micro models that provide projections for very small geographic areas. We conclude with a discussion of the California Urban Futures Model, a model that offers some new ideas and directions for urban systems models.

DRAM and EMPAL: Descendants of Lowry's Gravity Model

Lowry (1964) developed two gravity models that linked the spatial distributions of basic employment (e.g., manufacturing and transportation), population-serving employment (e.g., retail trade and services), and population. A number of models building on this framework soon followed (e.g., Crecine, 1968; Garin, 1966; Gouldner, Rosenthal, & Meredith, 1972). Many of the zonal land use and activity models in use today are direct descendants of Lowry's seminal work. The most successful and widely used are the Disaggregated Residential Allocation Model (DRAM) and Employment Allocation Model (EMPAL), models that have been updated and refined for more than 20 years (Putman, 1979, 1983, 1991). These models have been used in more policy and forecasting applications than any other land use and activity model (Wegener, 1994).

DRAM and EMPAL are production-constrained gravity models that have similar equation structures. Although separate, they are closely integrated with each other. DRAM produces residential projections (households, population, income) and EMPAL produces nonresidential projections (employment). Both incorporate three important modifications not found in previous models: (1) they use multivariate, multiparametric attractiveness functions; (2) they include consistent balanced zonal constraint procedures; and (3) they contain additive lagged terms.

DRAM can allocate households into as many as eight income categories. In most applications, however, three to five categories are used. DRAM allows up to nine variables in its attractiveness function, which can be customized to fit any application (e.g., Watterson, 1990). A constraint procedure permits the analyst to investigate the effects of policies that impose limits on growth, such as restricting the development of steep hillsides or habitat-preserve areas. The inclusion of a lagged household variable converts DRAM from a purely cross-sectional model into a quasi-dynamic model of household location (Wegener, 1994).

EMPAL projects the geographic location of four to eight categories of employment. Unlike its predecessors, EMPAL does not require the exogenous location of basic employment as a starting point. Rather, basic employment location in this model interacts directly with the location of other activities within the region. It is believed that this formulation reflects economic interactions more accurately than previous formulations, especially in light of the declining economic impact of heavy manufacturing industries (Prastacos, 1986a).

POLIS: An Optimization Model

Land use and activity models based on linear programming and optimization techniques first appeared in the 1960s (e.g., Harris, 1965; Herbert & Stevens, 1960). They gained prominence from the discovery that gravity models could be reformulated as convex programming problems and embedded within activity allocation frameworks (e.g., Wilson, Coelho, Macgill, & Williams, 1981). From this and other research, linear programming and optimization models emerged as alternatives to gravity models (Southworth, 1995).

The Bay Area Association of Governments developed the Projective Optimization Land Use Information System (POLIS) in the early 1980s. We believe that its planning and forecasting applications exceed those of any other optimization model. Prastacos (1986a, b) and Caindec and Prastacos (1995) describe the theoretical concepts, mathematical details, and calibration procedures that underlie POLIS.

POLIS determines residential and employment locations from a set of competing alternatives by selecting the alternative that optimally satisfies a set of objectives, subject to certain constraints. These objectives are based on random utility theory. Essentially, POLIS uses economic principles of individual decision making to determine the location of activities rather than relying on statistical descriptions of aggregate trends, as do gravity models.

Unlike gravity models, POLIS does not solve its equations through an iterative process. Rather, it uses complicated mathematical programming techniques to simultaneously determine employment location, residential selection, and travel patterns. This technique makes these factors consistent with each other and with the constraints. Consequently, there are no reallocations of activity overflows, which can distort projections if they occur in distant zones (Gouldner, 1980).

Land Pricing Models

Models such as DRAM, EMPAL, and POLIS have been criticized because they ignore the economics of urban land markets (e.g., Hunt, 1994b; Johnston & de la Barra, 1997; Southworth, 1995). Land prices influence the location of residential and nonresidential activities and are themselves influenced by changes in the supply and demand for land. Land pricing models thus incorporate a fiscal dimension that is missing in other land use and activity models. This dimension is very useful for policy evaluations and cost–benefit analyses.

Land pricing models, however, are difficult to implement. The simulation of land markets requires price data for land, housing, and other kinds of development. These data are difficult to obtain, especially for small geographic areas and numerous points in time. In addition, land pricing models are complex and time-consuming to calibrate (Hunt, 1994a). In many agencies, the expertise and resources needed to implement these models are not available. DRAM, EMPAL, and POLIS, on the other hand, rely on more accessible data sources and are among the easiest land use and activity models to implement.

The idea that land and transportation systems might be viewed as markets with endogenously determined prices and costs is rooted in urban economics (e.g., Alonso, 1964; Mills, 1967; Wingo, 1961). Because land and housing prices tend to decline as the distance from nonresidential activities increases, lower housing costs generally entail higher commuting costs. Under this framework, people select their optimal residential locations based on the trade-off between housing prices and transportation costs. A "bid-rent" function reflects this trade-off. A number of land use and activity models now include bid-rent functions or some other type of land pricing mechanism (e.g., Anas, 1992; de la Barra, 1989;

Echenique, 1994; Hunt, 1994b; Kim, 1989; Waddell, 2000; Wegener, Mackett & Simmons, 1991).

MEPLAN and TRANUS are widely used land use and activity models that incorporate land prices, but most of their applications have been outside the United States. This may now be changing. DRAM and EMPAL have been reformulated into a structure capable of including a land market component (Putman, 1995). In Sacramento, work is underway comparing DRAM, EMPAL, TRANUS, and MEPLAN (Johnston & de la Barra, 1997). The Baltimore Area Council of Governments recently began to implement TRANUS. UrbanSim— another model that incorporates land prices—has been applied in Lane County, Oregon, Honolulu, and the Salt Lake City Metropolitan Area (Waddell, 2000).

We use MEPLAN to illustrate some of the concepts that underlie a land pricing model. This model is the culmination of a long-term urban modeling effort started at Cambridge University in 1967. Hunt (1994a, b) and Hunt and Simmons (1993) describe some of the model's theoretical underpinnings, structural details, and calibration procedures.

MEPLAN treats land and transportation as two parallel and interacting markets. The land market model uses input-output modeling, demand functions, and random utility theory to determine the spatial distribution of residential and nonresidential activities, land use, and land prices. Commuting costs affect the demand for housing and employment; conversely, changes in housing and employment affect travel patterns. The framework is fully integrated. The transportation system responds quickly to changes in the land market (i.e., within the same period). Land use and housing, however, respond more slowly to changes in transportation conditions. Like many other land use and activity models, MEPLAN uses transportation factors from the previous period to influence current land use and activity locations.

Microgeographic Land Use and Activity Models

Many transportation models in the United States are based on much smaller geographic zones than those used in the models just described. Consequently, there is a need for land use and activity models that can produce projections for very small geographic areas such as blocks, traffic analysis zones, and grid cells. We refer to these as *micro models*. This section describes two micro models currently in use. One is a component of the Urban Development Model (UDM) used in San Diego (San Diego Association of Governments, 1998) and the other is the Subarea Allocation Model (SAM) used in Phoenix (Maricopa Association of Governments, 1996).

Urban Development Model. UDM is an integrated land use and activity modeling system that provides projections for micro areas (e.g., blocks and portions of blocks). It produces a wide range of demographic and economic information and, to our knowledge, its 85 land use categories are the most of any model used today. UDM is a two-stage nested allocation model. In the first stage, gravity models are used to project employment and residential activities for 208 zones. In the second stage, a micro model allocates these zonal projections to some 30,000 micro areas.

UDM is based on travel times, land use characteristics, and the spatial distribution of housing and employment activities. Two factors determine the allocation of zonal housing projections to micro areas. One is an accessibility weight that measures the number of housing units and jobs within 10 minutes of a micro area. Larger weights indicate higher levels of accessibility. The second is the micro area's capacity to accommodate new growth. This factor is based on the available supply of land and projected densities (i.e., housing units per acre). The available land supply includes both vacant land and land that can be redeveloped (e.g., multifamily units that replace single-family units).

The first step in the allocation process is to locate the most accessible micro area that has a capacity to add new houses. UDM allocates housing units to that area until the zonal housing projection is completely allocated or the area reaches its housing capacity, whichever comes first. If that area reaches its housing capacity before the zonal projection is completely allocated, UDM identifies the next most accessible micro area that has housing capacity and allocates housing units to it. This process is repeated until the zonal housing projection is completely allocated to micro areas. The model also removes housing units in micro areas in which redevelopment is occurring. Finally, housing projections for micro areas are converted into population projections by using the housing unit method (Tayman, 1996a).

Subarea Allocation Model (SAM).   SAM allocates zonal housing and employment projections produced using the DRAM and EMPAL models to micro areas within the metropolitan area. The micro areas are developed by using a GIS procedure that allocates polygon-based information—such as land use—into many small grid cells. The number of grid cells in a particular application is a function of the desired spatial resolution. For example, an application covering all of Maricopa County requires approximately 33 million ¼-acre grid cells and 2 million 1-acre grid cells.

SAM identifies grid cells that have the capacity to accommodate additional growth. Capacity is based on two conditions: (1) that vacant land is available, and (2) that the county's long-range plans permit additional growth. The model assigns each grid cell a locational preference score based on the availability of infrastructure and its proximity to roads, freeways, and areas that have already been developed. A high score indicates a high degree of locational preference.

SAM's allocation procedure is similar to the procedure followed by UDM. It allocates zonal projections to the grid cell that has vacant land and the highest locational preference score. When that grid cell is fully developed, the cell that has the next highest locational preference score is identified. The allocation process is repeated until all of the projected zonal growth has been allocated to grid cells. The model also has a feature that permits it to simulate the growth of subdivisions by combining related grid cells into a larger cluster.

California Urban Futures Model

The California Urban Futures (CUF) model exemplifies a new direction in land use and activity models. It was designed to provide planners, elected officials, and private citizens with a tool to create and compare alternative land use policies. Applications of this model generally focus on policy simulations rather than on the development of a specific projection or forecast.

The initial version of the CUF model, released in the mid-1990s, differed in several significant ways from other land use and activity models (Landis, 1994, 1995). It was the first model to incorporate interactions between private developers and local governments regarding the location, size, and density of proposed developments. It provided a detailed model of the supply side and profit potential of residential housing markets and made innovative use of GIS for spatial analysis, data preparation, and map display. Finally, it allocated growth to individual sites rather than to larger geographic areas such as census tracts or TAZs.

Despite these advantages, the initial version of the CUF model had some notable shortcomings. It was limited to residential land use and ignored the competition for land between residential and nonresidential development. Housing prices—which play an important role in the allocation process—were exogenous to the model and were unaffected by shifts in the supply and demand for residential units. As a result, residential demand that exceeded available supply simply went to other jurisdictions (e.g., cities) rather than feeding back into the allocation process through higher prices. Finally, the procedures for allocating residential development to specific sites were based primarily on developer profits rather than on historical patterns.

The second version of the model (CUF-2) attempted to remedy these shortcomings (Landis & Zhang, 1998a, b). Unlike the earlier version, the newer version includes multiple land uses, allows different uses to compete for preferred locations, is calibrated to historical data, and includes a pricing mechanism. CUF-2 has four main components: (1) an activity projection, (2) a spatial database, (3) land use change, and (4) a simulation engine.

The first component consists of several econometric models used to project population, households, and employment by industrial sector for counties, cities, and unincorporated areas at 10-year intervals. The employment projections are new to the model. Landis (1994) and Landis, Zhang, and Zook (1998) describe the specifications and calibration of the projection equations.

The second component is a database that covers the spatial geometry and characteristics of developable areas (i.e., sites that have the potential to develop or redevelop). These areas are defined as grid cells measuring 100 meters × 100 meters. The database includes digital map layers that describe several important factors related to the location of residential activities (e.g., land use, zoning, density, transportation networks, and environmental constraints). The CUF-2 model is far more data-intensive than its predecessor.

The third component uses a series of equations to project changes in land use. More than two dozen variables—for example, population, employment, availability of land, site-level environmental characteristics, and proximity to job centers, freeways, and transit stations—are included. Seven types of land use changes are modeled, including both vacant land and land that has a potential for redevelopment. In addition, the CUF-2 model predicts which vacant areas will not be developed and which developed areas will not change their uses. Landis and Zhang (1998b) discuss the specification and calibration of these models.

The simulation engine—the final component of the CUF-2 model—allocates projected households and jobs for a city or unincorporated area to its individual grid cells. The simulation engine calculates probabilities from the statistical results produced by the equations described before. These probabilities represent bid scores for development or redevelopment that are used in the allocation process. Bid scores vary by grid cell and by potential land use and allow for competition among uses as well as among sites. Land use and activity spillover from one area to another are also picked up by the model. Spillover can occur when there are not sufficient sites to accommodate the expected demand for residential and nonresidential activities within a particular city or unincorporated area.

CONCLUSIONS

Urban systems models differ from economic–demographic models in several fundamental ways. They focus on very small geographic areas; develop ties among the location of residential and nonresidential activities, land use characteristics, and the transportation system; have very high resource requirements; and rely heavily on GIS. Although they are similar to economic–demographic models in some ways, they represent a distinct class of structural models.

Urban systems models require a huge amount of data and a substantial degree of technical expertise. Consequently, they are expensive to develop and implement, although the recently released Transportation, Economic, and Land Use System (TELUS) may be an important step toward reducing costs (Pignataro &

Epling, 2000). In addition—given their small geographic scale—the forecast accuracy of their projections is not likely to be high and is often related to the stability of land use plans and policies. Yet, urban systems models are used more frequently today than ever before.

Technological advances (e.g., greater computing power, the development of desktop GIS software, and the proliferation of machine-readable data) and a greater understanding of the urban development process have greatly facilitated the development of these models. The growth of GIS has played a particularly important role, revolutionizing the ways in which data are reported, collected, manipulated, and accessed. Virtually all database development now uses this technology. Somewhat surprisingly, the development of GIS has not yet led to any significant advances in projection methodology (Putman, 1994; Wegener, 1994). We expect that will change in the coming years.

There have also been substantial increases in the demand for the types of projections that these models can produce. Federal transportation and clean air legislation has mandated that transportation plans and air quality analyses account for interactions among land use, residential and nonresidential activities, and the transportation system. Local planners and policy makers are under increasing pressure to understand and meet the challenges posed by traffic congestion, housing shortages, threats to water and air quality, and deteriorating physical infrastructure. The business community seeks economic and demographic information for ever smaller levels of geography. Urban systems models are well suited to meet these demands. We believe that technological advances and increasing demand will lead to further increases in developing and using these models.

CHAPTER 11

Special Adjustments

Cohort-component, trend extrapolation, and structural models can be applied in many situations without considering any factors beyond those discussed in previous chapters. In some situations, however, the basic projection model must be adjusted to account for special circumstances. Three of the most common adjustments are for international migration, special populations, and census enumeration errors. In addition, there are circumstances in which a set of projections must be controlled to an independent projection or adjusted to provide additional temporal or age detail. In this chapter, we discuss the circumstances in which unadjusted projections might provide unacceptable results and describe several ways for making the necessary adjustments. We also describe techniques for controlling to independent projections and interpolating within age groups or between target years. The adjustments described in this chapter increase the complexity of the projection process but greatly enhance the usefulness of the projections for some purposes.

INTERNATIONAL MIGRATION

The migration module described in Chapter 7 did not distinguish between internal and international migration. However, international migrants often have characteristics that differ from the characteristics of internal migrants and are often influenced by different factors. Consequently, if international immigration is an important component of growth for a particular state or local area, it may be useful to project it separately from internal migration.

How can international migration be projected? One approach is to take the net international migration projected for the nation as a whole and distribute it to a state or local area according to its proportion of the national foreign-born population who reported in the decennial census that they had been living abroad five or

10 years earlier (e.g., U.S. Bureau of the Census, 1966). For example, 8.7 million foreign-born persons counted in the 1990 Census reported that they had entered the country between 1980 and 1990. Of these, 718,000 were living in Texas in 1990 (U.S. Bureau of the Census, 1993a). Thus, Texas' share of net national immigration could be projected as

$$718,000 / 8,700,000 = 0.083$$

A similar approach is to base an area's projected share of total international migration on its historical share (e.g., Center for the Continuing Study of the California Economy, 1997; San Diego Association of Governments, 1999). The Immigration and Naturalization Service (INS) collects annual data on legal international migrants, refugees, and those seeking asylum; these data are available for states, metropolitan areas, and most counties. In addition, some state demographic agencies develop annual estimates of internal and international migration for the counties in their states. Using one of these sources, the analyst can develop trend data on the area's share of the nation's net international migration and base projections on the extrapolation of those trends.

Foreign immigration and emigration can also be projected separately (e.g., Wetrogan, 1988). Projected national immigration can be distributed to an area in a manner similar to that described above for net international migration. Because emigrants are drawn primarily from the foreign-born population, projections of emigration can be based on each area's share of the national foreign-born population. For example, 19.8 million foreign-born persons were counted in the 1990 Census, including 6.5 million living in California (U.S. Bureau of the Census, 1993a). Thus, California's share of emigrants from the United States could be projected as

$$6,500,000 / 19,800,000 = 0.328$$

These techniques refer to the number of international migrants. How can their demographic characteristics be projected? The simplest approach is to assume that international migrants for a particular area have the same characteristics as international migrants for the nation as a whole. This will be a reasonable assumption if the mix of countries that sends migrants to an area is similar to the mix for the United States and if the analyst believes that mix will continue. INS data can be used to compare the origins of international migrants at the national, state, and local levels.

Another approach is to assume that future international migrants into an area will have the same characteristics as those reported in the most recent decennial census or in recent INS data. Although the INS no longer publishes information on the demographic characteristics of international migrants for states or local areas, it still provides such information at the national level. One way to develop a demographic profile for a specific state or local area is to combine state or local data on the country of origin of international migrants with national data on their demographic characteristics. Then, the demographic profile of international immigrants for a particular area can be calculated as a weighted average of the U.S. demographic profile by country of origin, using the area's distribution by country of origin as weights. For example, if international migrants to an area come mostly from Mexico, the area's immigration profile can be made to reflect the national demographic profile for immigrants from Mexico. If an area's international migrants come primarily from two countries, its immigration profile can be made to reflect the average of the national characteristics of immigrants from those two countries.

SPECIAL POPULATIONS

A special population is defined as a group of persons located in an area because of an administrative or legislative action (Pittenger, 1976, p. 205). Common types include college students, prison inmates, residents of nursing homes, and military personal and their dependents. Special populations complicate the projection process because their growth and decline are not determined by the same factors that affect fertility, mortality, and migration rates; consequently, they often follow trends that differ from the rest of the population.

Special populations often have different demographic characteristics as well. Military personnel and college students are concentrated primarily in the young adult ages, residents of nursing homes are concentrated primarily in the older ages, and the prison population often has a high concentration of males and racial minorities. These differences can have a substantial impact on the mortality, fertility, and migration rates used in population projections.

Another confounding characteristic of special populations is that they often do not age in place like other population groups. Instead, their age structure may remain fairly stable over time. For example, a college town sees a large inflow of people aged 17–19 and a large outflow of people aged 21–23 every year. Consequently, a substantial proportion of the town's young adult population replaces itself repeatedly rather than aging in place.

Special populations do not cause any particular problems for population projections if they comprise only a small proportion of the population or if their growth rates are similar to the rest of the population. In these circumstances no special adjustments are needed. When special populations follow different trends and account for a substantial proportion of the total population, however, adjustments must be made.

Unfortunately, there is no rule of thumb that defines "different" or "substantial." Consequently, the analyst must evaluate each situation separately by focusing on the special population's growth trends, demographic composition, and— perhaps most important—its share of total population. It is a good bet that special populations will have a significant impact on projections in areas where a large prison, military installation, or university are located. Nursing homes, boarding schools, and mental institutions may also be important. The impact of special populations is generally greater in small areas than in large areas. For example, a prison may have little impact on the population of a large county but may comprise the entire population of a particular census tract.

Incorporating Special Populations into the Projection

Several steps can be followed to account for special populations when making population projections (Figure 11.1). The first is to create estimates of the "regular" population in the historical base data by subtracting the special population from the total population. The second is to project the regular population by using the methods described in earlier chapters. The third is to project the special population itself by using one of the approaches described later. The final step is to add the projection of the special population to the projection of the regular population.

How can special populations be projected? One approach is to develop a cohort-component model for the special population itself by using fertility, mortality, and migration rates that pertain specifically to that population (e.g., Pittenger, 1976). This approach will be useful if the special population accounts for a large proportion of the total population, if the necessary data are available, and if reasonable assumptions about future trends in fertility, mortality, and migration rates for the special population can be made. Data limitations often make this approach impractical, especially for small areas.

A second approach is to hold the special population constant over the course of the projection horizon. This approach will be useful if the size and demographic composition of the special population has been relatively stable over time and is expected to remain so in the future. It will also be useful if the direction and magnitude of future changes are completely unpredictable at the time the projection is made. This is the approach followed by the Census Bureau in several recent sets of projections (e.g., Campbell, 1996; Day, 1996a).

Finally, projections of special populations can be based on information collected from the administrators of facilities such as colleges, prisons, or nursing homes. Administrators often have information on planned changes in the facility's capacity and enrollment. This information can be used in conjunction with the analyst's judgment regarding future population trends (e.g., Smith & Nogle, 2000). Combinations of the various approaches can also be used, such as holding the demographic composition of the special population constant but allowing for changes in its total size.

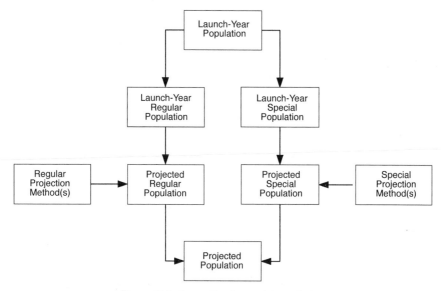

Figure 11.1. Accounting for special populations

Assessing Data for Special Populations

Data availability is a serious problem in developing projections of special populations. County-level population and migration data for some special population groups can be obtained from the decennial census by using either summary files or PUMS data. Data are available for persons who reside in group quarters facilities such as military barracks, prisons, and college dormitories but are not available for military dependents and for college students who do not live in dormitories. Estimates for these groups can sometimes be made by using PUMS data that identify all households headed by military personnel or students, but these estimates are far from perfect. Because PUMS data are available only for counties and subcounty areas that have 100,000 residents or more, they cannot be used for most small areas.

Migration data present a problem even for the types of group quarters explicitly covered by the decennial census. The data refer to migration during a five-year period, but group quarters status is noted only for the census year. For example, the 2000 Census provides migration information for persons who resided in group quarters facilities in 2000, but not for persons who resided in group quarters facilities in 1995. This creates a problem for constructing out-migration rates because group quarters status is as important in the earlier year as the later year.

It is also difficult to obtain mortality and fertility data specific to special populations. Birth and death certificates in some counties identify persons associated with the military, but this is the exception rather than the rule. Fertility and mortality data pertaining to other special populations are even more difficult to obtain. In these instances, the analyst may simply have to make an educated guess. Fortunately, the development of fertility and mortality rates specific to a special population is generally unnecessary, either because the special population's contribution to local births and deaths is very small or because rates for the special population are similar to rates for the population as a whole. The military population (including dependents) may be an exception. Fertility rates for this group are often higher than for the nonmilitary population. If the military population is a substantial portion of the total population, it may be advisable to account separately for the fertility of this population (see Box 11.1).

Obtaining special population data is even more difficult for subcounty areas than for counties. The decennial census provides subcounty-level data on total numbers for some types of special populations, but generally does not provide data on their demographic characteristics. One approach to this problem is to identify a small area—such as a census tract or individual block—in which all or most of the population belongs to the special population group. In instances like this, data that define the total population also define the special population. Then, those data can be used to estimate the characteristics of similar special populations in nearby areas. For example, suppose that projections are to be made for a census tract that contains a prison. Suppose further that two particular blocks within that census tract contain solely prison inmates. The demographic characteristics of those two blocks can be used as an estimate of the characteristics of the prison population for the entire census tract.

The presence of a special population in an area is a red flag warning the analyst to pay special attention to the data and assumptions used in the projection model. If the special population is large enough and differs significantly from the rest of the population in its demographic characteristics and growth rates, it must be accounted for explicitly. Sometimes the necessary data are available for adjusting mortality, fertility, and migration rates for special populations, and sometimes they are not. When the necessary data are not directly available, the analyst will have to become particularly creative. Developing reasonable adjustments for special populations is essential for developing reasonable population projections.

We have found that the use of proxy rates from similar areas or similar subgroups is very helpful. For example, if the opening of a prison had a substantial impact on the age-specific migration rates observed for males in a particular county, it would not be a good idea to use those rates for constructing population projections for that county. One solution to this problem is to use age-specific migration rates for females for projections of both males and females because

---

Box 11.1

Estimating Military Birth Rates

Birth certificates in some counties report the military status of parents, thereby providing an excellent source of data on military births. In most counties, however, birth certificates do not include this information. Administrators of military installations sometimes maintain birth data, but gaining access to those data is often difficult. When complete data on military births are available, the analyst can develop estimates of military birth rates and incorporate them directly into the projection process. Unfortunately, such data are seldom available. How can estimates of military birth rates be developed for places lacking these types of data?

One possibility is to use PUMS data from the most recent census to calculate child-woman ratios (CWRs) for the military and total populations. An adjustment factor can be developed by forming a ratio of the military CWR to the total CWR. Then, estimates of military age-specific birth rates (ASBRs) can be made by applying this adjustment factor to the ASBRs calculated for the entire population. This approach assumes—perhaps incorrectly—that the pattern of ASBRs is the same for military and nonmilitary populations.

Another possibility is to obtain information on births that occurred in military hospitals. This information can be obtained directly from the hospital or from birth certificates. Although useful, this information excludes data on births to military families that did *not* occur in military hospitals. When direct information on military births is not available, the analyst may have to combine information from a variety of sources to estimate military birth rates.

---

migration rates for males and females are often similar. Alternatively, age-specific rates for males in nearby counties that have similar demographic characteristics could be used. Solutions like these may not be perfect, but they are better than making no adjustments at all.

Illustrating the Impact of a Special Population

We use San Diego County, California, to illustrate the impact of a special population on a projection. San Diego has one of the largest concentrations of military personnel in the United States. In 1995, uniformed military personnel (101,000) and their dependents (90,000) accounted for about 7% of the county's total population and for about 15% of the county's births. This population is heavily concentrated in the young adult ages. Given their numbers and age distribution, the military population is likely to have a substantial impact on the county's population projections.

We developed two alternative sets of projections for white males in San Diego County, using 1995 as a launch year and 2000 as a target year. One set used a basic (i.e., unadjusted) cohort-component model and the other used an adjusted model that separated uniformed military personnel from the civilian population. In the basic model, net migration rates were based on the *total* population. In the adjusted model, net migration rates were based on the *civilian* population and were applied solely to that population. Then, an independent projection was made for the uniformed military population. Other than for net migration, the assumptions used in the two models were identical.

Table 11.1 shows projections of white males from the basic and adjusted models. The two projections of total population are similar, differing by just over 3,300 persons (0.4%). However, there are significant differences in some age groups. For ages 15–19 and 20–24, projections from the basic model *exceed* those from the adjusted model by 15% and 19%, respectively; for ages 25–29, projections from the basic model are 13% *less than* those from the adjusted model. In general, the basic model tends to overstate the population aged 15–24 and understate the population aged 25–54. Differences between the two models were very small for ages 55 and above. The impact of these adjustments would be even greater if the projections were extended further into the future.

CENSUS ENUMERATION ERRORS

Census enumeration errors can also have an impact on population projections, particularly those made by age and other demographic characteristics (e.g., Pittenger, 1976; Swanson, Vaidya, Yehya, Bennett, & Prevost, 1989). The Census Bureau began adjusting its national projections for census enumeration errors in the mid-1970s by using the *inflation–deflation* method. The objective of this method is to yield a census-like projection that incorporates adjustments for certain types of errors. These adjustments are made because the census tends to undercount some age, race, and ethnic groups more than others. Failing to account for this differential undercount could lead to persistently low projections for these groups as they age over time.

It must be emphasized that the inflation–deflation method does *not* account for all census enumeration errors. In particular, it does not adjust for the overall level of population undercount (or, in some instances, population overcount). The inflation–deflation method adjusts the demographic composition of the population, but not its total number.

The application of the inflation–deflation method is relatively straightfor-

Table 11.1. Alternative Projections for White Males,
San Diego County, 2000

Age	Cohort component model		Difference[c]	
	Basic[a]	Adjusted[b]	Number	Percent
0–4	50,680	50,680	0	0.0
5–9	52,133	52,133	0	0.0
10–14	53,542	53,542	0	0.0
15–19	59,580	50,826	8,754	14.7
20–24	71,993	58,512	13,481	18.7
25–29	65,709	74,360	−8,651	−13.2
30–34	75,102	79,013	−3,911	−5.2
35–39	82,750	84,521	−1,771	−2.1
40–44	84,998	87,524	−2,526	−3.0
45–49	71,359	72,789	−1,430	−2.0
50–54	61,243	61,785	−542	−0.9
55–59	42,677	42,752	−75	−0.2
60–64	32,203	32,217	−14	0.0
65–69	29,166	29,180	−14	0.0
70–74	25,527	25,512	15	0.1
75–79	21,558	21,558	0	0.0
80–84	13,473	13,473	0	0.0
85+	7,865	7,865	0	0.0
Total	901,558	898,242	3,316	

[a]No adjustments for uniformed military population.
[b]Separate projections for uniformed military and civilian populations.
[c]Difference = basic − adjusted.

ward. Each age, sex, and race group in the launch-year population is inflated by dividing it by the appropriate coverage rate. Coverage rates are less than 1.0 for an undercount and greater than 1.0 for an overcount; therefore, this adjustment increases the size of undercounted groups and decreases the size of overcounted groups. If special populations are involved, they are subtracted from the population after the adjustment is made. Then, the projection method is applied by using the usual procedures and the projection of the special population is added back in. Finally, each population subgroup is deflated by multiplying it by the same coverage rate originally used to inflate it.

Because the inflation–deflation method applies the final adjustment procedure to the target year population, the components of change may not exactly match the population change implied by the adjusted projection. A common way to resolve this discrepancy—which is usually quite small—is to adjust net migration, the major component of population change for many states and local areas.

The Census Bureau uses the inflation–deflation method only to adjust its national projections, not its state projections. The main problem in applying the method is determining the appropriate coverage rates, which vary considerably from place to place and are difficult to measure precisely (Pittenger, 1976). Unfortunately, age-sex-race-specific coverage rates are not available for states and local areas; consequently, the inflation–deflation method is seldom used for state and local projections. We are not aware of any studies of the impact of inflation–deflation adjustments on population forecast accuracy. Except in very unusual circumstances, we do not believe it is necessary to apply this method to state and local population projections.

The Census Bureau develops estimates of coverage rates for each decennial census. It uses two methods, one based on demographic analysis, the other on post-enumeration surveys. At this point, it is not known how those methods will be used to modify the results of the 2000 Census and to adjust future population estimates and projections. These issues have been hotly debated in Congress and will most likely have to be resolved in the courts. If the 2000 Census provides population counts for small areas that have been adjusted for census coverage errors, there will be no need to apply the inflation–deflation method when constructing population projections.

CONTROLLING

Analysts who make population projections often face two distinct but related problems. One is how to make projections of demographic composition (e.g., age, sex, race) match an independent projection of total population or migration. The second is how to make projections for a number of geographic areas add to an independent projection for a larger area (e.g., how to make the sum of census tract projections match a county projection). *Controlling* is the term we use to describe this adjustment process; *raking* is another commonly used term.

There are several reasons for controlling one set of projections to another. One is the requirement that a set of projections be consistent with an "official" projection that has been developed, adopted, or sanctioned by a governmental body or some other decision-making unit. Another is to tie projections of demographic composition or geographic distribution from an older set of projections to a projection of migration or total population from a more recent set. Perhaps most important, controlling facilitates the construction of a set of projections that is consistent across demographic subgroups and geographic areas: It ensures that projections of demographic characteristics will sum to projections of total population, and that projections for small geographic areas will sum to projections for larger geographic areas.

Controlling to Independent Projections

In this section, we describe methods for controlling projections of demographic characteristics to independent population or migration projections. We illustrate these methods using the projections of white females in Broward County shown in Chapter 7. In the first two illustrations, we control the age distribution from the original projection for 2000 to a new projection of total population. In the third illustration, we control the original in- and out-migration projections to a new projection of net migration. We refer to these new projections as "independent" projections.

Projections of Total Population. The simplest method for controlling demographic characteristics from one projection to the total population from another projection is to use a raking procedure based on a single adjustment factor. This factor can be computed by dividing the total population from the independent projection by the total population from the original projection. Then, the original projections for each demographic subgroup are adjusted by multiplying each one by the adjustment factor. The general equations are as follows:

$$\text{FACTOR}_t = \text{CNTLP}_t / P_t$$
$$_n\text{CP}_{x,t} = {_n}P_{x,t} \times \text{FACTOR}_t$$

where P_t is the original projection of total population in the target year, CNTLP_t is the new projection of the total population in the target year (i.e., the control total), FACTOR_t is the adjustment factor for the target year, $_nP_{x,t}$ is the original projection for each age group, and $_n\text{CP}_{x,t}$ is the controlled projection for each age group.

Table 11.2 shows the original and controlled projections for the white female population of Broward County in 2000. The original (uncontrolled) projection is 593,699, and the independent control total is 612,300, yielding an adjustment factor of 612,300 / 593,699 = 1.03133. The adjusted (controlled) projection for each age group is computed by multiplying the original projection by 1.03133. The sum of the adjusted age groups is equal to the new total of 612,300.

In this example, projections of age groups for white females were controlled to an independent projection of the total white female population. The same method can be applied using different demographic subgroups and different control populations. For example, projections of males and females could be based either on a different adjustment factor for each sex or on a single adjustment factor for both sexes. If there are two race categories (black and white), projections for black males, black females, white males, and white females could be based on four separate adjustment factors (one calculated specifically for each race-sex group), on two separate adjustment factors (one for each race or sex), or on a single

Table 11.2. The Raking Method: Controlling
to an Independent Projection of White Females,
Broward County, 2000

| Age | 2000 projection | | |
	Original[a]	Controlled[b]	Difference[c]
0–4	27,866	28,739	873
5–9	32,522	33,541	1,019
10–14	29,705	30,636	931
15–19	24,784	25,560	776
20–24	24,183	24,941	758
25–29	29,933	30,871	938
30–34	36,056	37,186	1,130
35–39	45,825	47,261	1,436
40–44	48,104	49,611	1,507
45–49	44,231	45,617	1,386
50–54	40,702	41,977	1,275
55–59	31,638	32,629	991
60–64	26,209	27,030	821
65–69	27,537	28,400	863
70–74	32,247	33,257	1,010
75–79	33,922	34,985	1,063
80–84	29,248	30,164	916
85+	28,987	29,895	908
Total	593,699	612,300	18,601
Original population projection			593,699
Independent population projection			612,300
Adjustment factor[d]			1.03133

[a]Projections from Table 7.4.
[b]Controlled projection = original projection × adjustment factor.
[c]Difference = controlled projection − original projection.
[d]Adjustment factor = independent population projection / original population projection.

adjustment factor based on the controlled and uncontrolled projections of the total population.

The choice of the appropriate control group will depend on the availability and reliability of independent projections for various demographic groups. For demographic groups that have similar growth characteristics, it is generally not necessary to develop separate control totals and adjustment factors. The main thing to remember when applying this method is that the sum of the demographic subgroups for which adjustments are made must equal the control total used in computing the adjustment factor (within rounding error).

Projections of Population Change. The first approach to controlling works well when the adjustments are moderate or small. When adjustments are large, it may produce unsatisfactory results because some demographic subgroups may be adjusted by a larger amount than warranted. In these circumstances, a method that focuses on population change over the projection interval rather than the population in the target year may produce better results.

The basic procedure is simple: Changes in total population over the projection interval are calculated for both the independent projection and the original projection. A ratio of the changes from the two projections is formed and applied to the change originally projected for each demographic subgroup, producing a set of adjusted changes. Then, these adjusted changes are added to the launch-year population for each demographic subgroup to provide a controlled projection for the target year.

Although this procedure is simple, its implementation becomes complicated when some demographic subgroups increase and others decline. To illustrate this point, suppose that there are only two subgroups. The projection for one increases by 100 and the other declines by 50, implying a total population change of 50. Suppose further that the projected change for the independent projection (i.e., the control total) is 55. These numbers produce an adjustment factor of $55 / 50 = 1.1$. Applying this factor to the changes originally projected for the two subgroups (100 and -50) produces adjusted changes of 110 and -55. These sum to 55, which is consistent with the change in the independent projection. However, the adjustment makes the subgroup that loses population lose even more than originally projected. Given that projected growth for the entire population has been adjusted upward, this is probably not a reasonable outcome.

This points to an important problem with using a simple raking procedure: When some demographic subgroups are projected to increase and others are projected to decline, a raking procedure based on a single upward adjustment factor makes both population gains and losses larger. A better procedure would account for population gains and losses separately.

This can be accomplished by calculating two separate adjustment factors—one for subgroups that are projected to grow and one for subgroups that are projected to decline. This adjustment procedure is known as the *plus-minus* method (Shryock & Siegel, 1973). The equations for this method are as follows:

$$CNTLPCHG = CNTLP_t - P_l$$
$$_nPCHG_x = {}_nP_{x,t} - {}_nP_{x,l}$$
$$ABSUM = \Sigma |{}_nPCHG_x|$$
$$SUM = \Sigma {}_nPCHG_x$$
$$POSFACTOR = [ABSUM + (CNTLPCHG - SUM)] / ABSUM$$
$$NEGFACTOR = [ABSUM - (CNTLPCHG - SUM)] / ABSUM$$

If $_n\text{PCHG}_x > 0$, then $_n\text{CP}_{x,t} = _nP_{x,l} + (_n\text{PCHG}_x \times \text{POSFACTOR})$

If $_n\text{PCHG}_x < 0$, then $_n\text{CP}_{x,t} = _nP_{x,l} + (_n\text{PCHG}_x \times \text{NEGFACTOR})$

where CNTLP is the independent projection of total population (i.e., the control projection), P_l is the total population in the launch year, CNTLPCHG is the population change between launch year l and target year t for the independent projection, PCHG is the population change implied by the original (uncontrolled) projection, ABSUM is the sum of the absolute values of uncontrolled population changes for each demographic subgroup, SUM is the sum of the uncontrolled population changes for each demographic subgroup, POSFACTOR is the adjustment factor for subgroups projected to increase, NEGFACTOR is the adjustment factor for subgroups projected to decline, CP is the controlled population projection for each demographic subgroup, x is the youngest age in an age group, n is the numbers of years in the age group, and Σ represents the sum over all demographic subgroups.

As these equations show, the formulas for positive and negative adjustment factors are similar, differing only by a single sign in the numerator. In fact, if projected changes for all demographic subgroups have the same sign, the plus-minus method produces the same adjustment factor as the single-factor raking procedure. It should also be noted that the sum of the two adjustment factors is equal to 2.0.

Table 11.3 shows the application of the plus-minus method to the projection of white females in Broward County. The adjustment factors indicate that population changes for age groups that gain population are increased by just over 23% (1.23682) and population changes for age groups and lose population are reduced by the same percentage (0.76318). Comparing the controlled and uncontrolled columns, we see that the adjustment process works as expected. The gains become larger for age groups projected to increase and the losses become smaller for age groups projected to decline. We believe that the plus-minus method will be particularly useful when some demographic subgroups are increasing and others are declining.

One weakness of the plus-minus method appears when the difference between the control total (CNTLPCHG) and the sum of the uncontrolled projections (SUM) exceeds the sum of the absolute values of the uncontrolled projections (ABSUM). Under this condition, one of the adjustment factors must be negative, reversing the signs of the projected changes. One solution to this problem is to transform the distribution of projected changes by adding or subtracting a fixed constant to each value before computing the adjustment factors (Shryock & Siegel, 1973). The control total also must be modified by the total amount added to or subtracted from the distribution. After the factors are applied, the controlled values are transformed back to the original scale by the amount of the fixed constant. It is relatively easy to find a transformation value that results in positive values for both adjustment factors (e.g., San Diego Association of Governments, 1998).

A comparison of Tables 11.2 and 11.3 shows the impact of the controlling method on the final projections. In terms of the projections themselves, the differences are not particularly large. Compared to the projected changes between 1995 and 2000, however, there are some notable differences. The 60–64 age group provides a good illustration. In the original projections, this group increased by only 254 between 1995 and 2000. In the application of the simple raking method (Table 11.2), it increased by 1,075 (254 + 821). In the application of the plus-minus method (Table 11.3), it increased by 314 (254 + 60). Thus, the plus-minus method produced an increase that was much more in line with the original projection than the simple raking method did.

Projections of Migration. The examples in Tables 11.2 and 11.3 show how to control population characteristics from one projection to the total population from another projection. In some instances, however, the application may call for controlling to an independent projection of migration rather than to an independent projection of total population. This may occur when migration (rather than total population) is the variable of interest or when the focus is on components of change rather than population *per se*. It should be noted that controlling to an independent projection of total population or population change makes projections of the components of growth inconsistent with projections of total change.

Consider Table 11.3, for example. The controlled projection for the population aged 0–4 in 2000 is 1,172 higher than the projection based exclusively on births and infant deaths (i.e., the original, uncontrolled projection). How many of these additions were the result of a larger number of births? How many were migrants? How many died during the projection interval? There is no satisfactory answer to these questions.

When the focus is on components of change, the most satisfactory solution to this problem may be to control to a projection of migration. A new migration projection can be obtained by rearranging the terms in the demographic balancing equation described in Chapter 2. This defines net migration as total population change minus births plus deaths. For white females in Broward County, for example, the level of net migration consistent with the population change implied by the independent (control) projection is

$$(612,300 - 574,142) - 28,058 + 39,243 = 49,343$$

In the example shown in Table 11.4, we use the plus-minus method to control in- and out-migration projections to a new net migration total. For these computations, we treat out-migrants as the group that has negative changes and in-migrants as the group that has positive changes. Because the independent projection of net migration is larger than the original projection of net migration, we want the adjustment to increase the number of in-migrants and decrease the number of out-migrants. Therefore, a positive adjustment factor is applied to in-migrants and a

Table 11.3. The Plus–Minus Method: Controlling to an Independent Projection of Population Change for White Females, Broward County, 2000

Age	1995 population	2000 projection			1995–2000 change		Absolute value original change
		Original[a]	Controlled[d]	Difference[e]	Original[b]	Controlled[c]	
0–4	32,814	27,866	29,038	1,172	-4,948	-3,776	4,948
5–9	28,867	32,522	33,388	866	3,655	4,521	3,655
10–14	26,044	29,705	30,572	867	3,661	4,528	3,661
15–19	22,733	24,784	25,270	486	2,051	2,537	2,051
20–24	25,987	24,183	24,610	427	-1,804	-1,377	1,804
25–29	34,929	29,933	31,116	1,183	-4,996	-3,813	4,996
30–34	43,345	36,056	37,782	1,726	-7,289	-5,563	7,289
35–39	45,137	45,825	45,988	163	688	851	688
40–44	42,035	48,104	49,541	1,437	6,069	7,506	6,069
45–49	37,618	44,231	45,797	1,566	6,613	8,179	6,613
50–54	30,660	40,702	43,079	2,377	10,042	12,419	10,042
55–59	24,778	31,638	33,263	1,625	6,860	8,485	6,860
60–64	25,955	26,209	26,269	60	254	314	254
65–69	32,125	27,537	28,624	1,087	-4,588	-3,501	4,588
70–74	36,920	32,247	33,354	1,107	-4,673	-3,566	4,673
75–79	35,118	33,922	34,205	283	-1,196	-913	1,196
80–84	28,941	29,248	29,321	73	307	380	307
85+	20,136	28,987	31,083	2,096	8,851	10,947	8,851
Total	574,142	593,699	612,300	18,601	19,557	38,158	78,545

Calculation of plus–minus adjustment factors:

Sum of original changes (SUM)	19,557
Sum of absolute values of original changes (ABSUM)	78,545
1995 population	574,142
Independent projection for 2000	612,300
1995–2000 population change[f] (CNTLPCHG)	38,158
Positive adjustment factor[g]	1.23682
Negative adjustment factor[h]	0.76318
Sum of adjustment factors	2

[a]Projections from Table 7.4.
[b]Original population change = original population projection − 1995 population.
[c]Controlled population change = positive original population change × positive adjustment factor or negative original population change × negative adjustment factor.
[d]Controlled population projection = 1995 population + controlled population change.
[e]Difference = controlled population projection − original population projection.
[f]Independent projection of population change = independent population projection − 1995 population.
[g]Positive adjustment factor = [ABSUM + (CNTLPCHG − SUM)]/ABSUM.
[h]Negative adjustment factor = [ABSUM − (CNTLPCHG − SUM)]/ABSUM.

Table 11.4. The Plus–Minus Method: Controlling to an Independent Projection of Migration for White Females, Broward County, 2000

Age	1995	2000 surviving population[a]	1995–2000 Original migrants[b]		1995–2000 Controlled migrants		2000 population projection Original[e]	Controlled[f]	Difference[g]
			In	Out	In[c]	Out[d]			
0–4	32,814	27,866	0	0	0	0	27,866	27,866	0
5–9	28,867	32,672	7,493	7,643	8,085	7,039	32,522	33,718	1,196
10–14	26,044	28,841	6,946	6,082	7,495	5,601	29,705	30,735	1,030
15–19	22,733	25,998	6,491	7,705	7,004	7,096	24,784	25,906	1,122
20–24	25,987	22,674	10,360	8,851	11,179	8,151	24,183	25,702	1,519
25–29	34,929	25,912	14,008	9,987	15,115	9,198	29,923	31,829	1,896
30–34	43,345	34,799	12,792	11,535	13,802	10,624	36,056	37,977	1,921
35–39	45,137	43,132	12,078	9,385	13,033	8,643	48,825	47,522	1,697
40–44	42,035	44,855	10,687	7,438	11,532	6,850	48,104	49,537	1,433
45–49	37,618	41,634	8,260	5,663	8,913	5,215	44,231	45,332	1,101
50–54	30,660	37,054	7,973	4,325	8,603	3,983	40,702	41,674	972
55–59	24,778	29,943	6,535	4,840	7,052	4,457	31,638	32,538	900
60–64	25,955	23,891	6,700	4,382	7,230	4,036	26,209	27,085	876
65–69	32,125	24,602	7,090	4,155	7,650	3,827	27,537	28,425	888
70–74	36,920	29,663	5,498	2,914	5,933	2,684	32,247	32,912	665
75–79	35,118	32,563	4,465	3,106	4,818	2,861	33,922	34,520	598
80–84	28,941	28,524	2,793	2,069	3,014	1,905	29,248	29,633	385
85+	20,136	28,334	2,868	2,215	3,095	2,040	28,987	29,389	402
Total	574,142	562,957	133,037	102,295	143,553	94,210	593,699	612,300	18,601

Calculation of plus–minus adjustment factors:

Sum of original migrants[h] (SUM)	30,742
Sum of absolute values of original migrants[i] (ABSUM)	235,322
1995 population	574,142
Independent population projection for 2000	612,300
1995–2000 population change[j]	38,158
1995–2000 births	28,058
1995–2000 deaths	39,243
1995–2000 net migration[k] (CONTROL)	49,343
Positive adjustment factor[l]	1.07904
Negative adjustment factor[m]	0.92096
Sum of adjustment factors	2

[a]Projections of surviving population from Table 7.4.

[b]Migration projections from Table 7.4.

[c]Controlled in-migrants = original in-migrants × positive adjustment factor.

[d]Controlled out-migrants = original out-migrants × negative adjustment factor.

[e]Original population projection = surviving population + original in-migrants − original out-migrants.

[f]Controlled population projection = surviving population + controlled in-migrants − controlled out-migrants.

[g]Difference = controlled population projection − original population projection.

[h]SUM = total original in-migrants − total original out-migrants.

[i]ABSUM = total original in-migrants + total original out-migrants.

[j]Independent projection of population change = independent population projection − 1995 population.

[k]Independent projection of net migration = independent projection of population change − births + deaths.

[l]Positive adjustment factor = [ABSUM + (CONTROL − SUM)]/ABSUM.

[m]Negative adjustment factor = [ABSUM − (CONTROL − SUM)]/ABSUM.

negative adjustment factor is applied to out-migrants. The adjustment factors indicate that the number of projected in-migrants in each age group is increased by almost 8% (POSFACTOR = 1.07904) and the projected number of out-migrants is reduced by the same percentage (NEGFACTOR = 0.92096). If separate projections were available for in- and out-migration, one could bypass the plus-minus method by computing two raking factors, one for in-migrants and one for out-migrants.

Controlling to Projections of Larger Geographic Areas

Population projections are often prepared for a number of geographic areas and for the geographic composite of those areas (e.g., all counties within a state, all census tracts within a county). In this section, we discuss methods for reconciling projections at different levels of geography. These methods make the sum of the projections for smaller geographic areas equal the projection for a larger area (within rounding error).

The simplest way to achieve this reconciliation is to use a bottom-up approach in which the projection for the larger area is simply calculated as the sum of the projections for the smaller areas. A bottom-up approach is most useful when the sum of the projections for the smaller areas is not significantly different from the projection for the larger area.

If the differences are large—or if there are other reasons for holding the projections for the larger area constant—a bottom-up approach should not be used. In these circumstances, some type of controlling procedure must be used. If controlling involves only one variable—a single dimension—we can use one of the controlling methods described in the previous section. If controlling involves several variables, however, more complicated procedures must be used.

Single-Dimensional Controlling. As an example of single-dimensional controlling, consider a set of projections for females in several census tracts in San Diego County (Table 11.5). These projections were made by using the Hamilton–Perry method and cover four census tracts and one subregional area (SRA). There are 42 SRAs in San Diego County, each composed of one or more census tracts. Because the SRA shown in Table 11.5 contains these four census tracts, consistency requires that the sum of the projections for the census tracts be equal to the SRA projection.

We illustrate single-dimensional raking by controlling age projections for the four census tracts to an independent projection of the SRA population by age. The procedure is very simple. First, SRA projections are tabulated by age group. Second, census tract projections are added together for each age group. Third, adjustment factors are computed for each age group by dividing the SRA projections by the sum of the census tract projections. Finally, these adjustment factors are multiplied by the original projections of each census tract by age group, providing a set of controlled projections.

Table 11.5. The Raking Method: Controlling to a Projection
of Females by Age for Selected Census Tracts in San Diego County, 2000

	Original census tract projections				Subregional area		Adjustment factor[c]
Age	188.00	189.01	189.02	190.00	Sum of tracts[a]	Control[b]	
0–4	565	572	1,037	379	2,553	2,488	0.97454
5–9	748	574	831	483	2,636	2,581	0.97914
10–14	765	470	595	423	2,253	2,203	0.97781
15–19	655	585	995	393	2,628	2,579	0.98135
20–24	346	648	1,153	166	2,313	2,173	0.93947
25–29	307	570	650	235	1,762	1,724	0.97843
30–34	664	497	722	282	2,165	2,165	1.00000
35–39	912	567	676	386	2,541	2,522	0.99252
40–44	868	506	660	422	2,456	2,474	1.00733
45–49	824	420	437	495	2,176	2,187	1.00506
50–54	771	286	344	431	1,832	1,814	0.99017
55–59	704	258	192	264	1,418	1,385	0.97673
60–64	685	152	216	262	2,315	1,250	0.95057
65–69	499	166	183	222	1,070	1,057	0.98785
70–74	441	189	158	202	990	991	1.00101
75–79	344	186	138	129	797	799	1.00251
80–84	240	153	105	97	595	596	1.00168
85+	271	196	79	71	617	602	0.97569
Total	10,609	6,995	9,171	5,342	32,117	31,590	

	Controlled census tract projections				Subregional area		
Age	188.00	189.01	189.02	190.00	Sum of tracts[a]	Difference[d]	
0–4	551	557	1,011	369	2,488	0	
5–9	732	562	814	473	2,581	0	
10–14	748	460	582	414	2,204	−1	
15–19	643	574	976	386	2,579	0	
20–24	325	609	1,083	156	2,173	0	
25–29	300	558	637	230	1,725	−1	
30–34	664	497	722	282	2,165	0	
35–39	905	563	672	383	2,523	−1	
40–44	874	510	665	425	2,474	0	
45–49	827	422	439	498	2,186	1	
50–54	763	283	341	427	1,814	0	
55–59	688	252	188	258	1,386	−1	
60–64	651	144	205	248	1,248	2	
65–69	493	164	181	219	1,057	0	
70–74	440	189	158	202	989	2	
75–79	345	186	138	129	798	1	
80–84	240	153	105	97	595	1	
85+	264	191	77	69	601	1	
Total	10,453	6,874	8,994	5,265	31,586	4	

[a]Sum of the population projections in each census tract.
[b]Independent population projections of the subregional area.
[c]Adjustment factor = subregional area control / sum of the population projections in each census tract.
[d]Difference = subregional area control − sum of the controlled population projections in each census tract.

The adjustment factors shown in Table 11.5 are quite small for most age groups. However, these projections cover only a five-year projection horizon. For projections that cover longer horizons (e.g., 20–30 years), the adjustments would be much larger.

The projections for the census tracts now sum to the SRA projection in every age group (except for small differences due to rounding). However, the projections of total population for each tract are now different from what they were originally. This is the major problem of single-dimensional controlling: making projections consistent across one dimension makes them inconsistent across another. If the projections focused on a single factor such as total population, of course, this would not be problem.

N-Dimensional Controlling.   What if we wanted to make the census tract age projections consistent with the SRA age projections but preserve the original population totals for each census tract? The methods discussed so far cannot handle this situation. Rather, a procedure is needed that can control across several dimensions simultaneously; this is called *n-dimensional controlling*.

N-dimensional controlling can be accomplished using the iterative proportions (IP) method, which approximates a least squares solution to obtain convergence in all *n* dimensions (Deming, 1943; Shryock & Siegel, 1973). This method can handle a wide range of situations.

There are three main conditions for applying the IP method. First, all projections must be greater than or equal to zero. Second, there must be projections for the totals of each controlling dimension (e.g., age and total population). If we are controlling census tract projections to a county projection by age group, for example, this condition requires that we have projections of total population for each census tract and an independent projection of the county's age distribution. The third condition is that the sum of all projections over all dimensions must be equal; for example, the sum of the age group projections for the county must be equal to the sum of the total population projections for the census tracts.

To illustrate *n*-dimensional controlling, we use two dimensions. One represents a demographic characteristic (age) and one represents the total population of several geographic regions (four census tracts and a SRA). The IP method begins with an initial matrix, whose body contains the population projection by age for each census tract in the SRA. The row total is the population projection by age for the larger geographic area (the SRA) and the column total is the total population projection for each census tract. These row and column totals are often called *marginals*. The goal of the IP method is to adjust the matrix so that, when summed horizontally, census tract projections equal SRA projections for each age group and, when summed vertically, census tract projections by age equal the total population for each census tract.

We achieve this goal by applying a single-factor raking procedure to the original matrix, alternating sequentially between rows and columns. Starting with the rows, we apply a row-specific raking factor to each cell in each row; we repeat this process for all rows. After this step, the sum across the rows matches the SRA population projection for each age group. However, the sum down the columns (i.e., all age groups within one census tract) no longer matches the total population projection for that tract. Then, we apply a column-specific raking factor to each cell in each column. After this step, the sum of the cells in each column matches the total population of that column, but the sum of the cells in each row no longer matches the SRA population in that age group. By continuing this sequence of adjustments, eventually we arrive at a convergence in which cells in both rows and columns sum to the marginal totals (except for small differences due to rounding).

The rate of convergence is relatively fast, typically requiring only two cycles of horizontal and vertical adjustments to achieve complete agreement in one dimension and close agreement in the other (Shryock & Siegel, 1973). It does not matter whether one begins the process by adjusting rows or columns; the results are essentially the same. One can refine the IP method to handle both positive and negative adjustments by using the plus-minus method described earlier to compute two separate adjustment factors to use in the iterative process.

We illustrate the IP method by using the original projections of the female population shown in Table 11.5. The objective is to make the age-group projections for the census tracts consistent with the age-group projections for the SRA *and* with the projections of total female population for each census tract.

Table 11.6 shows the mechanics of the IP method. To save space, we illustrate the method using 10-year age groups; the wider age groups have no impact on the way the method is carried out. The first panel ("Beginning Matrix, First Iteration") shows the initial conditions and the elements needed to apply the IP method. The main body of the matrix is contained in columns 2–5 and the rows for age groups 0–9 through 80+; the cells of this matrix show the uncontrolled projections produced using the Hamilton–Perry method. Column 6 shows the sum of the census tract projections for each age group and column 7 (Control) shows the row marginals (i.e., the independent projections by age for the SRA). The row labeled "Sum of Ages" shows the uncontrolled projections of total population for each census tract, and the row labeled "Tract Control" shows the column marginals (i.e., the independent projections of total population for each census tract).

The population control totals are all less than the sum of the uncontrolled projections. This is true both for columns (census tract populations) and rows (age groups). The two numbers in bold type ($-8,818$) are particularly important. They represent the total amount of the adjustment required in the rows and columns to make the projections consistent in both dimensions; they must be equal for the IP method presented here to work properly. The row and column adjustment factors are computed as the ratio of the control total to the sum of the corresponding cells; they are computed separately for each age group and each census tract. These are

Table 11.6. The Iterative Proportions Method: Controlling a Projection for Females in Two Dimensions, Selected Census Tracts, San Diego County, 2000

Beginning Matrix, First Iteration

| | Census tract | | | | Subregional area | | | |
Age	188.00	189.01	189.02	190.00	Sum of tracts[a]	Control	Adjustment[b]	Row factor[c]
0–9	1,313	1,146	1,868	862	5,189	3,737	−1,452	0.72018
10–19	1,420	1,055	1,590	816	4,881	3,527	−1,354	0.72260
20–29	653	1,218	1,803	401	4,075	2,875	−1,200	0.70552
30–39	1,576	1,064	1,398	668	4,706	3,457	−1,249	0.73459
40–49	1,692	926	1,097	917	4,632	3,438	−1,194	0.74223
50–59	1,475	544	536	695	3,250	2,359	−891	0.72585
60–69	1,184	318	399	484	2,385	1,702	−683	0.71363
70–79	785	375	296	331	1,787	1,320	−467	0.73867
80+	511	349	184	168	1,212	884	−328	0.72937
Sum of ages[d]	10,609	6,995	9,171	5,342				
Tract control	8,060	5,651	5,455	4,133				
Adjustment[e]	−2,549	−1,344	−3,716	−1,209	−8,818		−8,818	
Column factor[f]	0.75973	0.80786	0.59481	0.77368				

Rows Adjusted, First Iteration

| | Census tract | | | | Subregional area | | | |
Age	188.00	189.01	189.02	190.00	Sum of tracts[a]	Control	Adjustment[b]	Row factor[c]
0–9	946	825	1,345	621	3,737	3,737	0	1.00000
10–19	1,026	762	1,149	590	3,527	3,527	0	1.00000
20–29	461	859	1,272	283	2,875	2,875	0	1.00000

				Sum of tracts	Control	Adjustment[b]	Row factor[c]
30–39	1,158	782	1,027	491	3,458	3,457	−1	0.99971
40–49	1,256	687	814	681	3,438	3,438	0	1.00000
50–59	1,071	395	389	504	2,359	2,359	0	1.00000
60–69	845	227	285	345	1,702	1,702	0	1.00000
70–79	580	277	219	244	1,320	1,320	0	1.00000
80+	373	255	134	123	885	884	−1	0.99887
Sum of ages[d]	7,716	5,069	6,634	3,882			−2	
Tract control[e]	8,060	5,651	5,455	4,133				
Adjustment[e]	344	582	−1,179	251	−2			
Column factor[f]	1.04458	1.11482	0.82228	1.06466				

Columns Adjusted, First Iteration

	Census tract				Subregional area			
Age	188.00	189.01	189.02	190.00	Sum of tracts[a]	Control	Adjustment[b]	Row factor[c]
0–9	988	920	1,106	661	3,675	3,737	62	1.01687
10–19	1,072	849	945	628	3,494	3,527	33	1.00944
20–29	482	958	1,046	301	2,787	2,875	88	1.03158
30–39	1,120	872	844	523	3,449	3,457	8	1.00232
40–49	1,312	766	669	725	3,472	3,438	−34	0.99021
50–59	1,119	440	320	537	2,416	2,359	−57	0.97641
60–69	883	253	234	367	1,737	1,702	−35	0.97985
70–79	606	309	180	260	1,355	1,320	−35	0.97417
80+	390	284	110	131	915	884	−31	0.96612
Sum of ages[d]	8,062	5,651	5,454	4,133			−1	
Tract control[e]	8,060	5,651	5,455	4,133				
Adjustment[e]	−2	0	1	0	−1			
Column factor[f]	0.99975	1.00000	1.00018	1.00000				

continued

Table 11.6. (Continued)

Columns Adjusted, Third Iteration

	Census tract				Subregional area			
Age	188.00	189.01	189.02	190.00	Sum of tracts[a]	Control	Adjustment[b]	Row factor[c]
0–9	1,010	934	1,118	674	3,736	3,737	1	1.00027
10–19	1,087	855	948	636	3,526	3,527	1	1.00028
20–29	500	987	1,075	313	2,875	2,875	0	1.00000
30–39	1,219	872	841	525	3,457	3,457	0	1.00000
40–49	1,305	756	658	720	3,439	3,438	−1	0.99971
50–59	1,097	428	310	524	2,359	2,359	0	1.00000
60–69	867	247	228	360	1,702	1,702	0	1.00000
70–79	593	300	174	254	1,321	1,320	−1	0.99924
80+	379	273	105	127	884	884	0	1.00000
Sum of ages[d]	8,057	5,652	5,457	4,133				
Tract control	8,060	5,651	5,455	4,133				
Adjustment[e]	3	−1	−2	0	**0**		**0**	
Column factor[f]	1.00037	0.99982	0.99963	1.00000				

[a] Sum of the population in each census tract.
[b] Adjustment = subregional area control − sum of the population projections in each census tract.
[c] Row factor = subregional area control/sum of the population control in each census tract.
[d] Sum of the population projections for each age.
[e] Adjustment = census tract control − sum of the population projections for each age.
[f] Column factor = census tract control/sum of the population projections for each age.

the factors that are applied to each cell during the controlling process. All of the adjustment factors in the first panel are less than 1.0, which indicates that downward adjustments are necessary.

The second panel of Table 11.6 ("Rows Adjusted, First Iteration") shows the population projections by age for each census tract after we adjusted them to match the row control totals. For example, the adjusted population aged 0–9 in census tract 188.00 is

$$1,313 \times 0.72018 = 946$$

After the first set of adjustments, all of the row adjustment factors are 1.0 (or very close to 1.0), which indicates convergence to the projections by age for the SRA. In addition, the total amount of adjustment required is close to zero (-2 for both the sum of ages and the sum of census tracts). However, for individual census tracts, the sum of ages is still inconsistent with the control totals. In fact, three of the four column adjustment factors have changed from values less than 1.0 in panel 1 to values of greater than 1.0 in panel 2.

The third panel ("Columns Adjusted, First Iteration") shows the population projections by age for each census tract after we adjusted them to match the column control totals. For example, the adjusted population aged 0–9 in census tract 188.00 is now

$$946 \times 1.04458 = 988$$

Now, all of the column adjustment factors are 1.0 (or very close to 1.0), which indicates convergence to the total population projections for each census tract, but the column adjustments have made the sum of the age-group projections for census tracts inconsistent with the age-group projections for the SRA. However, the differences are much smaller than they were before; none is greater than 88 (in absolute value). This comparison shows that significant convergence to both marginal totals has occurred after only one full iteration of the process.

The last panel of Table 11.6 ("Columns Adjusted, Third Iteration") shows the results after three full iterations. The census tract projections by age have now converged (within rounding error) to both the SRA population projections by age and the projections of total population for each census tract.

The IP method—and other controlling methods—may not always come as close to the independent (control) projections as the examples shown here. Raising the level of demographic detail and reducing the geographic scale can make multiplicative adjustment routines lose their efficiency because the computations may not change the original values as much as is needed to produce complete convergence. For example, integer values less than five will not change unless the adjustment is at least 10% (e.g., 5×1.09 still equals five after rounding to the nearest integer). If this occurs with enough frequency, controlling falls short of its intended objective.

To handle circumstances where multiplicative adjustments are not adequate, alternative mathematical controlling strategies have been developed (e.g. San Diego Association of Governments, 1998). These involve probabilistic assignment routines and/or iterative schemes that apply small additive adjustments to the uncontrolled observations. Private data vendors do some of the most innovative work in this area but—for obvious reasons—are reluctant to reveal their trade secrets.

PROVIDING ADDITIONAL TEMPORAL AND AGE DETAIL

As mentioned earlier, many (perhaps most) applications of the cohort-component method use five-year age groups and produce projections in five-year intervals. With a launch year of 2000, for example, projections might be made for 2005, 2010, 2015, and 2020 for the age groups 0–4, 5–9, 10–14, ... 85+. What if somebody needs a projection for 2008 instead of 2010 or for ages 15–17 rather than 15–19? In this section, we discuss some methods for disaggregating projections by age and interpolating between target years.

The methods discussed in this section are approximate. They do not reflect all of the subtleties in the age distribution or pick up temporal growth patterns as well as a single-year cohort-component model. They are very useful, however, and offer a reasonable and cost-effective alternative to a single-year projection model.

Adding Temporal Detail

We treat the creation of projections for intermediate years as an interpolation problem. The goal is to develop a consistent set of annual projections for years between two target years. We describe three interpolation methods. The first assumes that change occurs linearly over the projection interval, which means that numerical changes are the same for each year. The second assumes that change occurs geometrically over the projection interval, which means that percent changes are the same for each year. The third uses osculatory interpolation techniques that incorporate information from several different projection intervals.

These methods can be applied to projections of the total population or to specific subgroups of that population. We illustrate them by using a set of age, sex, and race/ethnic projections for San Diego County produced by the San Diego Association of Governments in 1998. For each method, we develop annual interpolations for the projected white female population between target years 2000 and 2005. Because they exhibit sharply contrasting growth patterns, we use a different age group to illustrate each interpolation method.

Linear Interpolation.   For linear interpolation, the average annual absolute change between launch year l and target year t is calculated as the difference between the populations at the beginning and end of the interval, divided by the number of years in the interval:

$$AAAC = (P_t - P_l)/z$$

where P_t is the population in the target year, P_l is the population in the launch year, z is the number of years in the projection interval, and AAAC is the average annual population change during the projection interval. It should be noted that P_l refers to the start of any projection interval. For example, a projection for 2005 may be made using 2000 as the initial launch year. Then, the 2005 projection serves as the launch year for the 2010 projection.

We can compute the average annual change for total population or for any demographic characteristic (e.g., an age group). The annual change is successively added (or subtracted, in the case of a population loss) to the population at the beginning of the projection interval to obtain a projection for the intermediate years:

$$P_{l+w} = P_l + (AAAC \times w)$$

where w is the number of years from the beginning of the projection interval to the intermediate year. To illustrate linear interpolation, we develop annual population projections for white females aged 85+:

Population in 2000: 23,617

Population in 2005: 27,876

AAAC: $(27,876 - 23,617)/5 = 852$

2001 projection: $23,617 + 852 = 24,469$

2002 projection: $23,617 + (852 \times 2) = 25,321$

2003 projection: $23,617 + (852 \times 3) = 26,173$

2004 projection: $23,617 + (852 \times 4) = 27,025$

Geometric Interpolation.   Geometric interpolation assumes that the annual percent change is the same for each year in the projection interval. It uses the average annual growth rate to determine an intermediate year projection, following the geometric formula described in Chapter 2:

$$GF = (P_t/P_l)^{1/z}$$

GF is a growth factor that represents 1.0 plus the average annual population growth rate over the projection interval. This factor is between 0 and 1.0 for population losses and is greater than 1.0 for population gains. A growth factor of 1.0 indicates no change in the population. Much like calculating the balance in a bank account by using compound interest, the population at the beginning of the projection interval is compounded by the average annual rate of population growth to obtain a projection for the intermediate years:

$$P_{l+w} = P_l \times GF^w$$

To illustrate geometric interpolation, we develop annual population projections for white females aged 60–64:

Population in 2000: 34,025

Population in 2005: 42,989

Growth factor: $(42,989 / 34,025)^{(1/5)} = 1.04788$

2001 projection: $34,025 \times 1.04788 = 35,654$

2002 projection: $34,025 \times 1.04788^2 = 37,361$

2003 projection: $34,025 \times 1.04788^3 = 39,150$

2004 projection: $34,025 \times 1.04788^4 = 41,025$

Osculatory Interpolation. The third method—osculatory interpolation—produces smoother interpolations than either linear or geometric interpolation because it incorporates more information about population changes over time. Specifically, this method incorporates information on population change during the time intervals immediately before and after the projection interval, as well as information from the projection interval itself. For example, interpolations for 2000–2005 use information from 1995–2000 and 2005–2010 as well as from 2000–2005.

Osculatory interpolations are based on equations that combine two overlapping polynomial functions (Shryock & Siegel, 1973). Although the construction of these equations is somewhat complex, the application of the coefficients derived from those equations is simple. Shryock and Siegel (1973) present coefficients for the most commonly used osculatory interpolation techniques. To illustrate, we use Karup–King coefficients to develop interpolations for each year within a five-year interval.

There are three types of Karup–King coefficients. Middle-interval coefficients are used for most intervals in the projection horizon; they incorporate information from time intervals immediately before and after the projection interval. Last-interval coefficients are used for the final interval in the projection horizon (e.g., 2015–2020 for projections through 2020). They are not as reliable as middle-interval coefficients because they use information from only one side of the projection interval (i.e., the preceding interval). First-interval coefficients are also available but are seldom needed because historical data are usually available. For example, if the first projection interval were 1995–2000, we could use the 1990 Census and 1995 estimates to provide data for the period immediately preceding the 1995 launch year.

Using Karup–King middle-interval coefficients, we created annual interpolations for 2001–2004 for white females aged 40–44 (Table 11.7). There are four coefficients for each year in the interval for which interpolations are to be made. These coefficients correspond to the following years:

Table 11.7. Karup–King Interpolations between Target Years
2000 and 2005 for White Females Aged 40–44, San Diego County

	1995	2000	2005	2010
Population[a]	67,286	76,819	79,470	71,059

Middle-interval coefficients

Year	G1 (1995)	G2 (2000)	G3 (2005)	G4 (2010)
2001	−0.064	0.912	0.168	−0.016
2002	−0.072	0.696	0.424	−0.048
2003	−0.048	0.424	0.696	−0.072
2004	−0.016	0.168	0.912	−0.064

Intermediate calculations[b]

Year	G1 (1995)	G2 (2000)	G3 (2005)	G4 (2010)	Population projection[c]
2001	−4,306	70,059	13,351	−1,137	77,967
2002	−4,845	53,466	33,695	−3,411	78,905
2003	−3,230	32,571	55,311	−5,116	79,536
2004	−1,077	12,906	72,477	−4,548	79,758

[a]*Source*: San Diego Association of Governments, 2020 Regional Growth Forecast, July, 1998.
[b]Intermediate calculation = coefficient × population in the appropriate year.
[c]Projected population = sum of the intermediate calculations for each year.

1. Five years before the beginning of the interval (i.e., 1995)
2. The beginning of the interval (i.e., 2000)
3. The end of the interval (i.e., 2005)
4. Five years after the end of the interval (i.e., 2010)

For the intermediate computations, we multiply the coefficients by the population projection for the appropriate year. For example, for 2001 the intermediate calculations are as follows:

$$-0.064 \times 67,286 = -4,306$$
$$0.912 \times 76,819 = 70,059$$
$$0.168 \times 79,470 = 13,351$$
$$-0.016 \times 71,059 = -1,137$$

To obtain the population projection for an intervening year, we sum the four intermediate calculations for that year. For 2001, the projection is as follows:

$$-4,306 + 70,059 + 13,351 - 1,137 = 77,967$$

Interpolations based on the Karup–King method are essentially weighted averages of projected (or estimated) populations. For each interpolation year, the weights sum to 1.0. For the years defining the beginning and end of the projection interval, the weights are directly related to the length of time between the interpolation year and the beginning and ending years; that is, the closer the interpolation year is to the projection year, the greater the weight assigned to that projection year. In this example, the 2000 population is assigned a weight of 0.912 for the 2001 interpolation; the weight drops to 0.696 for 2002, 0.424 for 2003, and 0.168 for 2004. Conversely, the weight assigned to the 2005 population increases steadily throughout the interval. Thus, interpolations from the Karup–King method fit more smoothly with population changes observed for earlier and later periods than either of the other interpolation methods.

Evaluation.   Which of these interpolation methods is best? The linear and geometric methods require the least data and are the simplest to apply. The computations for osculatory interpolation are more extensive, require external multipliers, and usually require a projection for intervals outside of the interval being interpolated. However, the osculatory method provides interpolations that are more consistent across projection intervals. The choice of method will depend on the characteristics of projected population change and the intended use of the interpolated data.

Figure 11.2 shows the projected population change in five-year intervals from 1995 to 2010 for the three age groups used in our illustrations. The growth patterns suggest which interpolation method(s) might be most appropriate for each age group. Linear interpolation seems suitable for ages 85+ because the projections show relatively constant numerical change in each of the three periods. The steadily increasing numerical changes shown for ages 60–64 suggest the use of geometric interpolation. Finally, osculatory interpolation is likely to be the best approach when the growth pattern changes direction and has clearly defined turning points, such as shown for ages 40–44.

To judge the suitability of alternative interpolation methods in different population growth scenarios, we examined the accuracy of each method for each age group. We measured accuracy by comparing the interpolations with results derived from an independent set of annual forecasts for 2001, 2002, 2003, and 2004. These independent forecasts were produced by using a single-year cohort-component projection model. We call the differences between the interpolations and the forecasts *errors*, assuming that a single-year projection model will provide more reliable forecasts for intermediate years than an interpolation method. We calculated the mean absolute percent error for the four interpolated years for each interpolation method and each age group.

Figure 11.3 shows the results of these comparisons. The results generally support our suggestion that the choice of interpolation method should be based on

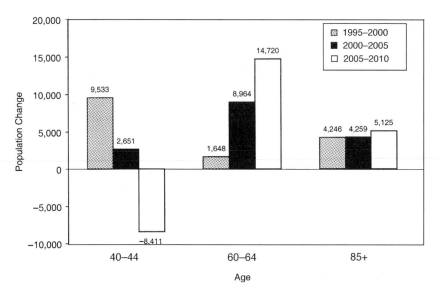

Figure 11.2. Projected changes for white females, selected ages, San Diego County, 1995–2010.

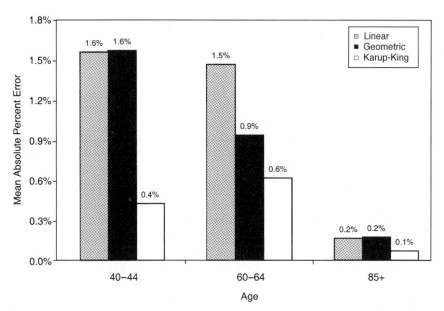

Figure 11.3. A comparison of interpolation errors for white females, selected ages, San Diego County, 2001–2004.

projected patterns of change. Osculatory interpolation performed much better than the other two methods for the 40–44 age group. Clearly, neither linear nor geometric interpolation was able to pick up the effects of the dramatic shifts in population growth observed for this age group. Linear interpolation performed poorly for both the 40–44 and 60–64 age groups, which were characterized by rapidly declining and increasing numerical changes, respectively. Geometric interpolation performed considerably better than linear interpolation for the 60–64 age group (although not quite as well as osculatory interpolation). For ages 85+, which exhibited an approximately linear growth trend over the three 5-year periods, errors were very low for all three methods.

The Karup–King osculatory interpolation method produced the lowest errors for every age group, which reflects the advantage of incorporating information from several time periods into the interpolation procedure. We believe that it is the preferred method when substantial shifts in growth patterns are projected. However, it was not much more accurate than geometric interpolation for the scenario of steadily increasing growth, and was only slightly more accurate than either linear or geometric interpolation for the scenario of roughly constant numerical increases. In the latter two scenarios, the simpler interpolation methods provide an acceptable alternative to the more complex osculatory method.

We do not deal with the full range of interpolation issues in this chapter. Our illustrations focused on the same *age groups* at two points in time (e.g., white females aged 40–44 in 2000 and 2005). Another approach is to focus on changes in *cohorts* over time (e.g., white females aged 40–44 in 2000 who become 45–49 in 2005). The cohort approach may be more useful than the period approach described here for some purposes, but is considerably more complicated because it requires dealing with a variety of nonstandard age groups. For example, persons aged 40–44 in 2000 are 43–47 in 2003, forming parts of two five-year are groups (40–44 and 45–49). We believe the procedures described here—although approximate—will be adequate for most purposes.

In addition, we did not consider approaches for making interpolations for all demographic subgroups of the population (e.g., the full age, sex, and race composition of the popuation). When this level of detail is needed, annual interpolations can be made by (1) developing annual interpolations of the total population of the area by using one of the methods described here; (2) developing annual interpolations for each demographic subgroup by using one of the methods described here; and (3) controlling the interpolations for the demographic subgroups to the interpolation for the total population for each year in the interval. A more detailed discussion of interpolation procedures can be found in Shryock and Siegel (1973).

Adding Age Detail

The disaggregation of broad age groups into smaller age groups can be achieved by using a number of different methods (e.g., Shryock & Siegel, 1973).

We discuss three of the most commonly used: rectangular distribution, osculatory interpolation, and historical patterns. The rectangular distribution and osculatory interpolation methods require only information directly available from the projection model, whereas the historical patterns method requires data external to the projection model.

Rectangular Distribution Method. The simplest method for splitting age groups into smaller categories is to apply a rectangular distribution, which assumes that all the smaller groups (e.g., the one-year age groups) have identical shares of the larger group (e.g., the five-year age group). The numbers in each of the smaller groups can be computed by dividing the population of the larger group by the number of subdivisions desired. For example, the projected white female population aged 15–19 in Broward County in 2000 is 24,784. Using the rectangular distribution method, the population within each one-year age group is computed as follows:

$$24,784 / 5 = 4,957$$

We can create a projection for any age group within the 15–19 group by multiplying the one-year number by the width of the age group desired. For example, a population projection for white females aged 18–19 is computed as follows:

$$4,957 \times 2 = 9,914$$

Osculatory Interpolation Method. The osculatory interpolation method for splitting age groups is similar to that described before for interpolating between target years. However, instead of using data for earlier and later time periods, it uses data for younger and older age groups. This allows the osculatory interpolation method to pick up some of the effects of past fertility and migration patterns. Shryock and Siegel (1973) present coefficients that can be used for splitting age groups. Although the concepts are similar, these are *not* the same coefficients used for interpolating between two points in time.

Again, we use Karup–King coefficients to illustrate osculatory interpolation. Several sets of coefficients are available, corresponding to different sizes of age groups. We use the coefficients needed for dividing five-year age groups into one-year age groups.

To illustrate this method, we divide the projected white female population aged 15–19 in Broward County in 2000 into single years of age (Table 11.8). As before, we use middle-interval coefficients. First- and last-interval coefficients have been constructed for the youngest (e.g., 0–4) and oldest (e.g., 85+) groups, but interpolations for these groups are not as reliable as interpolations for other groups because they use information from only one side of the relevant group (Shryock & Siegel, 1973).

Table 11.8. Karup–King Interpolations within an Age Group
for White Females Aged 15–19, Broward County, 2000

Age group	10–14	15–19	20–24
Population[a]	29,705	24,784	24,183

Middle-interval coefficients

Age	G1 (10–14)	G2 (15–19)	G3 (20–24)
15	0.064	0.152	−0.016
16	0.008	0.224	−0.032
17	−0.024	0.248	−0.024
18	−0.032	0.224	0.008
19	−0.016	0.152	0.064

Intermediate calculations[b]

Age	G1 (10–14)	G2 (15–19)	G3 (20–24)	Population projection[c]
15	1,901	3,767	−387	5,281
16	238	5,552	−774	5,016
17	−713	6,146	−580	4,853
18	−951	5,552	193	4,794
19	−475	3,767	1,548	4,840

[a]Projections from Table 7.4.
[b]Intermediate calculation = coefficient × population in the appropriate 5-year age group.
[c]Projected population = sum of the intermediate calculations for each age.

There are three coefficients for each single year of age: one corresponds to the group to be subdivided (15–19), one corresponds to the next younger group (10–14), and one corresponds to the next older group (20–24). For the intermediate calculations, we multiply the coefficients by the population in the appropriate 5-year age group. For example, the intermediate calculations for age 18 are

$$-0.032 \times 29,705 = -951$$
$$0.224 \times 24,784 = 5,552$$
$$0.008 \times 24,183 = 193$$

To obtain the projection for each year of age, we sum the three intermediate calculations for each age group. For age 18, the projection is

$$-951 + 5,552 + 193 = 4,794$$

Karup–King and other osculatory interpolation methods are self-normalizing in the sense that the sum of the population projections for each single year of age is within rounding error of the projected population in the larger age group. This occurs because the weights for the age group to be interpolated (e.g., 15–19) sum to 1.0, whereas the weights for the immediately younger and older age groups sum to zero.

Historical Patterns Method. The rectangular distribution and osculatory interpolation methods both assume that patterns in the distribution of broader age groups provide a reasonable representation of the distributions *within* broader age groups (Shryock & Siegel, 1973). If the distribution within the age group has a special or distinctive pattern not reflected by the grouped data, however, these methods yield poor results. For example, places that have a college or university will have a relatively large number of persons aged 18–19, throwing off the typical distribution within the 15–19 age group. Similar effects may occur in places that have large military installations or high levels of retiree migration.

One way to handle these unique situations—or, for that matter, the entire disaggregation process—is to split broader age groups by using historical data on the distribution of persons within those groups. We call this the *historical patterns* method. It requires compiling and integrating data beyond that needed by the other two methods. The additional level of effort required depends on the number of age groups involved, the number of historical points used, and the number of strata involved (e.g., two sexes and five race/ethnic groups).

Despite the additional data requirements, the application of the historical patterns method is not complicated. To illustrate this method, we split Broward County's projected white female population aged 15–19 in 2000 into two groups: 15–17 and 18–19. We show two approaches based on the number of females aged 18–19 as a share of the number aged 15–19. In the last two censuses (1980 and 1990), these shares were 0.395 and 0.419, respectively. The projection for females aged 15–17 is made simply by subtracting the projection of females aged 18–19 from the projection of females aged 15–19.

The first approach assumes that the most recent population share does not change. For example, we can project the population aged 18–19 in 2000 by multiplying the 1990 share by the 2000 projection for the population aged 15–19:

$$24{,}784 \times 0.419 = 10{,}384$$

The second approach is based on the continuation of recent trends in population shares. In our example, that share increased during the 1980s. This approach assumes that the share will increase by the same amount during the 1990s. To project the 2000 share, we subtract the 1980 share from the 1990 share and add that result to the 1990 share. Then, we apply the projected share to the projected white female population aged 15–19 in 2000:

$$0.419 + (0.419 - 0.395) = 0.443$$
$$24,784 \times 0.443 = 10,979$$

One must be careful when specifying the assumptions used in the historical patterns method. The most widely used assumption is that the share from the most recent census will remain constant. This assumption avoids the error of extrapolating short-term changes that prove to be temporary and is particularly useful for places that have institutional populations that retain a relatively constant age structure year after year (e.g., colleges and universities). On the other hand, the extrapolation of recent trends will be appropriate if there is evidence that recent changes in shares reflect a long-term trend.

Evaluation. Figure 11.4 compares the results from these methods for projections of white females aged 18–19 in 2000 for two Florida counties. Broward County is a large metropolitan county in south Florida. Although the projections generated by the four alternative methods in Broward County differ from each other, the differences are relatively modest and fall within a normal range of errors for county-level projections of narrow age groups. The difference between the lowest and highest projection is only 14%. Excluding the trended share method, the other three projections fall within a range of only 8%.

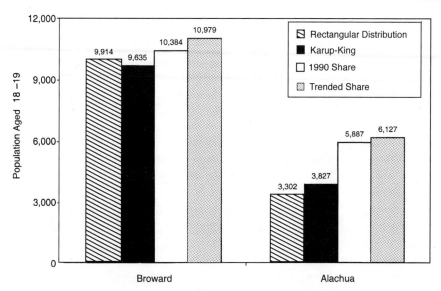

Figure 11.4. Alternative projections for white females aged 18–19, Broward and Alachua Counties, 2000.

Alachua County, a medium-sized county in north-central Florida, is home to the University of Florida. For Alachua County, projections from the Karup–King and rectangular distribution methods differ dramatically from projections from the two historical patterns methods. Because of the presence of the university, Alachua County has disproportionately more females aged 18–19 than most areas. In situations like this, the rectangular distribution and osculatory methods do not provide satisfactory results. These examples emphasize the importance of considering the unique characteristics of an area before deciding on an interpolation strategy (or, more generally, before deciding on the entire projection strategy).

CONCLUSIONS

In some circumstances, the projection methods described in this book cannot be applied in a simple, straightforward way. Missing or inaccurate data, complicated population dynamics, and unique events sometimes mean that a basic projection model must be adjusted before reasonable projections can be made. In addition, the needs of data users may require that adjustments be made to the resulting projections.

In this chapter, we described several circumstances in which the straightforward application of a projection method might produce inconsistent or misleading results. We then presented a number of procedures for dealing with the problems created by these circumstances. We also described a number of interpolation procedures for creating additional age and temporal detail and for controlling population characteristics from one projection to a total population (or migration) value from another projection. These adjustment procedures will not solve every problem that might be encountered, of course, but they will help the analyst deal with some of the special circumstances affecting the production of state and local population projections.

CHAPTER 12

Evaluating Projections

In previous chapters we discussed the three major approaches to population projection: the cohort–component method, trend extrapolation methods, and structural models. All three approaches include a wide variety of models, techniques, assumptions, special adjustments, and types of data that can be used to produce the desired projections. Given all the possibilities, how does one go about choosing the specific models, techniques, assumptions, and data sources for a particular set of projections? Is there a single "best" approach, or at least some that are better than others? Are some approaches better under some circumstances, whereas others are better under other circumstances? How can we even go about answering these questions?

In this chapter, we describe a number of criteria that can be used to evaluate population projections. We begin with a discussion of the criteria that we believe are most important: provision of necessary detail, face validity, plausibility, costs of production, timeliness, ease of application and explanation, usefulness as an analytical tool, political acceptability, and forecast accuracy. Then, we consider how they must be balanced against each other when choosing the appropriate methodology for any given situation. We close with an assessment of the way different methods stack up according to these criteria. Forecast accuracy is such an important criterion that we devote an entire chapter to its measurement and evaluation (Chapter 13). Further discussion of the criteria for evaluating population projections and other types of forecasts may be found in Ahlburg (1995), Armstrong (2001b), Keilman (1990), Long (1995), Murdock, et al. (1991), and Yokum and Armstrong (1995).

We distinguish between two types of projections. *General-purpose projections* are those produced without reference to a specific use or data user. Examples include projections produced by the Census Bureau for all states in the United States, projections produced by state demographers for all counties in their states, and projections produced by private companies for all census tracts in the United

States. *Customized projections* are those produced for a particular data user or for a particular purpose. Examples include population projections by block group for developing a county's transportation plan, birth projections by market area for evaluating a hospital's need for obstetrical services, and school enrollment projections by grade level for evaluating a school district's construction plans.

EVALUATION CRITERIA

Provision of Necessary Detail

Perhaps the most fundamental criterion for evaluating population projections is whether they provide the level of geographic, demographic, and temporal detail required by the data user. State projections are of little use to someone who needs county projections; projections of total population are of little use to someone who needs projections by age and sex; and projections for 2010 are of little use to someone who needs projections for 2025.

It is easy to determine whether projections will provide the necessary level of detail when they are made for a specific client or for a particular purpose. In fact, the producer can make sure of this. When general-purpose projections are made by a government agency, university, research institute, or private company, however, it is much more difficult to determine whether they meet user needs. Which geographic areas should be covered? Which demographic characteristics? Which projection intervals and time horizons? It is virtually impossible to meet all user needs with one set of projections because needs vary so much across data users.

Geographic Detail. Many data users need population projections for states and counties. These needs can be met relatively easily because the geographic boundaries for states and counties generally remain stable over time and many types of data are routinely available at the state and county levels. In addition, the number of states and counties is finite and relatively manageable. There are approximately 3,100 counties or county equivalents nationwide; the largest numbers are in Texas (254) and Georgia (159). Most states have fewer than 100 counties or county equivalents.

For subcounty areas, however, the number of potential areas—and even the ways in which those areas might be defined—is virtually endless. Possibilities include cities, townships, census tracts, block groups, blocks, school districts, traffic analysis zones, and many types of market or service areas. Projections that meet the needs for geographic detail for all (or almost all) data users would have to be made at least at the block level. Then, those projections could be aggregated to fit the geographic region required by each individual data user. Such a process, of course, would be extremely expensive and fraught with problems of data availability and reliability.

Demographic Detail. The need for demographic detail also varies from user to user. Some require only total population numbers whereas others require breakdowns by age, sex, race, and ethnicity. Some need age data in single-year groups; for others, five- or 10-year age groups are sufficient. Some require projections of specific population subgroups such as college students, military personnel, seasonal residents, and persons who are disabled. Others require projections by income, education, occupation, poverty status, or other socioeconomic and demographic characteristics. Again, the potential for variation in user needs is virtually unlimited.

Temporal Detail. By *temporal detail*, we mean the length of the projection horizon and the length of time interval between projection dates. Some data users need projections for one- or two-year horizons, some need projections for five- or 10-year horizons, and a few need projections for horizons stretching 50 years and beyond. Some need projections in five- or 10-year intervals, some need annual projections, and a few need quarterly or even monthly projections. The longer the horizon and the shorter the interval, the greater the potential usefulness of a set of projections to a broad range of data users. However, data and techniques that are satisfactory for short-range projections may be unsatisfactory for long-range projections. In addition, data and techniques suitable for projections made in five- or 10-year intervals may be unsuitable for monthly, quarterly, or annual projections.

Meeting User Needs. The needs of the largest number of potential data users can be met (at least theoretically) by making projections that are highly disaggregated by geographic area and demographic characteristic and that cover long time horizons in frequent intervals. Armed with these building blocks, data users can put together projections that cover the specific geographic areas, demographic characteristics, and projection horizons they need. For example, block projections can be added together to provide projections of school districts, traffic analysis zones, and market areas within a county, using whatever demographic categories and time horizons might be required.

The greater the degree of disaggregation, however, the greater the data requirements, the lower the data reliability, the higher the production costs, and the lower the forecast accuracy for each detailed category. These are strong incentives *against* producing highly disaggregated projections. As a result, most producers of general-purpose projections provide projections that cover only a limited number of geographic areas, demographic categories, and time horizons.

The Census Bureau, for example, makes projections by age, sex, race, and Hispanic origin for the nation and each state, but does not make projections for any substate areas. Many state demographic agencies make projections by age and sex (and sometimes by race and/or Hispanic origin) for counties within their state, but few make projections for subcounty areas. Some local governments make projections (with widely varying degrees of demographic detail) for census tracts, traffic analysis zones, or other subcounty areas within their boundaries. Several private companies make projections for all counties in the United States, and a few break them down into a variety of subcounty areas.

Projection horizons are often longer for national projections than for state and local projections. Some projections are made in one-year intervals, but five-year intervals are the most frequently used. As discussed in Chapter 11, interpolation procedures can be used to transform projections made in five- or 10-year intervals into projections for intervening years.

The most basic criterion for judging the potential usefulness of a set of population projections, then, is whether those projections provide the level of geographic, demographic, and temporal detail needed for any particular purpose. If the projections cannot at least come close to meeting those requirements, they will not be very useful regardless of how well they do with respect to other evaluation criteria. General-purpose projections will be able to meet the needs of many data users for many purposes, but some projects will require creating projections specifically for the purposes at hand.

Face Validity

By *face validity*, we mean the extent to which a projection uses the best methods for a particular purpose, is based on reliable data and reasonable assumptions, and accounts for the effects of relevant factors. Because of the effects of population and geographic size, evaluating face validity is considerably more complex and time-consuming for small areas than for large areas.

Choice of Methods. The face validity of a method depends primarily on the purposes for which the projections will be used. All of the methods discussed in this book can be used for projecting total population; even simple extrapolations of recent trends will be acceptable for many purposes. For projections by age group, the analyst must account for shifts in age structure over time; this implies the use of some variant of the cohort-component method. For projections of the components of growth, the model must distinguish among the effects of fertility, mortality, and migration. Projections that incorporate interactions between economic and demographic variables require the use of structural models.

Does the degree of complexity or sophistication affect the face validity of a projection method? We believe that it does, but only to the extent to which complexity or sophistication is required to accomplish the purposes for which the projections will be used. For projections used strictly as forecasts of total population, neither a sophisticated structural model nor a complex multiregional model is necessarily better than simple extrapolations of recent trends. For projections that trace out the implications of alternative economic or demographic scenarios, however, structural models or relatively complex cohort–component models will be required. The face validity of a particular model or technique cannot be generalized; rather, it is conditional upon the specific purposes for which the projections will be used.

Data and Assumptions. Face validity is affected by the quality of the data and assumptions used to create the projections. Although they are not perfect, data from the short form of the decennial census are generally quite accurate, especially for larger geographic regions. Long-form data are somewhat less accurate (especially for small areas) but are still reliable in most instances. Postcensal population estimates are somewhat less accurate than decennial census data, especially for rapidly growing or declining areas. Vital statistics data are highly accurate for states and counties but are less accurate for subcounty areas. The quality of data used in structural models varies by type and by geographic area. An important part of assessing the face validity of population projections, then, is evaluating the quality of the input data and making adjustments when necessary to correct for errors.

The timeliness of input data may also affect face validity. Demographic data vary considerably in terms of time lags and frequency of release. Birth and death data are typically available at least annually, but migration data from the decennial census are available only once every 10 years. In addition, migration data from the decennial census are not tabulated and released until three or four years after the census is conducted. Consequently, projections made in 2002 and 2003 may still be using migration data from 1985–1990. These data may no longer provide a reliable basis for projecting future migration flows.

The application of any projection method requires making certain assumptions. Cohort-component projections require assumptions regarding future levels of fertility, mortality, and migration. Structural models require assumptions regarding the form of the model, the choice of independent variables, and the estimation of parameters. Even simple extrapolation methods are based on assumptions regarding the length of the base period and adjustments for special circumstances. Assessing the "reasonableness" of the underlying assumptions is an important aspect of evaluating face validity.

Accounting for Relevant Factors. Face validity is also determined by the extent to which the projection methodology accounts for the impact of factors that affect population change. Drawing on the discussion by Murdock et al. (1991), we suggest that the following factors may impact small-area projections:

1. Physical features such as the size of an area and the prevalence of potentially growth-constraining factors such as flood plains, lakes, mountains,
   and environmentally protected areas.
2. Location characteristics such as distances from recreational areas, major
   employers, and shopping centers.
3. Land use patterns and policies, including population density, land use
   plans, and zoning or regulatory restrictions.
4. Housing characteristics such as housing density, household size, and
   housing units by type (e.g., single family, multifamily, mobile home).
5. Transportation characteristics such as current and likely future access to
   major highways, airports, railways, and other modes of transportation.
6. Socioeconomic characteristics such as income, education, occupation,
   and poverty status.
7. Population characteristics such as population size, rate of change, distribution within the area, and composition (e.g., age, sex, race, ethnicity).
8. Demographic processes (mortality, fertility, migration).
9. Special populations such as persons who reside in prisons, college dormitories, and military barracks.

We are not suggesting that all these factors must be accounted for in every set
of projections, of course. In many instances, the information contained in a simple
trend extrapolation or cohort-component model will be sufficient to produce
reasonable projections. In other instances, structural models may be able to
account for the relevant factors. For states and most counties, additional adjustments to the basic projection model are not likely to be needed. For some counties
and subcounty areas, however, such adjustments may be necessary, especially to
account for the impact of unique events, special populations, and potential growth
constraints.

Two examples illustrate how these factors might be accounted for. Suppose
that county projections are made in five-year intervals from 2000 to 2020 based on
historical data from 1995–2000. Suppose further that a state prison that houses
1,000 inmates was built in a small county in 1997. If the addition of those inmates
is not explicitly accounted for in the base data, the projections would in effect
assume that 1,000 inmates will be added to the population every five years
between 2000 and 2020. This is probably not a reasonable assumption. This effect
can be accounted for by taking the inmates out of the base data, making the
projections based on the remaining data, and adding independent projections of
inmates as a final step in the projection process (see Chapter 11 for further details).

As a second example, suppose that projections for census tracts (CT) are
based on population trends from 1990–2000. Suppose further that CT 123 grew
very rapidly during that period, from 1,000 persons living in 400 housing units to

5,000 persons living in 2,000 housing units. If no adjustments were made, it would be projected that the tract would continue growing rapidly in future decades. This will not be a realistic assumption if no more vacant residential land is available in CT 123. Situations like this can be dealt with by introducing constraints imposed by the availability of vacant land, zoning restrictions, topographical features, and the like.

Accounting for relevant factors is not always easy. In many instances, the necessary data are not available. In other instances, the data are out-of-date or otherwise unreliable. Even when reliable data are available, models that can use those data to improve the quality of population projections may not exist: Simply knowing that a variable has a particular effect on population growth does not mean that such knowledge can be used to improve the basic projection model. As shown in Chapter 13, population projections based on simple methods that incorporate relatively little data are at least as accurate (on average) as projections based on complex methods that incorporate much larger amounts of data.

Plausibility

By *plausibility*, we mean the extent to which a projection is consistent with historical trends, with the assumptions inherent in the model, and with projections for other areas. Plausibility is closely related to face validity; in fact, the two may be thought of as opposite sides of the same coin. Face validity focuses on the inputs into the projection process, whereas plausibility focuses on the outcomes. If a projection is not based on valid data, techniques, and assumptions, it is not likely to provide plausible results.

Plausibility, of course, is a subjective concept. Just as beauty is in the eye of the beholder, so too is plausibility. A trend that appears eminently plausible to one observer may seem totally implausible to another. How can plausibility be evaluated?

There are a number of ways. One is to compare numerical tables that summarize historical and projected values of key variables. For example, suppose that we want to evaluate a set of county projections. We could construct a table that shows the average annual change in total population projected for each county for future time periods (e.g., 2000–2005, 2005–2010, 2010–2015 ...) and compare those changes with the changes observed during several historical periods (e.g., 1990–1995 and 1995–2000). We could construct another table that compares the age, sex, and race distributions projected for future years (e.g., 2010 and 2020) with those observed in the past (e.g., 1990 and 2000). Are the projections consistent with the underlying assumptions? Are projected changes consistent with those observed in the past? If not, what are the reasons for the differences? Is one of the assumptions invalid? Have some special circumstances been overlooked? Were

---

Box 12.1

Plausibility and Assumption Drag

Assumption drag, defined as "the continued use of assumptions long after their validity has been contradicted by the data" (Ascher, 1978, p. 53), is a common problem in forecasting. It may be caused by several factors: First is the socialization of experts, determined by their education, training, and association with other experts. The "received wisdom" in any field is not often questioned. Second, if recent data contradict longstanding assumptions, they are often viewed as temporary deviations from trends rather than as changes in the trends themselves. Third, there are often delays in collecting and disseminating data that lead to substantial lags between the point at which trends start to change and the point at which those changes are finally incorporated into forecasting models.

Assumption drag means that projections that are consistent with past population trends may not provide the best forecasts of future population change. Analysts must consider this possibility when evaluating the plausibility of population projections, especially projections used as forecasts. All objective projection methods—even cohort-component and structural models—are extrapolations of one type or another. The critical question is, how can we tell when the model's underlying trends have started (or will start) to change? This is the most difficult question in population forecasting (or any other type of forecasting, for that matter).

---

errors made while entering data or writing computer programs? Answering questions like these provides one type of "plausibility check."

Plausibility can also be evaluated by comparing projections for one area with those for another. For example, trends for one county can be compared with trends for another county or for the state as a whole. Are the changes in total population size projected for one area consistent with those projected for another? What about projected changes in the age, sex, and race distribution? Can reasonable explanations be given for diverging trends?

Checking for consistency between projected and historical values and comparing projections for one area with those for another requires a substantial investment of time and effort, but has a potentially large payoff. Given their subjective nature, however, plausibility checks must be viewed as suggestive rather than conclusive. They provide hints and clues but cannot "prove" that one set of projections is better than another. In particular, relying too much on comparisons with past trends might cause the analyst to miss the beginning of a new trend (see Box 12.1).

Costs of Production

The costs of production for a set of population projections are determined primarily by labor costs. A great deal of time must be spent considering all of the relevant details involved in producing a set of projections; collecting, verifying, and cleaning up the input data; putting together a projection model; and evaluating the plausibility of the results. Other costs (e.g., computer hardware and software, purchases of proprietary data) are usually small in comparison.

Very little research has focused on the costs of producing population projections. Just how high are those costs and how do they vary by method, level of geographic and demographic detail, and frequency of application? Logic and personal experience suggest that costs increase with the degree of methodological complexity, with the level of geographic and demographic detail, and with the attention paid to special populations and unique events. However, costs can be expected to decline with the number of times a specific application is repeated; it takes more time to produce a set of projections the first time than to repeat the process additional times.

Can economies of scale reduce costs of production; that is, can projections for a large number of places be made for a lower average cost than projections for a small number of places? One study reported that producing a set of cohort-component projections for counties required about 2,000 person-hours in Ohio and 1,000 person-hours in Washington (Swanson & Tayman, 1995). Because Ohio has about twice as many counties as Washington (88 compared to 39), these results suggest that economies of scale have little impact on the costs of production. If projections are made simply by feeding data into a projection model and spitting out the results, economies of scale will have a large impact on the costs of production. If attention is paid to the reliability of the input data, the potential impact of local characteristics, and the plausibility of the projection results, however, the benefits of economies of scale are likely to be small.

Further research on the costs of production would be very helpful. Other things being equal, lower costs are preferable to higher costs. Other things, however, are rarely equal. Trade-offs must be made between costs of production and other attributes of population projections. Assessing the costs of production—and their relationship to other projection attributes—is central to the evaluation process.

Timeliness

There are several aspects to the concept of timeliness. We covered one in the discussion of face validity; namely, the frequency with which input data are released and the time lag between the reference date and the date when the data actually become available. Another is the frequency with which projections are produced. Since the mid-1950s, the Census Bureau has produced two or three sets of state projections each decade. Demographers in some states produce state and county projections annually, some produce them every other year, and some produce them at irregular intervals (Judson, 1997). Frequent revisions are particularly important for small areas because of the volatility in their growth patterns.

A third aspect of timeliness is the amount of time needed to construct the projections. This is determined by the scope of the project and by the number of analysts available to work on it. Production time takes on particular importance when a set of customized projections is created for a specific client. The client (who may be someone within the same organization as the analyst) may require that the projections be completed within a short (perhaps unreasonably short) period. In some circumstances, production time is a major factor in the choice of projection methods.

Ease of Application and Explanation

Ease of application is determined by the amount of time and the level of expertise needed to collect, verify, and adjust the input data; develop a projection model; and generate the desired projections. This criterion will be particularly important for analysts who have limited training or expertise in producing population projections or who face severe time or budget constraints. At present, no widely available projection software package can be implemented quickly and easily. Instead, the analyst generally must develop a set of computer algorithms specifically for the project at hand. We expand on this point in Chapter 14.

Ease of explanation refers to the extent to which data users can be provided with a clear description of the data sources, assumptions, and techniques used to produce the projections. For some data users, this criterion is irrelevant. They are interested only in the projections themselves, not in the way they were produced. Other data users, however, can truly evaluate (and properly use) a set of projections only if they understand how those projections were made. Indeed, some may have little or no use for projections based on unknown methods or "black-box" models. For those data users, the clearer and more complete the description of the methodology, the more valuable the projections (Rainford & Masser, 1987).

Usefulness as an Analytical Tool

Population projections are used most frequently as forecasts of future trends in population size, distribution, and composition. However—as discussed in Chapter 1—they are also used to analyze the components of growth, trace out the effects of recent trends or specific changes in those trends, demonstrate the sensitivity of population growth to particular variables or assumptions, and relate changes in demographic variables to changes in economic or other variables. In some circumstances, the extent to which projections can be used for these purposes is the main determinant of their usefulness.

Projections can answer a wide variety of questions. What impact would a 10% decline in the birth rate have on future population size and composition? What would be the impact of eliminating a particular cause of death? How would the expansion of a major employer affect an area's migration rates? What would the continuation of current growth rates mean for future water consumption? How is population aging likely to affect the inflow and outflow of funds in the Social Security System? The answers to these and similar questions can teach us a great deal about the determinants and consequences of population growth and demographic change.

Political Acceptability

Population projections are never produced in a vacuum. They are influenced by the context in which they are produced and by the perspectives of those who produce (or approve) them. Cohort-component models are based on decisions regarding future mortality, fertility, and migration rates. Structural models are based on decisions regarding which variables to include and how to estimate the parameters. Even simple extrapolation methods involve choices regarding data and technique(s), length of base period, and adjustments for special populations, unique events, and potential growth constraints. All projections are political statements in the sense that they are based on a particular view of population change.

The political acceptability of population projections can be interpreted in several ways. One is the extent to which projections are acceptable to the persons or agencies sponsoring the projections. Sponsoring parties are often federal, state, or local government agencies but can also be businesses or nongovernmental organizations. These parties may have strong vested interests in the projected numbers and may seek to influence those numbers. Another interpretation of political acceptability is the extent to which projections are accepted as unbiased, reasonable, or authoritative by data users and members of the general public. This will be determined by the reputation and track record of the analyst or agency that produces the projections. Both interpretations are useful, but we focus primarily on the first in this discussion.

As Moen (1984) pointed out, population growth and distribution are deeply embedded in politics. A county government might want projections to show that the county has a rapidly growing population attractive to new businesses. An environmental group might want them to show that there is a need for growth restrictions or more stringent pollution controls. A real estate developer might want them to show that there is a need for additional housing and roads. A school board or parent group might want them to show that there is a need for another elementary school. In these examples, projections are meant to play an active rather than a passive role; that is, they are meant to *influence* future growth rather than simply *chart the likely course* of that growth.

Political considerations can create problems for analysts who construct projections. Dennis (1987) described a case in which three governmental agencies were given the task of constructing projections of vehicle miles traveled in the Denver metropolitan area. It was anticipated that these projections would have a major impact on the funding and planning activities of each agency. In the early stages of the projection process, the projections were based on parameter estimates developed jointly by the technical staffs of the three agencies. These estimates were developed in a "veil of ignorance," or without knowing their impact on the final projections. This approach is considered an excellent way to minimize bias and ensure objectivity. However, an impasse developed when the results were politically unacceptable to one of the agencies. Eventually, the impasse was broken only after decision-making power was passed to a different governmental agency, and the final decision was based on political considerations rather than technical merits.

Tayman and Swanson (1996) cited an example from Detroit, where the technical staff of the planning department produced a set of projections that showed a population loss between 1980 and 1990. Those projections were consistent with previous trends but were unacceptable to political decision makers; consequently, they were revised upward to show a population increase. This revision was based strictly on political considerations; all of the technical factors pointed to a population decline. (Later it was found that the original projections were considerably more accurate than the revised projections as forecasts of the future population.)

McKibben (1996) described a situation in which political opposition was successfully overcome. A small county in Indiana had been experiencing steady declines in school enrollment for years. The school board formulated a long-term building and consolidation plan that called for closing three small elementary schools located in rural areas of the county. This plan was strongly opposed by a sizable segment of the rural population and by several members of the school board, who believed the widely publicized "echo of the baby boom" would produce enough enrollment growth to justify renovating those schools and keeping them open. A group of consultants was hired by the school board to make 10-year projections of school enrollment by grade. Their projections showed further declines in school enrollment and were initially met with widespread disbelief and fervent opposition. However, by providing a clear description of the data, techniques, and assumptions used in developing the projections, the consultants were able to convince most of the skeptics that future declines in school enrollment were more likely than future increases.

The potential conflict between political and technical considerations raises a number of difficult questions. What role should political considerations play in the projection process? How can the (perhaps conflicting) viewpoints of a variety of interest groups be incorporated? At what point does the incorporation of political considerations cause the projections (and the analyst) to lose their credibility? What are analysts to do when the person who signs their paychecks wants to change the projections for purely political reasons?

In our experience, major conflicts between political and technical considerations occur relatively rarely. When they do occur, however, they create thorny ethical and procedural dilemmas. In these circumstances, balancing political acceptability with technical legitimacy may be the most difficult part of the projection process. In the long run, we believe that the negative consequences of producing politically motivated projections far outweigh whatever short-run benefits they might provide.

Political influences are not uniformly negative, however. There are circumstances in which such influences have a *positive* impact on the validity and usefulness of population projections. For example, this may occur when projections are produced as part of a comprehensive urban plan. Tayman (1996b) described the interplay between the construction of population projections and the development of a growth management strategy in San Diego County. This strategy incorporated policies related to housing, public transportation, commuting, employment, government services, environmental considerations, and alternative land uses. The development of several sets of hypothetical projections allowed public officials to observe the potential effects of different policies and to choose the policies expected to be most beneficial for the county. The projections ultimately adopted were those consistent with the policies that were to be implemented.

The Tayman study illustrates the "active" approach to population forecasting. Under this approach, political decision makers first decide which future outcomes are the most desirable and then design policies to achieve those outcomes. If the policies prove successful, projections consistent with those policies will be more realistic than projections that ignore the political context in which they were made. In circumstances like these, incorporating the influence of political considerations improves forecast accuracy and enhances the overall usefulness of the projections.

As we have shown, political factors can have either a positive or negative impact on the validity and usefulness of population projections. When evaluating projections, then, data users must be aware of the context in which they were made. Who made the projections? Why were they made? What roles were they expected to play? Did the producers have a vested interest in the projection results? Did they provide a clear description of the methodology and a convincing explanation for using particular methods and making particular assumptions? The answers to these questions will provide important information for judging the validity of the projections.

Forecast Accuracy

The final criterion for evaluating population projections is forecast accuracy. For many data users, this will be the most important criterion. A recent study of forecasting methods in a variety of fields found that accuracy was more important to data users and producers than any other criterion (Yokum & Armstrong, 1995). Other studies have reported similar results (e.g., Carbone & Armstrong, 1982; Mentzer & Kahn, 1995).

To summarize the empirical evidence briefly, we can say that forecast errors are generally larger for small places than for large places; are generally larger for places that have very high or negative growth rates than for places that have moderate, positive growth rates; generally increase with the length of the projection horizon; and vary from one launch year to another. The degree of complexity or sophistication of the methodology, however, seems to have no consistent impact on forecast accuracy, at least for projections of total population. Given its importance, we take an in-depth look at forecast accuracy in the next chapter.

A BALANCING ACT

All of the criteria discussed before are potentially important for choosing the data, techniques, and assumptions that will be used in constructing a set of population projections or for evaluating a set produced by someone else. The relative importance of each criterion, however, varies according to the purposes for which the projections will be used.

The provision of necessary detail is essential for all purposes. If data for the relevant geographic areas, demographic categories, and time periods are not available, the projections clearly will not be very useful. Face validity, plausibility, and timeliness are of almost universal importance; exceptions might occur when projections are used simply to illustrate the outcomes of various hypothetical scenarios or to push a particular political agenda. Ease of application and costs of production generally do not matter to the data user but are important to the producer. In fact, these criteria may drive methodological decision making when time is limited or budgets are tight. Ease of explanation is unimportant for some data users but critical for others. Political acceptability and analytical usefulness are essential in some circumstances but irrelevant in others. Forecast accuracy may be the most important criterion when projections are used to guide decision making but is irrelevant when projections are used for simulations or as political propaganda. Indeed, there are circumstances in which planning and intervention may be intended to *prevent* projections from providing accurate forecasts (Isserman, 1984).

Choosing the relevant criteria for evaluating a set of projections is clearly a balancing act. Some criteria may be much more important than others, and decisions based on one criterion may be inconsistent with decisions based on another. Choices must be made regarding which criteria are most important for a particular set of projections and—when they conflict with each other—which to rank ahead of the other. An optimal projection strategy can be chosen only after weighing the relative importance of each of the evaluation criteria.

COMPARING METHODS

Once the relevant evaluation criteria have been chosen, a second type of balancing act occurs as the specific data, methods, procedures, and assumptions used to create the projections are chosen. How do various projection methods stack up according to the criteria discussed in this chapter? Table 12.1 summarizes our views regarding the characteristics of the projection methods covered in this book. These rankings are somewhat imprecise because of the potential variability in the ways each method can be applied. For example, the cohort–component method can be applied by using simplified Hamilton–Perry procedures or by

Table 12.1. Ranking of Projection Methods

Evaluation criteria	Simple extrapolation	Complex extrapolation	Cohort-component	Structural models
Geographic detail	****	***	***	****
Demographic detail	*	*	****	***
Temporal detail	****	****	****	****
Face validity	—	—	—	—
Plausibility	—	—	—	—
Cost of production	****	***	**	*
Timeliness	****	***	**	*
Ease of application	****	***	**	*
Ease of explanation	****	**	***	*
Usefulness as analytical tool	*	*	***	****
Political acceptability	—	—	—	—
Forecast accuracy	—	—	—	—

Note: **** Top ranking (performs well); *** second ranking; ** third ranking; * lowest ranking (performs poorly); — cannot generalize.

using complex multiregional models. However, these rankings will give the reader a quick overview of the strengths and weaknesses of the different approaches to constructing population projections.

Provision of Detail

Geographic Detail. Trend extrapolation methods have the smallest data requirements of all of the methods that can be used for projecting total population. The simplest methods require data from only one or two points in time; more complex methods (e.g., time series models) require data from a number of points. Because total population data are readily available (or can be developed) for many different levels of geography and for many different points in time, trend extrapolation methods perform very well in providing projections for a wide variety of geographic areas, including very small areas.

Some cohort-component and structural models do not meet this criterion quite as well. Complete cohort-component models require mortality, fertility, and migration data, as well as population data by age, sex, and perhaps other characteristics as well. Structural models require historical and projected data for the independent variables. These data are often unavailable or of questionable reliability for subcounty areas. However, the Hamilton–Perry method and urban systems models have been applied to very small geographic areas. Adjustments may have to be made, but cohort-component and structural models can be used for projections that cover a wide variety of geographic areas.

Demographic Detail. One of the major advantages of the cohort-component method is that it can provide projections by age, sex, race, and other population characteristics. Structural models can also be used to project population characteristics but are generally used in conjunction with a cohort-component model. Both types of models perform very well in providing projections of demographic characteristics.

Trend extrapolation methods do not meet this criterion nearly as well. They are generally used only for projections of total population or for projecting birth, death, or migration rates; consequently, they generally provide no projections of demographic detail. Could this shortcoming be overcome? Could trend extrapolation methods be applied to subgroups of the population, rather than to the population as a whole? For groups *not* differentiated by age (e.g., females, blacks, Hispanics), trend extrapolation methods could potentially be applied because people generally do not change from one group to another (e.g., female to male, black to white). We are aware of only a few instances in which trend extrapolation methods have been used in this manner (e.g., Leach, 1981; Smith & Nogle, 1999), but there are circumstances in which this approach might be useful (e.g., projections of race groups for very small areas). For age projections, however, some type of cohort approach must be used. Trend extrapolation methods applied to specific age groups are not likely to provide acceptable results (e.g., Long, 1995).

Temporal Detail. Trend extrapolation, cohort-component, and structural models are about equal in their ability to produce projections that cover specific intervals and time horizons. Some methods that are acceptable for short projection horizons, however, may not be valid for long projection horizons. Chapter 13 discusses why this might be true.

Face Validity and Plausibility

Many of the factors that affect face validity are the same for trend extrapolation, cohort-component, and structural models. All three approaches require paying attention to the quality of input data, to the reasonableness of assumptions, and to factors such as physical features, growth constraints, and the impact of special populations or unique events. However, face validity is also affected by the purposes for which projections will be used. All three approaches can provide valid projections of total population. Some type of cohort approach should be used for projecting age groups. Structural models can be used for projections that incorporate interactions between demographic and other variables. The face validity of a projection method cannot be properly judged without considering the purposes for which the projections will be used.

It is also impossible to generalize regarding the plausibility of trend extrapolation, cohort-component, and structural models. All three approaches can produce either plausible or implausible results, depending on the particular assumptions employed. For example, the extrapolation of recent growth rates may provide plausible 20-year projections for a county that has been growing by 1% per year, but implausible projections for a county that has been growing by 10% per year. Similarly, the specific assumptions used for mortality, fertility, and migration rates will determine the plausibility of cohort-component projections. The plausibility of projections from structural models will be determined by the structure of the models themselves and the nature of the assumptions regarding future values of the independent variables. The plausibility of all three approaches can be evaluated only after comparing projected trends with those observed in the past and those projected for other areas.

Costs and Timeliness

Costs of production—which are determined primarily by labor costs—vary tremendously by projection method. Simple extrapolation methods have the smallest data requirements and take the least time to apply. Consequently, they are the least expensive of all of the projection methods. More complex extrapolation methods have larger data requirements and take more time to apply but are still relatively inexpensive. Cohort-component methods require considerably more time for model development and data collection than trend extrapolation methods. Structural models are very time-intensive, requiring a large investment in data collection, model building, and testing. Urban systems models are particularly expensive to develop and implement.

Increasing the level of methodological complexity and sophistication is likely to increase the level of expertise needed to produce a set of projections. Persons who have higher skill levels can command higher wages than persons who have lower skill levels; this is likely to lead to higher costs as well. Incorporating the potential impact of special populations, unique area-specific events, and potential growth constraints adds considerably to the costs of production, regardless of the methodology employed.

There are several aspects to the concept of timeliness. It can refer to the recency of the input data, the frequency with which projections are produced, or the amount of time required by the production process. Trend extrapolation methods perform better than cohort-component and structural models in all three of these aspects. Due to their smaller data requirements and relatively simple mathematical structures, trend extrapolation methods can generally incorporate more recent data, be applied more frequently, and be produced in less time than either cohort-component or structural models. Specific applications of cohort-component and structural models differ considerably from each other, depending on the level of complexity and degree of sophistication employed. In general, the simpler the method, the more timely the projection is likely to be.

Ease of Application and Explanation

Simple extrapolation methods are the easiest to apply and to explain to data users because they have the simplest mathematical forms, the smallest data requirements, and the least amount of disaggregation. More complex extrapolation methods are more difficult to apply and to describe clearly. The cohort-component method is also more difficult to apply because of its large data requirements and complex set of interrelationships. Although the basic concepts underlying this method can be explained easily, a full description of all the data sources, techniques, and assumptions requires a lengthy discussion. Structural models are the most difficult to apply and to explain clearly in terms of their technical details, especially when they involve large numbers of variables and simultaneously determined equations. Interpreting the results of structural models can also be a difficult task.

Usefulness as an Analytical Tool

Simple extrapolation methods are not useful for most analytical purposes. They are not directly related to any theories of population change or to any variables that affect change, and are seldom appied to individual components of population change. Complex extrapolation methods are only slightly more useful. The logistic method can be related to a theory of growth in which the population first grows slowly, then enters a period of rapid growth, and eventually levels off (e.g., Romaniuc, 1990). Time series models can provide prediction intervals that indicate the degree of uncertainty surrounding specific projections (e.g., De Beer, 1993; Lee, 1993). For most analytical purposes, however, neither simple nor complex extrapolation methods are very useful.

The cohort-component method, on the other hand, is very useful. It can determine the proportion of population change caused by each individual component. It can trace out changes in the demographic composition of the population. It can demonstrate the sensitivity of population projections to specific changes in individual components of growth. These analyses raise our understanding of population dynamics and improve our ability to plan for the future.

Structural models are even more useful. Models can be developed that investigate the effects of a variety of economic, social, cultural, and other factors that affect fertility, mortality, migration, or total population change. These models can be constructed to cover both the determinants and consequences of population growth and demographic change. They can be used to create population projections that are consistent with a variety of economic, land use, and transportation projections. Regardless of the accuracy of their forecasts, structural models are extremely useful for analytical purposes.

Political Acceptability

It is impossible to generalize about the political acceptability of population projection methods. In some instances, a method may be unacceptable simply because it cannot produce the type of projections required. For example, simple extrapolation methods will not be acceptable when an interrelated set of economic-demographic projections are needed. In other instances, any method may be politically acceptable if the analyst or agency that produces the projections has a good reputation and a proven track record.

Sometimes political acceptability is determined by the outcome of the projection process rather than by the methods employed. From this perspective, any projection method is acceptable, as long as it provides the desired results. If it does not, it is not acceptable regardless of its technical merits. Consequently, any projection method may be politically acceptable in some circumstances and unacceptable in others.

One warning about the political acceptability of simple methods should be mentioned. Simplicity is sometimes interpreted as simple-mindedness. The use of simple methods may make analysts appear ignorant, unprepared, or incompetent, whereas the use of complex methods may make them appear knowledgeable, well-trained, and thorough. Perceptions may be more important than reality, especially when projections must be approved by an outside group or when they produce controversial results. In some circumstances, structural or multiregional cohort-component models may be more acceptable than other projection methods simply because they are more complex.

Forecast Accuracy

We will provide a detailed discussion of forecast accuracy in the next chapter. To preview that discussion, we can say that no single projection method provides consistently more accurate forecasts of total population than any other method. In specific circumstances, however, some methods perform better than others.

CONCLUSIONS

Evaluating population projections is a two-step process. The first step is to choose the criteria upon which the projections will be evaluated. Potential criteria include the provision of necessary detail, face validity, plausibility, costs of production, timeliness, ease of application and explanation, usefulness as an analytical tool, political acceptability, and forecast accuracy. The choice of criteria will depend on the purposes for which the projections will be used and the constraints imposed on the analyst who produces the projections. For any given purpose, some criteria may be very important, some may be moderately important, and a few may be completely unimportant.

The second step is to use these criteria to guide the selection of projection methods. Simple extrapolation methods are characterized by timeliness, ease of application and explanation, low costs of production, and applicability to very small areas; however, they cannot provide much demographic detail and have little usefulness as analytical tools. More complex extrapolation methods share many of these attributes but typically require more data and modeling expertise. Cohort-component methods are much more costly and less timely but are more useful as analytical tools and can provide a rich array of demographic detail. Structural models are the most data-intensive, time-consuming, and costly but can provide a variety of interrelated projections and offer the greatest analytical usefulness.

Again, we are left with a balancing act. The importance of each criterion must be weighed against the importance of all of the others, and the characteristics of each method must be weighed against the characteristics of all of the other methods. Typically, cost and timeliness must be traded off against analytical usefulness and richness of geographic and demographic detail. The most fundamental task facing the analyst is to choose the optimal bundle of characteristics, based on the resources available and the purposes for which the projections will be used. This choice will guide the analyst through the selection of projection methods, the collection of input data, and all of the other steps of the projection process.

CHAPTER 13

Forecast Accuracy and Bias

Demographers often claim they are not in the business of predicting the future. To emphasize that point, they typically call their calculations of future population "projections" rather than forecasts or predictions. They frequently produce several sets of projections rather than a single set, often without providing any indication of the relative likelihood of their occurrence. This reluctance to predict is not surprising, given the frequency with which past forecasts have been wide of the mark.

But data users want forecasts, not projections. They want the analyst's views of what will actually happen in the future, not some series of hypothetical scenarios or conditional probabilities. In fact, data users generally interpret projections as forecasts regardless of the analyst's intentions and whatever terminology or disclaimers might be used. A basic fact of life for demographers is that their projections become forecasts as soon as they reach the public.

Given the widespread use of population projections as forecasts—and the many planning decisions and funding allocations tied to those projections—it is essential to evaluate the forecast accuracy and bias of the most commonly used projection methods. This chapter provides such an evaluation. We start with a discussion of various statistics that can be used to measure forecast accuracy and bias. Then we provide an overview of the empirical evidence, focusing on the effects of differences in projection methodology, population size, growth rate, length of base period, length of forecast horizon, and launch year. We also consider the possibility of producing forecasts by combining several projections. We close with a discussion of ways to account for uncertainty in population projections. Throughout this chapter, we use the terms *projection* and *forecast* interchangeably because we are interpreting projections as if they were meant to be used as forecasts of future population.

MEASURING ACCURACY AND BIAS

Defining Forecast Error

We define *forecast error* (E) as the difference between the population forecast (F) for a particular geographic area in a particular target year (t) and the actual population (A) for the same area and year:

$$E_t = F_t - A_t$$

For example, if the population of a county had been projected at 55,000 in 2000, and the actual population turned out to be 50,000, the forecast error would be 5,000. If the population had been projected at 45,000, the forecast error would be $-5,000$.

Forecast errors are often expressed as percent differences rather than as absolute differences. This specification is useful when measures of relative error rather than absolute error are needed. The use of percent errors (PE) is particularly helpful when making comparisons across regions because—without adjustments for population size—errors for places with large populations would dominate the effects of errors for places with small populations:

$$PE_t = [(F_t - A_t)/A_t] \times 100$$

In the previous example, if the population of a county had been projected at 55,000 in 2000 and the actual population turned out to be 50,000, the percent error would be $(5,000/50,000) \times 100 = 10\%$. If the population had been projected to be 45,000, the percent error would be $(-5,000/50,000) \times 100 = -10\%$.

Population counts from the decennial census are often used as proxies for the "actual" population of an area. For postcensal or intercensal years, population estimates produced by the Census Bureau, state demographic agencies, or private data companies are typically used. These proxies are not perfect, of course. Census counts are subject to errors that may be substantial for some places or demographic groups; population estimates are subject to even larger errors.

Enumeration and estimation errors can either raise or lower forecast errors, depending on whether they reinforce or offset the differences between projected and actual populations. Because of these offsetting effects, the impact of enumeration/estimation errors on average forecast errors is not likely to be large, especially for longer projection horizons. Most empirical studies do not attempt to adjust for enumeration or estimation errors when evaluating population forecast accuracy.

Common Error Measures

Many error measures can be found in the general forecasting literature (e.g., Armstrong, 2001b; Armstrong & Collopy, 1992; Fildes, 1992; Mahmoud, 1987;

Makridakis, Wheelright, & Hyndman, 1998). We will describe a number of measures, including those most commonly used to evaluate population forecasts.

The first two measures refer to the average error for a set of n individual forecasts:

$$\text{Mean Error (ME)} = \frac{\Sigma E_t}{n}$$

$$\text{Mean Absolute Error (MAE)} = \frac{\Sigma |E_t|}{n}$$

The first measure takes account of the direction of errors; consequently, positive and negative errors offset each other. In fact, they could offset each other completely, resulting in a ME of zero even when individual errors are large. For example, three forecasts with errors of 400, 200, and -600 would yield a ME of zero.

The second measure ignores the direction of the error, so positive and negative errors do *not* offset each other. This measure—sometimes called the *mean absolute deviation*—shows the average difference between forecasted and actual populations, regardless of whether the forecasts were too high or too low. Using the example cited above, forecasts that have errors of 400, 200, and -600 would yield a MAE of 400.

These measures are based on the numerical differences between projected and actual populations; they do not account for differences in the size of the populations being projected. Yet a forecast error of 1,000 has a very different meaning for a place with 2,000 residents than a place with 200,000 residents. The next two measures account for population size by focusing on percent errors rather than numerical errors:

$$\text{Mean Algebraic Percent Error (MALPE)} = \frac{\Sigma \text{PE}_t}{n}$$

$$\text{Mean Absolute Percent Error (MAPE)} = \frac{\Sigma |\text{PE}_t|}{n}$$

The MALPE (often called the *mean percent error*) is a measure in which positive and negative values offset each other. Consequently, it is often used as a measure of bias (e.g., Keilman, 1999; Smith, 1987; Tayman & Swanson, 1996). A positive MALPE reflects a tendency for forecasts to be too high and a negative MALPE reflects a tendency for forecasts to be too low. The proportion of positive errors (%POS) or negative errors (%NEG) are other commonly used measures of bias (e.g., Smith & Sincich, 1992; Voss & Kale, 1985; White, 1954).

The MAPE, on the other hand, is a measure in which positive and negative values do *not* offset each other. It shows the average percent difference between forecasts and actual populations, regardless of whether the individual forecasts were too high or too low. The MAPE is a widely used measure of forecast accuracy in evaluations of population projections (e.g., Ahlburg, 1982; Isserman, 1977; Long, 1995; Smith & Sincich, 1992; Tayman & Swanson, 1996) and in the general forecasting literature (e.g., Armstrong, 1983; Ashton & Ashton, 1985; Mahmoud, 1984; Makridakis, Andersen, Carbone, Fildes, Hibon, Lewandowski, Newton, Parzen, & Winkler, 1982; Weller, 1990).

Sometimes it is important to use error measures that give more weight to large errors than to small errors; for example, when a large error has a disproportionately large impact on the cost of being wrong. In these situations, the following measures can be used:

$$\text{Mean Squared Error (MSE)} = \frac{\Sigma (E_t)^2}{n}$$

$$\text{Root Mean Squared Error (RMSE)} = \sqrt{\frac{\Sigma (E_t)^2}{n}}$$

Although these two measures are commonly used in general forecasting applications (e.g., Armstrong & Collopy, 1992; Mahmoud, 1987; Makridakis et al., 1998), they are less useful for evaluating population forecast errors because results for areas that have large populations swamp the results for areas that have small populations. This problem can be handled by using percent errors rather than absolute errors. A number of studies have used the Root Mean Squared Percent Error (RMSPE) to evaluate population forecasts (e.g., Keilman, 1990; Smith & Sincich, 1992):

$$\text{Root Mean Squared Percent Error (RMSPE)} = \sqrt{\frac{\Sigma (PE_t)^2}{n}}$$

Some accuracy measures focus on other aspects of the distribution of errors rather than the mean value. The median absolute percent error (MEDAPE) is the error which falls right in the middle of the distribution: half of the absolute percent errors are larger and half are smaller (e.g., Tayman, 1996a). This measure is useful when the objective is to highlight the "typical" error and ignore the effects of outliers. The 90th percentile error (90PE) is the absolute percent error that is larger than exactly 90% of all other absolute percent errors. This measure gives an indication of the range of errors and can be used to construct prediction intervals (e.g., Smith, 1987; Smith & Sincich, 1988; Tayman et al., 1998). We return to this topic later in this chapter.

Other error measures can also be used. Theil's U-statistic measures the difference between errors produced by a formal forecasting method and a naïve alternative, such as the assumption that no change will occur. This statistic squares the errors so that large errors are given heavier weights than small errors (Theil, 1966).

The proportionate reduction in error (PRE) also shows the extent to which a forecast can improve on the naïve assumption of no change, but without giving heavier weights to large errors (Tayman & Swanson, 1996). To adjust for the impact of outliers, some analysts have constructed mean errors based on statistical transformations of absolute percent errors (e.g., Mahmoud, 1987; Tayman & Swanson, 1999).

All of the measures discussed before focus on differences in population levels in the target year. This is the approach most commonly used to evaluate population forecast accuracy. An alternative approach focuses on differences between projected and actual *growth rates* rather than differences between projected and actual *population sizes*. Keyfitz (1981), Long (1995), and Stoto (1983) used this approach for evaluating national population projections; Tayman (1996a) used it for evaluating census tract projections. This approach is often used for evaluating short-run economic and business forecasts. We present several examples of this approach later in this chapter.

Selection Criteria

In the general forecasting literature, accuracy measures are used not only to show how well forecasts have performed over the projection horizon but also to show how well a particular model fit the data observed during the base period (e.g., Ascher, 1981; Makridakis, 1986; Pant & Starbuck, 1990). For population projections, however, accuracy measures are generally used solely to show how well (or poorly) projections have performed as forecasts. Given the many different statistics that can be used to measure forecast accuracy and bias, how can one go about choosing the most appropriate measure(s)?

A number of researchers have discussed criteria that might be used to select measures of forecast error (e.g., Ahlburg, 1995; Armstrong & Collopy, 1992; Makridakis, 1993). Several criteria are mentioned frequently. Error measures should be reliable; that is, repeated applications should yield similar results. They should be valid in the sense that they actually measure what they are purported to measure. They should convey as much information about forecast errors as possible and should be easy for the data user to understand. They should be sensitive to differences in error distributions but should not be unduly influenced by outliers.

It has also been noted that error measures should be related to loss functions that specify the cost of forecast errors to data users (e.g., Ahlburg, 1995; Fildes, 1992). For example, if the cost of forecast errors is linear in absolute terms, an error measure such as the MAE is appropriate. If the cost of errors is linear in percent terms, a measure such as the MAPE is appropriate. If the cost of large errors is disproportionately high, a measure that assigns larger weights to larger errors is appropriate (e.g., MSE, RMSE, or RMSPE). If the direction of error is important, measures such as ME, MALPE, or %POS are useful. The best error measure for any given data user, then, depends on the purposes for which the projections are to be used.

Unfortunately, data users rarely know the exact costs of forecast errors. Even if they did, loss functions would still be difficult to estimate because error distributions are usually unknown and rarely conform to standard statistical assumptions (Armstrong & Fildes, 1995). Perhaps more important, population projections are typically produced for general use rather than for a specific use or data user. Consequently, it is impossible to specify a unique loss function that will be best for all data users and for all purposes. For these reasons, loss functions are seldom used to evaluate population forecast accuracy.

A recent survey of the literature on population forecasting reported that the MAPE was used far more frequently than any other measure of error, followed in order by the RMSE, RMSPE, and Theil's U-statistic (Ahlburg, 1995). The MAPE is a good choice as a general accuracy measure because it is "a relative measure that incorporates the best characteristics among the various accuracy criteria" (Makridakis, 1993, p. 528). We believe that it provides a reasonable measure for evaluating accuracy under a wide variety of circumstances. Because of the impact of a few large errors, however, the MAPE may overstate the "typical" error in a set of projections (Tayman & Swanson, 1999).

The MALPE is widely used as a measure of bias (e.g., Keilman, 1990, 1999; Smith & Sincich, 1992; Tayman, 1996a). We believe that it provides a useful way to investigate the tendency for projections to be too high or too low. The next section discusses the empirical evidence on population forecast accuracy and bias, focusing on differences among methods and the effects of differences in population size, growth rate, length of base period, length of projection horizon, and launch year. Due to the frequency of their use in the literature, MAPE and MALPE are the measures we discuss most frequently.

Can valid conclusions be drawn when only a few error measures are analyzed? We believe that they can. Although different error measures provide different perspectives on forecast accuracy and bias, we have found that error patterns are stable across a variety of error measures; that is, the impact of factors such as population size, growth rate, and length of projection horizon on forecast accuracy is generally about the same regardless of which error measure is used. The same is true for alternative measures of bias. Because of these similarities, it is not necessary to analyze a wide variety of error measures to obtain valid conclusions.

When population projections are used to guide real-world decision making, however, the analyst should consider more than a single measure of error. In particular, it is important to consider the cost of being wrong. When population projections are used to plan the location of a retail outlet, construct a new electric power plant, or add a wing to a hospital, what are the implications of inaccurate forecasts? Will the cost of forecasting too little growth be considerably greater than the cost of forecasting too much growth? Will small errors have little impact on costs, but large errors have a disproportionately large impact? These are the types of questions the analyst must answer before using evaluations of population forecast errors to guide decision making.

FACTORS THAT AFFECT ACCURACY AND BIAS

Projection Method

Projection methods differ tremendously in data requirements, mathematical structure, degree of disaggregation, number of variables included, choice of assumptions, and modeling skills required. A common perception among both the producers and the users of population projections is that complex methods are more accurate than simple methods (e.g., Alho, 1997; Beaumont & Isserman, 1987; Irwin 1977; Keyfitz, 1981; Morrison, 1977; Pittenger, 1980). Other analysts have challenged this perception, claiming that it is not supported by empirical evidence (e.g., Ascher, 1978; Kale, Voss, Palit, & Krebs, 1981; Pflaumer, 1992; Smith & Sincich, 1992). Who is right? Do increases in methodological complexity— including the use of structural models—lead to smaller forecast errors? More generally, what does the evidence show about the accuracy of different population projection methods?

Classification of Methods.   Before we can answer these questions, we must develop a framework for evaluating the complexity of various projection methods. Smith and Sincich (1992) classified methods according to their mathematical and causal structures. Mathematical structures range from very simple (e.g., linear extrapolation) to very complex (e.g., multiregional cohort-component models). Causal structures may specify that population variables are affected solely by their own historical values (e.g., ARIMA time series models) or by other variables as well (e.g., structural models). Combining these two characteristics yields a 2 × 2 matrix containing four types of methods: simple extrapolative, simple structural, complex extrapolative, and complex structural.

This classification scheme can be enriched by considering several additional factors. Long (1995) highlighted three types of complexity: model specification, degree of disaggregation, and the selection of assumptions and alternative scenarios. The complexity of a model is determined not only by its mathematical structure, but also by the number of factors it takes into account and the ease with which it can be explained to data users. The degree of disaggregation refers to the level of demographic detail provided by the projections (e.g., age, sex, race). Selection complexity is determined by the manner in which assumptions are made and by the number of alternative scenarios provided. In cohort-component models, for example, assumptions about future fertility, mortality, and migration rates may be based on their most recent values, on time series extrapolations of historical values, or on structural models. The number of scenarios may also vary widely. Recent projections from the Census Bureau provided as many as thirty (Spencer, 1989) and as few as two alternative scenarios (Campbell, 1996).

Other factors could be considered as well, such as linear versus nonlinear models, the level of modeling skills required, and the degree of interaction among variables (Armstrong, 1985; Ascher, 1981). It is unlikely that a standard classification scheme could be developed that would fully cover all the possibilities. Furthermore, as A. Rogers (1995b) points out, the simple versus complex classification is really a continuum rather than a dichotomy. Methods should be defined in relative rather than absolute terms, or as "simpler versus more complex" rather than "simple versus complex." Any given method may be relatively simple compared to one method and relatively complex compared to another.

In this chapter, we classify projection methods as relatively simple or complex according to their mathematical structures, data requirements, degree of disaggregation, and level of modeling skills required. We classify them as structural or extrapolative according to whether they are affected by economic and other variables or solely by their own historical values. Other things being equal, we view structural models as more complex than strictly extrapolative methods. According to these criteria, we rank the projection methods described in this book as follows:

1. *Simple*: Linear extrapolation (LINE), exponential extrapolation (EXPO), constant-share, shift-share (SHIFT), and share-of-growth (SHARE) methods. These methods are mathematically simple and require little input data and few modeling skills.

2. *More complex*: Regression, logistic, and ARIMA time series models. These methods are considerably more complex mathematically and require more input data and modeling skills than simple methods. However, they rely primarily on highly aggregated data and do not account for the effects of other variables.

3. *Most complex*: Cohort-component models, including those that incorporate structural models. These models are mathematically complex, highly disaggregated, and require large amounts of input data. The assumptions used in these models can be relatively simple (e.g., fertility, mortality, and migration rates remain constant at current levels) or complex (e.g., fertility, mortality, and migration rates derived from structural or time series models).

The ranking of specific models and techniques is not always clear-cut. Cohort-component models have greater data requirements and a higher level of disaggregation than ARIMA time series models but require fewer statistical modeling skills. Specific applications of the cohort-component method may themselves vary considerably in simplicity or complexity. Structural models also vary a great deal, from simple recursive models that contain only a few variables to large simultaneous equation models that contain hundreds of variables and parameters.

Projections of Total Population. Table 13.1. summarizes the conclusions of a number of empirical studies comparing the forecast accuracy of two or more population projection methods. Five studies found projections of total population from cohort-component models to be no more accurate than projections from simple extrapolation methods (Kale et al., 1981; Long, 1995; Smith & Sincich, 1992; Stoto, 1983; White, 1954). Two studies found projections from cohort-component models to be no more accurate than projections from complex extrapolation methods (Leach, 1981; Pflaumer, 1992). Two studies found projections from complex extrapolation methods to be no more accurate than projections from simple extrapolation methods (Kale et al., 1981; Smith & Sincich, 1992). Three studies found projections from structural models to be no more accurate than projections based solely on the extrapolation of historical population trends (Kale et al., 1981; Murdock et al., 1984; Smith & Sincich, 1992). In some instances, in fact, projections from simpler methods were found to be *more* accurate than projections from more complex methods (Leach, 1981; Long, 1995; Stoto, 1983).

These studies covered a number of different geographic regions, historical time periods, and measures of forecast error. An example from Smith and Sincich (1992) illustrates typical results for states. They evaluated five sets of state projec-

Table 13.1. Accuracy of Alternative Population Projection Methods:
A Summary of Conclusions from Empirical Studies

Conclusion	Studies
1. Complex methods are no more accurate than simpler methods.	Kale et al., 1981; Leach, 1981; Long, 1995; Pflaumer, 1992; Smith & Sincich, 1992; Stoto, 1983; White, 1954
2. Complex methods are more accurate than simpler methods.	Keyfitz, 1981
3. Structural models are no more accurate than models based solely on historical population data.	Kale et al., 1981; Murdock et al., 1984; Smith & Sincich, 1992
4. Structural models are more accurate than models based solely on historical population data.	Sanderson, 1999(?)

tions; launch years ranged from the mid-1950s to the early 1980s, and projection horizons extended from five to 20 years. Their analysis covered four simple trend extrapolation methods (LINE, EXPO, SHIFT, SHARE); an ARIMA time series model; the Census Bureau's application of the cohort-component method; and employment-based structural models from the Bureau of Economic Analysis (BEA) and the National Planning Association (NPA). Table 13.2 summarizes the results, showing forecast errors averaged across all launch years. Errors for all series of Census Bureau projections have been averaged together and are depicted as CB.

Except for EXPO projections for longer forecast horizons (and, to a lesser extent, SHIFT projections), accuracy levels were similar for all projection methods. For 10-year horizons, MAPEs for all projections fell between 6 and 7%. For 20-year horizons, the range was 11–16% (11–13%, excluding EXPO). The EXPO projections had a consistent upward bias and the ARIMA projections had a consistent downward bias; the other methods displayed little tendency to consistently project too high or too low. Analyses by method, launch year, and forecast horizon found that differences in errors were small and statistically insignificant for almost every possible combination of method, launch year, and horizon. Smith and Sincich concluded that differences in the complexity or sophistication of projection methods did not lead to any consistent differences in forecast accuracy.

We know of only two empirical studies that have not shared this conclusion. Keyfitz (1981) compared national cohort-component projections published by the United Nations in the late 1950s with projections based on the extrapolation of 1950–1955 exponential growth rates. He found that the cohort-component projections had forecast errors smaller than the exponential extrapolations. It should be noted, however, that five-year base periods (such as those used by Keyfitz) often produce larger forecast errors than either 10- or 20-year base periods for long-range projections, especially for the exponential method and in rapidly growing areas (Smith & Sincich, 1990). We return to this point later in this chapter.

Sanderson (1999) evaluated projections of the world population and projections of birth and death rates for several countries. He compared results from standard cohort-component models with results from economic-demographic models and found that projections from the structural models were more accurate than cohort-component projections in the majority of the comparisons. Recognizing the limitations of his very small sample size, however, he stopped short of concluding that structural models *generally* provide more accurate forecasts than extrapolation methods. (This is why we have a question mark by this citation in Table 13.1.) Rather, he concluded that structural models can improve forecast accuracy when projections from structural models are averaged together with projections from other models. We return to the potential benefits of combining forecasts later in this chapter. It should also be noted that Sanderson's study did not consider migration, the most volatile component of population growth at the state and local level.

There is a substantial body of evidence, then, supporting the conclusion that

Table 13.2. Measures of Accuracy and Bias for State
Population Projctions: Averages Covering All Launch Years

Measure	Method	Length of projection horizon (years)			
		5	10	15	20
MAPE	LINE	3.5	6.0	8.0	11.3
	EXPO	3.9	7.0	10.6	16.2
	ARIMA	3.3	6.3	9.1	11.5
	SHIFT	3.8	6.4	9.2	13.4
	SHARE	3.6	6.0	8.2	11.7
	CB	3.7	6.1	8.3	12.4
	NPA[a]	4.3	6.8	8.4	—
	BEA	4.0	6.5	9.1	12.8
RMSPE	LINE	5.1	8.2	10.8	14.3
	EXPO	6.3	11.7	20.2	33.0
	ARIMA	4.6	8.2	11.7	14.8
	SHIFT	5.5	9.3	13.2	18.7
	SHARE	5.2	8.4	11.3	15.2
	CB	5.0	8.2	10.7	15.1
	NPA[a]	5.3	8.5	10.3	—
	BEA	5.8	8.8	11.6	15.2
90PE	LINE	7.7	11.8	26.4	22.3
	EXPO	8.6	13.6	21.3	32.0
	ARIMA	7.2	13.6	18.9	23.6
	SHIFT	8.1	13.1	19.5	27.7
	SHARE	7.8	12.1	17.9	23.4
	CB	8.1	13.2	17.5	24.7
	NPA[a]	8.3	13.4	17.9	—
	BEA	9.7	14.1	18.3	26.1
MALPE	LINE	0.1	−0.5	−1.1	−1.9
	EXPO	1.2	2.4	4.3	7.8
	ARIMA	−1.1	−2.8	−4.4	−6.0
	SHIFT	0.4	0.2	−0.2	−0.8
	SHARE	0.4	0.2	0.2	0.4
	CB	−0.7	−1.1	−0.4	2.4
	NPA[a]	−2.4	−0.9	−0.6	—
	BEA	1.7	−3.6	−2.6	−4.9
% POS	LINE	51.3	46.7	47.5	44.7
	EXPO	59.0	60.4	61.5	60.7
	ARIMA	40.6	40.0	36.0	34.7
	SHIFT	54.0	51.6	51.5	46.7
	SHARE	54.0	51.2	54.0	49.3
	CB	44.4	46.0	50.3	55.7
	NPA[a]	33.0	46.0	49.0	—
	BEA	64.0	34.0	43.0	40.0

[a]20-year projection horizon not available.

Source: S. Smith and T. Sincich (1992). Evaluating the forecast accuracy and bias of
alternative population projections for states, *International Journal of Forecasting*, 8,
495–508, reprinted with permission from Elsevier Science.

more complex methods—including structural models—do not produce more accurate forecasts of total population than can be achieved by using simple extrapolation methods. There is very little evidence to suggest that the opposite is true. We believe that the weight of the evidence is sufficient to conclude that—to date—neither the sophistication of structural models nor the complexity of time series and cohort-component models has led to greater forecast accuracy for projections of total population than can be achieved by simple extrapolation methods.

Why is it that complex methods do not produce more accurate forecasts than simpler methods? We believe that there is a certain irreducible level of uncertainty regarding the future. No projection method—no matter how complex or sophisticated—can consistently improve forecast accuracy beyond that level. Based on the evidence to date, it appears that the relatively small amount of historical information contained in simple trend and ratio extrapolation methods provides as much guidance to this uncertain future as the much larger amount of information contained in more complex and sophisticated methods.

Cohort-component projections are generally no more accurate than trend extrapolations because forecasting fertility, mortality, and migration rates is just as difficult as forecasting changes in total population (perhaps more difficult). This difficulty offsets the potential advantages offered by data disaggregation and the momentum of population aging. Would the application of time series methods or the development of new data sources improve the accuracy of cohort-component projections? We doubt it. The projections would still be based on the extrapolation of past trends, and those trends are likely to be highly correlated with those underlying the simple extrapolation methods. It is unlikely that more complex approaches to extrapolating past trends will lead to any significant improvements in forecast accuracy.

What about structural models? Those evaluated by Kale et al. (1981), Murdock et al. (1984), and Smith and Sincich (1992) were simple deterministic models that tied net migration to exogenous changes in employment. Would more sophisticated structural models lead to improvements in forecast accuracy? Again, we are doubtful. Knowledge regarding the determinants of population change is far from perfect. Consequently, it is impossible to construct models that realistically incorporate the effects of all of the factors that affect population change. Even if such models could be constructed, there is no certainty that past relationships between independent and dependent variables will remain constant in the future. More critical yet, even if those relationships were to remain constant, the future values for the independent variables themselves would still be unknown. Is there is any reason to believe that these variables can be projected more accurately than demographic variables? We do not believe so. In fact, given the relative stability of demographic processes, the opposite is more likely to be true.

We do not mean to imply that all relatively simple methods perform equally well under all circumstances. There are circumstances in which some simple methods produce less accurate or more biased forecasts than other simple methods or more complex methods. We investigate some of these possibilities later in this chapter. "Simple" should not be confused with "simplistic." Informed judgment is needed to determine when and how simple methods can best be applied.

One further caveat should be mentioned. Most empirical studies have focused on projection horizons that ranged between five and 25 years. Little evidence on population forecast accuracy exists for very short horizons (i.e., less than five years) or very long horizons (i.e., greater than 25 years). Thus, our conclusions regarding forecast accuracy refer to projections that span horizons of five to 25 years. We look at the relationship between forecast errors and length of forecast horizon later in this chapter.

Projections of Demographic Characteristics. Many studies have evaluated forecast accuracy for projections of total population, but few have considered the accuracy of projections of demographic characteristics such as age, sex, race, or ethnicity. In a study of national population projections for the Netherlands, Keilman (1990) found that errors were greatest for the youngest and oldest age groups. Projections were generally too high for the youngest age groups and too low for the oldest, which indicates that forecasters underestimated future declines in fertility rates and future increases in survival rates for the elderly. In a later study, Keilman (1999) evaluated projections made by the United Nations for seven regions in the world. Again, he found that errors were generally largest for the youngest and oldest age groups, although the direction of these errors was not the same in every region.

The results presented by Keilman (1990, 1999) were based on analyses of projections for nations and regions of the world, where migration typically plays a small role in overall population change. Consequently, errors in projecting age structure were caused primarily by errors in the mortality and fertility assumptions. What about projections for subnational areas, where migration plays a major role? Smith and Shahidullah (1995) evaluated 10-year projections for census tracts in Florida. They found MAPEs that ranged from 20 to 29% for individual age groups but did not find any consistent relationship between age and forecast errors: MAPEs were largest for ages 25–34 and 65+ and smallest for ages 45–54 and 55–64. On average, MAPEs for individual age groups were about 40% larger than the MAPE for the population as a whole.

Are these results typical? Would similar results be found in different places, time periods, and levels of geography? These results refer solely to differences by age group. What about differences by sex, race, or ethnicity? We simply do not know. More research must be done before we can answer these questions definitively.

One interesting question is whether differences in methodological complexity might have a greater impact on forecast accuracy for projections of demographic characteristics than for projections of total population. Long (1995) provided one answer to this question by comparing two sets of national population projections: the Census Bureau's cohort-component model and a simple model in which the exponential growth rate for each age group in the year immediately before the launch year was extrapolated into the future. Focusing on two age groups (15–19, 60–64), he found that errors for 10-year projections were much larger for the exponential extrapolations than for the cohort-component projections. He concluded that cohort-component models are more accurate than simpler methods for projecting population by age group.

This conclusion may yet prove to be true, but we believe that a better simple model can be developed by using 10 years rather than one year as the base period and by extrapolating by cohort rather than by age group (Hamilton & Perry, 1962). This model was discussed in Chapter 7. Although it breaks the population into age groups, it is still a simple model that requires only data by age (or age and sex) in two consecutive censuses rather than age-specific fertility, mortality, and migration rates and assumptions regarding future changes in those rates. Despite its simplicity, this model incorporates the effects of population aging and provides projections of the demographic characteristics of a population, thereby making up for two of the major shortcomings of simple extrapolation methods. A comparison of age group projections from the Hamilton–Perry method with those from more complex cohort-component models would provide an interesting and potentially useful test of simpler versus more complex methods.

Projections in Other Fields. The discussion so far has focused exclusively on the forecast accuracy of population projections. What about studies of forecast accuracy in fields such as business, economics, political science, sociology, psychology, and meteorology? What does the evidence show regarding the forecasting performance of different methods? In particular, how do simple methods compare with more complex methods and how do structural models compare with strictly extrapolative methods?

Studies that cover variables as diverse as GNP, employment, inflation, housing starts, company earnings, sales of particular products, market share, crime rates, psychiatric diagnoses, and annual rainfall have found results similar to those reported here for population projections. Many analysts have concluded that complex or statistically sophisticated forecasting methods are generally no more accurate than simpler methods (e.g., Armstrong, 1985; Ascher, 1981; Mahmoud, 1984; Makridakis, 1986; Pant & Starbuck, 1990; Schnaars, 1986). Others have concluded that structural models are generally no more accurate than trend extrapolation methods (e.g., Brodie & De Kluyver, 1987; LeSage, 1990; Makridakis, 1986; Pant & Starbuck, 1990). Some have even questioned the value of incorporating expert judgment in the forecasting process, at least beyond some minimal level (Pant & Starbuck, 1990).

There is not unanimity on these conclusions, however, particularly with respect to the benefits of structural models. A number of analysts have concluded that structural models *do* produce more accurate forecasts than strictly extrapolative methods, at least under certain circumstances (e.g., Armstrong, 1985; Batchelor & Dua, 1990; Clemen & Guerard, 1989; Fildes, 1985; Hagerty, 1987; Leitch & Tanner, 1995). For example, West and Fullerton (1996) forecasted a number of economic and demographic variables for metropolitan areas in Florida, using structural models and four different trend extrapolation methods. Their forecasts extended from one to 10 quarters. They found that, on average, the structural models performed better than the extrapolation methods. However, the relative performance of several of the extrapolation methods improved as the forecast horizon increased, so that by the ninth and tenth quarters errors for the extrapolation methods were virtually identical to errors for the structural models. They also found that differences in the complexity of extrapolation methods had no impact on forecasting performance: simple linear and exponential extrapolations performed just as well as more complex time series models.

Most of the studies cited before focused on forecasts of two years or less. This time frame is very different from that used in most studies of population projections, which typically focus on horizons of five years or longer. Consequently, it is important to note that several of the studies that found structural models to produce more accurate forecasts than trend extrapolation methods also found that the superiority of structural models declined as the forecast horizon increased. Clemen and Guerard (1989) and Leitch and Tanner (1995) found that the superiority of structural models disappeared within four quarters; West and Fullerton (1996) found the same result within ten quarters. Armstrong (1985), however, reported exactly the opposite: structural models are likely to be more accurate than strictly extrapolative models for long-term forecasts but not for short-term forecasts.

Evaluating the Evidence. A substantial body of evidence supports the conclusion that more complex methods do not lead to more accurate forecasts of total population than can be achieved with simpler methods. This evidence has been drawn from studies that cover a wide variety of methods, launch years, forecast horizons, and geographic regions. Studies from other fields have found similar results. Although one method may have greater accuracy than another for a particular set of projections, no single model or technique is *consistently* more accurate than all of the others. On the contrary, most models and techniques produce similar results when applied in similar circumstances (e.g., launch year, target year, level of geography).

This does not imply, of course, that complex methods should never be used. There are many purposes for which complex methods and structural models are very useful, such as evaluating components of growth, providing demographic detail, conducting simulations, evaluating alternative scenarios, accounting for the effects of economic variables, and connecting population projections to other types of projections. The political benefits of complex methods may also be important in some circumstances. Complex methods and structural models have several important advantages over simple extrapolation methods. Greater forecast accuracy, however, is not one of them.

Population Size

Many studies—covering a variety of projection methods, geographic regions, launch years, and time horizons—have found that forecast accuracy improves as population size increases (e.g., Isserman, 1977; Murdock et al., 1984; Smith & Sincich, 1988; Tayman, 1996a; White, 1954). This relationship has been found even when the effects of variables such as the population growth rate have been accounted for (e.g., Smith, 1987; Tayman et al., 1998). A number of studies, however, have found that this relationship weakens (or disappears completely) once a certain population size has been reached (e.g., Schmitt & Crosetti, 1951; Smith, 1987; Smith & Shahidullah, 1995; Tayman, 1996a; Tayman et al., 1998). The threshold level at which the relationship begins to weaken varies with the size of the geographic area under consideration. For example, the relationship begins to weaken at a smaller population size for projections of counties than for projections of states. It appears that population size matters and so does the relationship between population size and the size of the geographic area.

A clear relationship between forecast errors and population size is generally found only for measures of accuracy (e.g., MAPE), but not for measures of bias (e.g., MALPE). A number of studies have found no consistent relationship between population size and bias (e.g., Murdock et al., 1984; Smith & Sincich, 1988). Even when a relationship is found, it is often specific to a particular projection method or time period or is caused by a spurious correlation rather than by population size *per se* (e.g., Smith, 1987; Tayman et al., 1998). Although the evidence is not completely clear-cut, it appears that population size has no predictable impact on the tendency for population projections to be too high or too low.

Table 13.3 illustrates the relationship between population size and forecast errors. It shows errors for 10-year projections for counties in the United States (Smith, 1987). For all four projection methods, MAPEs declined steadily as population size increased. MAPEs for the smallest counties were approximately twice as large as MAPEs for the largest counties. Differences in MAPEs among the smallest size categories were quite large, but differences among the largest categories were very small. MALPEs also showed a strong positive relationship

Table 13.3. Errors for 1980 County Population Projections by Size

Population size in 1970	N	MAPE			
		LINE	EXPO	SHIFT	AVERAGE
<5,000	302	20.8	18.0	25.1	21.2
5,000–14,999	918	15.9	14.7	18.4	16.3
15,000–24,999	555	12.6	11.8	14.4	12.9
25,000–49,999	539	11.8	11.3	13.0	11.9
50,000–99,000	324	10.2	11.0	11.3	10.7
100,000+	333	9.3	11.7	10.5	10.3
All counties	2,971	13.7	13.1	15.7	14.1

Population size in 1970	N	MALPE			
		LINE	EXPO	SHIFT	AVERAGE
<5,000	302	−18.0	−14.2	−22.3	−18.2
5,000–14,999	918	−15.0	−13.5	−17.7	−15.4
15,000–24,999	555	−11.7	−10.5	13.5	−11.9
25,000–49,999	539	−9.1	−7.2	−10.0	−8.8
50,000–99,999	324	−5.6	−1.9	−5.7	−4.4
100,000+	333	2.7	7.4	3.3	4.5
All counties	2,971	−10.6	−8.3	−12.3	−10.4

Source: Unpublished tables to accompany S. Smith (1987). Tests of forecast accuracy and bias for county population projections, *Journal of the American Statistical Association*, 82, 991–1003.

with population size, but this relationship was spurious, having been caused by a strong positive correlation between population size and growth rate.

The effect of population size on forecast accuracy can also be seen by comparing errors for different types of geographic areas. For 10-year projection horizons, MAPEs for states generally average 4–8% (e.g., Kale et al., 1981; Smith & Sincich, 1988, 1992; White, 1954). For counties, they generally average 8–14% (e.g., Murdock et al., 1984; Smith, 1987). For census tracts and areas of similar size, they generally average 15–21% (e.g., Murdock et al., 1984; Smith & Shahidullah, 1995; Tayman, 1996a). For areas smaller than census tracts, errors are even larger (Tayman et al., 1998). Clearly, the larger the population of the area to be projected, the more accurate the forecast is likely to be.

Population Growth Rate

Population growth rates also have a strong impact on forecast errors. MAPEs are generally smallest for places that have small but positive growth rates and increase as growth rates deviate in either direction from those levels; that is, there is a U-shaped relationship between forecast accuracy and population growth rates. The largest errors are typically found for places that are either growing very rapidly or declining rapidly. A number of empirical studies have reported this result (e.g., Isserman, 1977; Murdock et al., 1984; Smith, 1987; Tayman, 1996a).

Some studies have evaluated forecast accuracy using growth rates observed during the projection horizon; others have used growth rates observed during the base period. Both approaches are valid for analytical purposes, but we believe that it is more useful to focus on growth rates during the base period because that information is available at the time a set of projections is made. Our discussion in the remainder of this section refers to growth rates during the base period.

Table 13.4 illustrates the relationship between growth rates and forecast errors using the same county projections as Table 13.3. There is a strong U-shaped relationship between growth rates and MAPEs for all four methods. MAPEs were

Table 13.4. Errors for 1980 County Population
Projections by Growth Rate

Population growth rate		MAPE			
1960–1970	N	LINE	EXPO	SHIFT	AVERAGE
<−10%	516	20.5	17.0	25.0	20.8
−10–0%	800	13.3	13.0	15.8	14.0
0–10%	766	10.6	10.5	12.0	10.9
10–25%	558	11.0	10.8	11.9	11.2
25–50%	243	13.9	13.9	14.5	13.9
50–100%	75	19.8	23.7	20.3	20.3
100%+	13	24.1	47.5	24.6	28.3
All counties	2,971	13.7	13.1	15.7	14.1

Population growth rate		MALPE			
1960–1970	N	LINE	EXPO	SHIFT	AVERAGE
<−10%	516	−20.3	−16.6	−24.9	−20.6
−10–0%	800	−12.5	−12.2	−15.2	−13.3
0–10%	766	−8.2	−8.0	−9.8	−8.7
10–25%	558	−4.9	−5.0	−5.3	−4.4
25–50%	243	−6.1	0.6	−4.6	−3.4
50–100%	75	−6.0	12.1	−1.3	1.6
100%+	13	−6.1	40.4	1.4	11.9
All counties	2,971	−10.6	−8.3	−12.3	−10.4

Source: Unpublished tables to accompany S. Smith (1987). Tests of forecast accuracy and bias for county population projections, *Journal of the American Statistical Association*, 82, 991–1003.

smallest for counties that had growth rates from 0 to 25%, were somewhat larger for counties that had growth rates of 25–50% or moderate negative growth rates, and were much larger for counties with growth rates greater than 50% or less than −10%. The effect of very high growth rates on errors was greatest for the EXPO method; the effect of negative growth rates was greatest for the SHIFT method.

Bias is also strongly affected by differences in population growth rates. The MALPEs shown in Table 13.4 have large negative values for counties that lost more than 10% of their populations during the base period, but those values increase steadily as the growth rate increases and eventually reach positive levels for three of the four projection methods. The SHIFT method had the strongest downward bias for counties that lost population and the EXPO method had the strongest upward bias for rapidly growing counties.

What caused the patterns shown in Table 13.4? We believe that they were caused by the tendency for extreme growth rates to regress toward the mean over time (Smith, 1987). As Table 13.5 shows, for 1950–1980 the vast majority of counties that lost population or grew very slowly during one decade grew more rapidly during the following decade (or lost less rapidly). Conversely, the vast majority of counties that grew very rapidly during one decade grew more slowly during the following decade. As a consequence, projections based on periods of very rapid growth are likely to be too high and projections based on periods of large population declines are likely to be too low. This phenomenon of "regression toward the mean" provides an explanation for the U-shaped relationship

Table 13.5. Comparison of Decade Growth Rates for Counties, 1950–1980

Growth rate in decade t	N	Counties with higher growth rate in decade $t+1$ than decade t	
		Number	Percent of counties
<−10%	1,271	1,125	88.5
−10–0%	1,532	1,221	79.7
0–10%	1,354	883	65.2
10–25%	1,031	487	47.2
25–50%	485	157	32.4
50–100%	213	42	19.7
100%+	56	3	5.4
All counties	5,942	3,918	65.9

Source: S. Smith (1987). Tests of forecast accuracy and bias for county population projections, *Journal of the American Statistical Association*, 82, 991–1003. Reprinted with permission from *The Journal of the American Statistical Association*. Copyright 1987 by the American Statistical Association. All rights reserved.

between growth rates and MAPEs and for the positive relationship between growth rates and MALPEs.

Why do growth rates tend to regress toward the mean over time? To answer this question, we must consider the components of growth. Migration is the demographic variable most responsible for differences in growth rates among states and local areas (e.g., Congdon, 1992; Shryock & Siegel, 1973; Smith & Ahmed, 1990). Mortality and fertility rates do not differ nearly as much from place to place as migration rates and do not change as rapidly over time. For states and local areas, then, migration is the demographic variable that usually contributes most to differences in population growth rates and to changes in those rates over time.

For rapidly growing areas to maintain high growth rates, levels of net in-migration must continue increasing year after year. Yet, if out-migration rates are based on the size of an area's population and in-migration rates are based on the size of the population outside that area, levels of net in-migration can increase only if in-migration rates go up or out-migration rates go down (Smith, 1986). Similarly, for areas losing population, levels of net out-migration must eventually decline because the source of out-migrants is growing more slowly than the source of in-migrants. As a result, it is unlikely that an area will maintain an extremely high or low rate of population growth for extended periods of time.

There are several other reasons for expecting extreme migration rates to become more moderate over time. Rapidly growing areas have increasing numbers of "migration-prone" residents who are likely to move again, whereas declining areas have declining numbers of such residents. Some in-migrants may become disenchanted with their new locations and return to their former places of residence. It could also be argued that migration itself is a "self-equilibrating mechanism" that causes the comparative advantage of one area over another to fade with time, thereby leading to more moderate rates of in- or out-migration (e.g., Greenwood, 1997; Hunt, 1993; Sjaastad, 1962).

In summary, theory and empirical evidence suggest that extreme growth rates are likely to regress toward the mean over time. Consequently, projections based on historical growth trends will often be too high for rapidly growing areas and too low for declining or very slowly growing areas. As we explain later in this chapter, this finding provides a basis for developing an alternative approach to constructing population projections.

Length of Horizon

On average, forecast accuracy declines as the projection horizon becomes longer. This result has been found in many studies of population projections (e.g., Keilman, 1990; Keyfitz, 1981; Schmitt & Crosetti, 1951; Smith & Sincich, 1992; White, 1954) and forecasts in other fields (e.g., Ascher, 1981; Batchelor & Dua, 1990; Makridakis, 1986; Schnaars, 1986; Zarnowitz, 1984). Such a result is not surprising, of course. The farther into the future a projection extends, the greater the opportunity for unforeseen events to occur and for the factors that affect population growth to diverge from their predicted trends.

What does the error-horizon relationship look like? A number of studies have found that MAPEs increase about linearly with the forecast horizon (e.g., Ascher, 1981; Kale et al., 1981; Schmitt & Crosetti, 1951; Schnaars, 1986; Smith, 1987; White, 1954). Table 13.2 (earlier in this chapter) provides information on this relationship for a number of projection methods and measures of forecast error.

Smith and Sincich (1991) used statistical tests to analyze the error-horizon relationship in detail. They made projections for states using four simple extrapolation methods (LINE, EXPO, SHIFT, SHARE), 10-year base periods, and launch years at 10-year intervals from 1910 to 1970. Projection horizons ranged from five to 50 years, in five-year intervals. For all methods, they found MAPEs to increase about linearly as the projection horizon increased to 35 years; after 35 years, MAPEs began to deviate from the linear trend (especially for the EXPO technique). Similar results were found for projections for each individual launch year and when states were divided into size and growth-rate categories. Statistical tests provided formal confirmation of this generally linear relationship.

The only exceptions were MAPEs produced by the EXPO projections, which grew at an increasing rate for states with high growth rates during the base period. An approximately linear relationship was found for several other measures of forecast accuracy as well (e.g., RMSPE, 90PE), but it was not as strong as the relationship for MAPE. The MAPE-horizon relationship is illustrated in Figure 13.1 for the LINE and EXPO methods.

Keyfitz (1981) and Stoto (1983) analyzed a large number of population projections for countries. Instead of using the MAPE as a measure of error, they focused on the difference between the projected rate of population increase and the actual rate realized over time. They concluded that this difference tends to remain constant over the entire length of the projection horizon; this conclusion is very similar to Smith and Sincich's conclusion that the MAPE grows linearly with the projection horizon.

Further research will add to our ability to generalize these results. Would a linear relationship be found for other measures of error? Would it be found for other commonly used projection methods? Would similar results be found for small geographic units, which often exhibit high levels of growth-rate volatility over time? Would a linear relationship be found for projection horizons shorter than five years? What about horizons of 50 or 100 years? There are still many gaps in our understanding of error-horizon relationship.

The empirical regularities discussed before refer only to measures of accuracy, not to measures of bias. Studies that also considered bias reported no clear, consistent relationship between forecast errors and the length of the projection horizon. Smith and Sincich (1991) found that MALPEs differed from one projec-

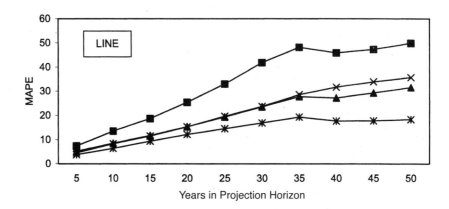

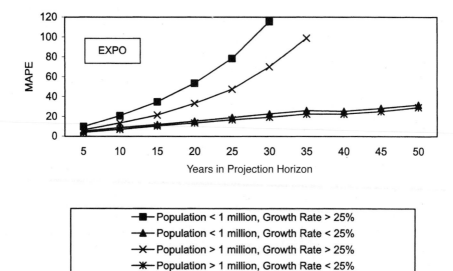

Figure 13.1. MAPEs for state population projections by length of horizon. (*Source*: S. Smith and T. Sincich [1991]. An empirical analysis of the effect of the length of the forecast horizon on population forecast errors, *Demography*, 28, 261–274. Reprinted with permission of the Population Association of America)

tion method to another, from one size-growth category to another, from one launch year to another, and over the length of the projection horizon. Table 13.2 also showed no clear pattern between measures of bias and the length of the projection horizon. We do not believe that the length of the projection horizon has any consistent impact on the tendency for projections to be too high or too low.

Length of Base Period

The length of the base period is one of the most fundamental decisions that must be made when producing population projections, but few studies have considered how this decision is made or what effect it has on population forecast errors. If simple extrapolation methods are used, what historical period should be used as the basis for those extrapolations? If a time series model is used, how many observations are needed? If a cohort-component model is used, what historical time period should be considered in choosing the appropriate mortality, fertility, and migration rates? How many data points are needed to construct reliable structural models?

A common rule of thumb for trend extrapolation methods is that the length of the base period should correspond to the length of the projection horizon (e.g., Alho & Spencer, 1997; Kale, Voss, & Krebs, 1985). For example, five years of base data may be sufficient for a five-year projection, but 20 years of data are needed for a 20-year projection. Is this valid? What does the empirical evidence show?

Smith and Sincich (1990) projected populations for states using three simple extrapolation methods (LINE, EXPO, SHIFT) and annual population data from 1900–1980. Projections were made for one-, five-, 10-, 20-, and 30-year horizons using base periods of one, five, 10, 20, 30, and 40 years. For all three methods they found that MAPEs for one-year horizons were virtually identical for base periods of one, five, 10, and 20 years; they were slightly larger for base periods of 30 and 40 years. For five-year horizons, there was a barely discernible U-shaped relationship between MAPEs and the length of the base period. For all three methods, projections derived from 10-year base periods had slightly smaller MAPEs than projections derived from either shorter or longer base periods. They concluded that differences in the length of the base period have very little impact on forecast accuracy for short projection horizons.

For longer horizons, however, differences in the length of the base period had a much larger impact (Figure 13.2). For LINE and SHIFT, MAPEs declined steadily as base periods increased from one to five to 10 years but remained very stable thereafter. This pattern was found for all three horizons but was most consistent for 20- and 30- year horizons. For EXPO, a similar pattern was found for the 10-year horizon, but for the 20- and 30-year horizons MAPEs continued to decline as the base period increased to 40 years. For increases beyond 20 years, however, the declines were very small.

Smith and Sincich refined their analysis by controlling for the effects of population size and growth rate (not shown here). They concluded that, except for EXPO and SHIFT projections of rapidly growing states, increasing the base period beyond 10 years did not lead to greater forecast accuracy. In fact, longer base periods sometimes led to larger forecast errors. It appears that too short a base period (e.g., one or five years) may incorrectly interpret short-run fluctuations as

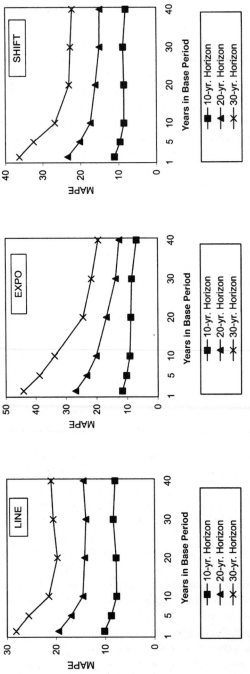

Figure 13.2. MAPEs for state population projections by length of base period and length of horizon. (*Source:* S. Smith and T. Sincich [1990]. The relationship between the length of base period and population forecast errors, *Journal of the American Statistical Association*, 85, 367–375. Reprinted with the permission of the American Statistical Association. All rights reserved.)

long-run trends, whereas too long a base period (e.g., 30 or 40 years) may reflect historical trends that are no longer valid. For EXPO and SHIFT projections of rapidly growing states, however, increasing the base period from 10 to 20 years led to substantial improvements in forecast accuracy, especially for the 30-year horizons. Increases beyond 20 years generally led to no further improvements.

Why does a longer base period improve forecasting performance for EXPO and SHIFT in rapidly growing places? Again, the explanation lies with the tendency for extreme growth rates to regress toward the mean over time. For rapidly growing places, the EXPO method projects a high, unchanging growth rate, and the SHIFT method projects an increasing share of the parent population. High growth rates, however, are generally not maintained for long periods. Consequently, using a longer base period helps reduce the large errors and strong upward bias often found in EXPO and SHIFT projections for places that grew rapidly during the 10 years immediately before the launch year.

Smith and Sincich (1990) also considered whether differences in the length of the base period have an impact on the tendency of population projections to be too low or too high. They found no consistent relationship between MALPEs and the length of the base period. They concluded that the degree of bias for simple extrapolation methods was not significantly affected by differences in the length of the base period.

How do the results reported by Smith and Sincich compare with those found in other studies of population projections? Using a ratio method similar to SHIFT, White (1954) found that a 60-year base period led to considerably larger errors in projections of state population than a 30-year base period. Beaumont and Isserman (1987) made LINE and EXPO projections for states that grew by more than 20% during the decade immediately before the launch year. For EXPO projections, they found that a 40-year base period produced smaller errors and less upward bias than a 10-year base period; for LINE, however, they found that a 40-year base period did not improve forecast accuracy and led to considerably more bias. Voss and Kale (1985) projected populations for Minor Civil Divisions in Wisconsin and found that a 30-year base period produced slightly more accurate forecasts than a 10-year base period for the EXPO technique. They also found that weighting more recent decades in the base period more heavily than more distant decades improved forecast accuracy. Although the length of base period was not the primary focus of these studies, the results were generally consistent with those found by Smith and Sincich.

At the beginning of this section, we cited a rule of thumb stating that—for trend extrapolation methods—the length of the base period should correspond to the length of the projection horizon. The evidence we have presented supports that "rule" to some extent, but not completely. For short-range projections, a short period of historical data appears to be sufficient. For long-range projections, at least 10 years of base data are needed to achieve the best possible results.

However, base periods longer than 10 years are generally not necessary and may even cause forecast accuracy to deteriorate. The only exceptions are EXPO and SHIFT projections of rapidly growing areas (and, to a smaller degree, SHIFT projections of declining areas), where increasing the base period to 20 years seems to improve forecast accuracy by moderating the impact of extreme growth rates.

The studies cited before used simple extrapolation methods to project total population. Little research on this topic has been done for other projection methods. In a study of cohort-component projections for the Netherlands, Keilman (1990) found that 10 years of base data led to more accurate forecasts of deaths than did five years of base data; however, he found no significant impact of differences in base period on forecasts of births. We are not aware of other studies on this topic.

There is no uniformity among practitioners regarding the choice of the base period for population projections. Simple extrapolation methods have been applied using anywhere between one and 60 years of base data. Time series models often use 50 or more observations, but some demographic applications have used as few as 11 (e.g., Voss & Kale, 1985). Some cohort-component models use only the most recent data for constructing mortality, fertility, and migration rates, whereas others use a long time series. For structural models, the length of the base period often depends primarily on the number of years for which relevant data series are available. Although the evidence is clear for simple extrapolation methods, more research is needed before we can draw firm conclusions regarding the optimal length of the base period for other projection methods.

Launch Year

Some time periods are characterized by a high degree of stability in demographic trends; others are characterized by sudden dramatic changes. Because all objective projection methods are extrapolations of one type or another, the degree of stability is likely to have an impact on forecast accuracy. Therefore, it might be expected that forecast errors would be larger for some time periods (i.e., launch years) than for others. In fact, a number of researchers have found this to be true (e.g., Keilman, 1990; Keyfitz, 1982; Long, 1995; Smith & Sincich, 1988; Stoto, 1983).

Long (1995) evaluated forecast errors for a number of national and state projections produced by the Census Bureau. At the national level, he evaluated projections made between the mid-1940s and mid-1980s, using the RMSE of the annual growth rate as a measure of error. For five-year horizons, he found that the RMSE varied from less than 0.1% to almost 1.0%. For 15-year horizons, it varied from less than 0.2% to 1.1%. Errors were highest for projections made during the mid-1940s, but showed no clear trend from the 1950s onward, bouncing up and down from one launch year to another.

For state projections, Long evaluated projections made between the mid-1960s and mid-1980s. For five-year horizons, he found that MAPEs varied between approximately 3% and 5%. For 15-year horizons, MAPEs varied from less than 6% to more than 12%. Again, there was no clear trend regarding changes in accuracy over time.

The differences in accuracy by launch year that Long noted were considerably smaller for state projections than for national projections. The most likely explanation for this finding is that errors for state projections were based on averages that covered all states, whereas errors for national projections were based on a single projection for each launch year. Generally, one would expect more instability in error characteristics for a single place than for an average that covers many places.

Smith and Sincich (1988) also evaluated the forecast accuracy of state projections, but covered a broader range of launch years, considered the direction of errors as well as their magnitude, and evaluated variances as well as means. Using 10-year base periods and 10- and 20-year horizons, they made projections for states using four trend extrapolation methods (LINE, EXPO, SHIFT, SHARE) and an average of the projections from all four methods (AVE). They used all census years from 1910 to 1980 as launch years. Table 13.6 shows the MAPEs and MALPEs for the 10-year projections by launch year.

MAPEs were largest for the 1910 launch year, especially for the EXPO method. Thereafter, MAPEs fluctuated within a range of 4%–8%, with most falling between 6% and 8%. After 1910, there was no apparent trend over time: MAPEs sometimes increased from one launch year to the next and sometimes declined. These data show some variation from one launch year to another, but not a great deal.

Results were considerably different for MALPEs, which ranged between −7% and +14% (most fell between −5% and +5%). All methods had positive errors for launch years 1910, 1930, 1960, and 1980 and negative errors for launch years 1940, 1950, and 1970; all except EXPO had negative errors for launch year 1920. These data do not show any consistent trends over time or any general tendency for projections to be too high or too low.

Smith and Sincich used statistical tests to evaluate stability over time in the means and variances of forecast errors. They concluded that there was a high degree of stability for both the means and variances of absolute percent errors, somewhat less stability for the variances of algebraic percent errors, and no stability at all for the means of algebraic percent errors. Because the MALPE is a measure of bias, this implies that the study of past forecast errors can tell us little (or nothing) about the likelihood that current projections will be too high or too low. However, the stability observed for the means and variances of absolute percent errors implies that the study of past errors can tell us something about the expected degree of accuracy of current projections.

Several other studies have compared forecast errors from different launch

Table 13.6. Errors for 10-Year Population Projections
for States by Launch Year, 1910–1980

	MAPE				
Launch year	LINE	EXPO	SHIFT	SHARE	AVERAGE
1910	8.4	15.6	9.5	11.0	10.9
1920	6.6	7.3	6.7	7.4	7.0
1930	6.5	8.6	7.1	7.8	7.5
1940	7.9	7.3	7.8	7.8	7.7
1950	6.9	5.6	6.6	6.7	6.4
1960	4.4	7.3	5.0	6.0	5.6
1970	8.4	7.9	8.3	8.5	8.2
1980	6.1	8.0	6.9	6.4	6.2

	MALPE				
Launch year	LINE	EXPO	SHIFT	SHARE	AVERAGE
1910	5.2	14.1	7.1	7.9	8.6
1920	−1.4	1.2	−0.6	−0.8	−0.4
1930	3.5	5.9	4.4	3.8	4.4
1940	−6.8	−5.9	−6.6	−6.6	−6.5
1950	−5.2	−2.6	−4.5	−4.7	−4.2
1960	1.3	5.4	2.4	2.0	2.8
1970	−3.7	−1.7	−3.1	−3.3	−2.9
1980	3.5	6.4	4.5	4.2	3.6

Sources: S. Smith and T. Sincich (1988). Stability over time in the distribution of population forecast errors, *Demography*, 25, 461–474, and unpublished data. Reprinted with the permission of the Population Association of America.

years. Kale et al. (1981) focused on projections for states, Smith (1987) focused on projections for counties, and Isserman (1977) focused on projections for townships. All found roughly the same results reported by Smith and Sincich (1988): Measures of accuracy were similar (but not identical) from one launch year to another, but measures of bias varied considerably. More research—covering a variety of projection methods, geographic regions, error measures, forecast horizons, and launch years—will help us refine these conclusions. Information about the stability of error distributions over time is crucial for constructing empirical prediction intervals, a subject we turn to later in this chapter.

COMBINING FORECASTS

A common finding in the general forecasting literature is that a combination of forecasts often leads to greater accuracy and less variability than can be achieved by using individual forecasts by themselves (e.g., Armstrong, 2001c; Clemen, 1989; Makridakis et al., 1998; Mahmoud, 1984; Webby & O'Connor, 1996). This result has been found for many types of forecasts, including gross national product, inflation, corporate earnings, stock prices, exchange rates, electricity demand, tourism, psychiatric diagnoses, political risk, rainfall, and sunspot cycles. Why might this be true?

Theoretically, the best model should provide the most accurate forecasts. However, statistical models are based on the assumption that data patterns and mathematical relationships remain constant over time. This is rarely the case. Consequently, models that fit the data well during the base period do not necessarily provide accurate forecasts, and models that provide the best forecasts for one time period do not necessarily provide the best forecasts for other periods (e.g., Armstrong, 2001b; Fildes, 1992; Makridakis, 1986; Pant & Starbuck, 1990).

Combinations of forecasts seem to work well because each individual method and data set provides potentially useful information; thus, using several forecasts increases the total amount of information going into the final forecast. In addition, offsetting errors tend to cancel each other out. Consequently, a combined forecast has a smaller risk of making a very large error than an individual forecast. A number of studies have concluded that combined forecasts are generally more accurate than most (sometimes all) of the individual forecasts that make up the combination (e.g., Armstrong, 2001c; Makridakis et al., 1982; McNees, 1992; Schnaars, 1986; Zarnowitz, 1984). Perhaps most important, even though combined forecasts do not outperform every individual forecast in every situation, there is no way to know in advance *which* of the individual forecasts (or forecasters) will perform better or worse than the combination.

Combining can be done in a number of ways. Individual forecasts can be based on different methods or different specifications of the same method. Combinations can incorporate the effects of many individual forecasts or only a few. They can be averages in which each individual forecast is weighted equally (i.e., a simple average) or averages in which some forecasts are weighted more heavily than others. For weighted averages, weights can be based on objective criteria such as previous forecasting performance or on subjective criteria such as the personal judgment of experts.

There is no consensus in the literature regarding the best way to combine forecasts. Bates and Granger (1969) suggested the use of weights based on the size of errors found in previous applications of the individual forecasting techniques. Although some researchers have found that weighted averages produce more accurate forecasts than simple averages (e.g., Ashton & Ashton, 1985), there is no guarantee that weights found optimal in the past will prove optimal in the future. Given this uncertainty, many researchers have concluded that a simple average will generally perform about as well as (or better than) more sophisticated weighted averages (e.g., Clemen, 1989; Diebold & Pauly, 1990; Pant & Starbuck, 1990; Schnaars, 1986).

How many individual forecasts should be used in combining? Some empirical investigations have used only two, whereas others have used 20 or more. The general conclusion seems to be that the greater the number of individual forecasts, the better (e.g., Granger, 1989; Makridakis & Winkler, 1983; Webby & O'Connor, 1996). However, a substantial part of the improvement in forecasting performance comes with the first four or five forecasts included in the combination; further additions result in smaller and smaller improvements. It appears that many of the benefits of combining can be achieved with a relatively small number of individual forecasts (e.g., Armstrong, 2001c; Ashton & Ashton, 1985; Clemen, 1989).

Although combining is a common practice for many types of forecasting (including population forecasts made by nondemographers), we know of only a few instances in which it has been used by demographers for population projections. Voss and Kale (1985) made projections for Minor Civil Divisions in Wisconsin using 11 extrapolation techniques. In evaluating forecast errors over a 10-year horizon, they found that the average of all 11 techniques had smaller errors than the majority of the individual techniques and that no single technique consistently outperformed the average. Isserman (1993) made cohort-component projections for counties in West Virginia, using migration rates based on an average of the rates from two decades. County projections in Florida are based on an average of projections from four simple trend extrapolation methods, each applied to several base periods (Smith & Nogle, 2000). All three of these studies used a simple average, sometimes after excluding the highest and lowest of the individual projections to reduce the impact of outliers (i.e., a "trimmed" mean).

One approach to combining that may prove particularly useful for population projections for small areas is the "composite" method suggested by Isserman (1977). This method is based on the assumption that some models or techniques perform substantially better (or worse) than others under particular circumstances or for places with particular characteristics. If consistent patterns can be observed with enough regularity to draw general conclusions, forecasts for particular places can be based solely on the models or techniques that are most accurate for places with those characteristics.

The empirical evidence discussed earlier in this chapter showed that the EXPO method performs poorly for rapidly growing places and the SHIFT method performs poorly for places losing population. This evidence—drawn from projections covering several time periods and different levels of geography—suggests that EXPO should not be used for projections of rapidly growing places and SHIFT should not used for projections of places losing population. One study tested the "composite" approach by constructing an average that excluded EXPO projections for rapidly growing places and SHIFT projections for slowly growing or declining places (Smith & Shahidullah, 1995). They found that this approach produced more accurate forecasts for census tracts than a simple average of all methods. We believe that looking for methods that work particularly well (or poorly) in particular situations will lead to greater improvements in forecast accuracy than can be achieved by looking for the method that performs best in all situations.

Why has combining so rarely been used for population projections? There are several possible explanations. One is the longstanding tradition in demography of producing sets of illustrative projections rather than a single forecast. If projections are intended simply to illustrate the outcomes of various combinations of assumptions, there is no need to consider various combinations of projections (or to be concerned with forecast accuracy, for that matter). Another explanation is that the dominance of the cohort-component method in official population projections has prevented demographers from investigating other methods. A third is that demographers have been searching for the one "true" model of population change and are unwilling to give up that quest. From a conventional point of view, using a combination rather than a single forecast amounts to an admission that the analyst has been unable to build a properly specified model (Clemen, 1989).

Whatever the explanation, we believe that combining offers the same potential benefits to population projections that have been achieved in other fields. Combining could be done by using projections from several methods, including cohort-component and structural models, as well as various extrapolation methods. It could also be done by using several alternative sets of assumptions for a given method (e.g., different combinations of mortality, fertility, and migration assumptions for cohort-component projections). Combining could be done by taking simple averages or by assigning weights based on historical observations or professional judgment. Multiple regression analysis could be used to uncover statistical regularities and to estimate the optimal weights to use in developing "composite" projections. Many approaches are possible.

There is some evidence that demographers have recently become more interested in the potential benefits of combining. Ahlburg (1999) found that combining projections of births from an economic-demographic model with projections from the Census Bureau's cohort-component model improved the forecast accuracy of the cohort-component model. Sanderson (1999) found that averaging projections from structural models with projections from extrapolation methods often led to more accurate forecasts than could be achieved by relying on a single model or method. We believe that further research on this topic will improve the forecasting performance of population projections.

ACCOUNTING FOR UNCERTAINTY

There is no uncertainty involved in producing illustrative population projections. Unless a mathematical error is made in the underlying calculations, illustrative projections are exact representations of the hypothetical future population.

For population projections used as forecasts, however, the story is very different. As the preceding discussion showed, population forecasts entail a tremendous amount of uncertainty, especially for long horizons and for places that have small or rapidly changing populations. About the only thing we know for sure is that our forecasts are going to be wrong. If we are good forecasters—and lucky—most errors will be relatively small. If we are not so good—or not so lucky—some errors may be huge.

How can we account for this uncertainty? Two basic approaches have been used in the past. One is to produce several alternative projections based on different sets of assumptions. The other is to develop prediction intervals based on statistical models.

Range of Projections

The traditional approach to dealing with uncertainty has been to construct a range of projections. A range could be based on the application of different projection methods, but the most common practice is to apply different assumptions using the cohort-component method. This approach has a long history (e.g., Thompson & Whelpton, 1933; Whelpton, Eldridge, & Siegel, 1947) and has been an integral part of the Census Bureau's state and national projections for many years (e.g., U.S. Bureau of the Census, 1950, 1957, 1966, 1972). It has been widely used by state and local demographers as well. Among the producers of "official" population projections, constructing a range is the most common way to deal with uncertainty.

A range of projections is typically developed using various combinations of mortality, fertility, and migration assumptions. For example, one set of U.S. projections used three assumptions for each of the three components of population growth, yielding 27 projection series (Spencer, 1989). Adding an additional assumption for immigration brought the total to 30. For the year 2080, these projections ranged from 185 million to 501 million.

Most sets of projections do not contain nearly as many alternative series. National projections generally have 10 or fewer series; some have had only three (e.g., U.S. Bureau of the Census, 1950). An early set of state projections had four alternative series, based on combinations of two fertility assumptions and two migration assumptions for each state (U.S. Bureau of the Census, 1966). A more recent set of state projections had only two alternative series, one based on a time series model of migration and the other based on an economic model of migration (Campbell, 1996). There are exceptions, but the most common practice is to construct two, three, or four series when producing a range of projections.

Two interpretations can be given to the individual series in a range of projections. One is that each series gives a reasonable view of future population change and that no particular series is any better than any other. This is the interpretation given to the Census Bureau's state projections from the 1950s to the early 1990s (e.g., U.S. Bureau of the Census, 1957, 1966, 1972, 1979; Wetrogan, 1990). Not only did Census Bureau analysts decline to designate a "preferred" or "most likely" series, but they took great pains to emphasize that the projections were not intended to be used as forecasts or predictions. Of course, most data users promptly disregarded those warnings and chose a particular series (typically the middle one) as the forecast.

The second interpretation is that although each series provides a reasonable view of the future, one particular series is preferred to all of the others. The designation of a "preferred" or "most likely" series can be based on an empirical investigation of past forecast errors or on a subjective evaluation of the assumptions used in producing each series. This is the interpretation given to the Census Bureau's most recent state projections (Campbell, 1996). In these projections, one particular series was designated as "preferred" and the other was simply an alternative.

There are also several possible interpretations of the range itself. One is that the high and low series provide a "reasonable" range around the preferred series. It is expected that the range will contain the future population observed in the target year, although no specific probability statements are made (e.g., Whelpton et al., 1947; U.S. Bureau of the Census, 1950). Another interpretation is that the series that make up the range simply provide several alternative views of the future. Under this interpretation there are no stated expectations that the range will contain the future population (e.g., Day, 1996a; U.S. Bureau of the Census, 1957).

The assumptions for mortality, fertility, and migration used in each projection series are typically based on historical values and the analyst's views regarding reasonable future changes in those values. However, the highest and lowest series in the range generally are *not* based on the highest and lowest levels each component of growth could feasibly reach. Consequently, there is no guarantee that the projected range will encompass the future population. In fact, several studies have reported that subsequent populations often fall outside the projected range (Alho & Spencer, 1997; Keyfitz, 1981; Stoto, 1983). It has also been found that the range of individual point forecasts produced by several different projection methods typically understates the true level of uncertainty, sometimes by a substantial margin (McNees, 1992).

Producing a range of projections has several benefits. One is that it makes it easy to observe the effects of differences in assumptions. For example, suppose that two series are based on identical mortality and migration assumptions, but one assumes that fertility rates will increase by 20% and the other assumes that fertility rates will fall by 20%. Differences in population size and age structure caused solely by differences in fertility rates can easily be determined by comparing these series. Outcomes from other combinations of assumptions can also be compared.

Another benefit is that a range gives the data user several options from which to choose. Because each series is based on clearly defined mortality, fertility, and migration assumptions, data users can make choices based on their own judgments regarding the validity of those assumptions. This will be particularly important when the data user has a high level of demographic expertise and knowledge of the area being projected. If the data user lacks this knowledge and expertise, of course, this benefit will be lost.

The primary limitation of producing a range of projections is that it does not provide an explicit measure of uncertainty. How likely is it that any particular series will provide an accurate forecast of future population change? How likely is it that the future population will fall within the range suggested by two alternative series? These questions cannot be answered simply by producing a range. Although a range may provide several reasonable views of the future, it does not provide a means for evaluating what is "reasonable." Even with a range, data users are left without a clear idea of the uncertainty surrounding a population forecast.

Prediction Intervals

The second approach focuses on statistical measures of uncertainty. Prediction intervals—based on statistical theory and data on error distributions—provide an explicit estimate of the probability that a given range will contain the future population. (Prediction intervals are sometimes called *forecast intervals*, *probability intervals*, *confidence intervals*, or *confidence limits*. We call them *prediction intervals* to distinguish them from traditional confidence intervals, which—strictly speaking—apply only to sample data.)

Two main types of prediction intervals have been used with population forecasts. The first is based on the development of statistical models of population growth. The second is based on empirical analyses of errors from past population projections. Both rely on the assumption that historical or simulated error distributions can be used to predict future error distributions.

Model-Based Intervals. Model-based prediction intervals are based on the stochastic (or random) nature of population processes. They can be developed in a number of ways. Past applications have included maximum likelihood estimators of population growth rates (Cohen, 1986), Monte Carlo simulations of fertility and migration rates (Pflaumer, 1988), regression-based projection models (Swanson & Beck, 1994), models based on the opinions of a group of experts (Lutz, Sanderson, & Scherbov, 1999), and a variety of time series models covering total population or specific components of growth (Alho & Spencer, 1990; Carter & Lee, 1992; De Beer, 1993; Pflaumer, 1992).

Time series models (especially ARIMA models) are the models most commonly used for prediction intervals for population projections. These models assume that the pattern (structure) of the data does not change over time, that errors are normally distributed with a mean of zero and a constant variance, and that errors are independent of each other (Makridakis, Hibon, Lusk, & Belhadjali, 1987). They require a fairly long series of historical observations and can be difficult to apply, especially when attempting to develop prediction intervals for all three components of growth and for various subgroups of the population.

Providing a detailed description of model-based prediction intervals is beyond the scope of this book. However, we can give several examples of the intervals produced by these models and compare them to subsequent population counts or to high and low projection series produced using the traditional approach. Lee and Tuljapurkar (1994) projected a population of 398 million for the United States in 2065, with a 95% prediction interval of 259–609 million. This interval is wider than the spread between the low and high projections produced by the Census Bureau at about the same time; those projections ranged from 276 to 507 million in 2050, and had a medium projection of 383 million (Day, 1992). The previous set of Census Bureau projections reported a medium projection of 300 million and a range of 230–414 million for 2050 (Spencer, 1989).

Pflaumer (1992) made two time series projections of the U.S. population, one based on population size and the other based on the natural logarithm of population size. The first model produced a medium projection of 402 million and a 95% prediction interval of 277–527 million in 2050. These numbers are similar to the Census Bureau's projections from the same time. The second model produced a medium projection of 557 million and a 95% prediction interval of 465–666 million. These numbers are much higher and provide a narrower range than the Census Bureau's projections (Day, 1992).

McNown, Rogers, and Little (1995) made time series projections of the components of growth for the U.S. population, as well as total population size. For 2050, they projected a total population of 373 million, with a 95% prediction interval of 243–736 million. The total fertility rate was projected at 2.46 in 2050, with a 95% prediction interval of 0.91–5.53. Life expectancy at birth for males was projected at 75.5, with a 95% prediction interval of 68.5–82.8. For fertility, these intervals were much larger than those found in the Census Bureau projections, which assumed that the total fertility rate would range only from 1.83 to 2.52 in 2050 (Day, 1992). For mortality, the intervals were not much different from those reported by the Census Bureau (life expectancy of 75.3–87.6 in 2050).

The projections produced by McNown et al. (1995) also had wider prediction intervals for the components of growth than those reported in several other studies using time series techniques. For example, Lee (1993) projected a total fertility rate of 2.3 in 2065 and a 95% prediction interval of 0.5–3.4. Lee and Carter (1992) projected life expectancy at birth at 86.1 in 2065 and a 95% prediction interval of 80.5 to 90.0 years.

Swanson and Beck (1994) developed a regression-based model for short-range county population projections in the state of Washington. They compared the 67% prediction intervals associated with this model to census counts of Washington's 39 counties in 1970, 1980, and 1990. They found that the prediction intervals contained the 1970 census count in 30 counties (77%), the 1980 census count in 24 counties (62%), and the 1990 census count in 31 counties (79%). These results suggest that Swanson and Beck's 67% prediction intervals provided a reasonably accurate view of forecast uncertainty.

Model-based prediction intervals are valid only to the extent that the assumptions underlying the models are valid. Despite their objective appearance, they are strongly influenced by the analyst's judgment. The models themselves are complex and require a substantial amount of base data. They are subject to errors in the base data, errors in specifying the model, errors in estimating the model's parameters, and future structural changes that invalidate the statistical relationships that underlie the model (Lee, 1992). In addition, many alternative forecasting models can be specified, each providing a different (perhaps dramatically different) set of intervals (Cohen, 1986; Lee, 1974). To our knowledge, model-based prediction intervals have not been used to produce official population projections.

Despite these problems, model-based prediction intervals offer one important benefit: they provide explicit probability statements to accompany point forecasts. The intervals are often very wide, exceeding the low and high projections produced by official statistical agencies. Given that many data users (and producers) tend to overestimate the accuracy of population projections, model-based prediction intervals provide an important reality check.

Empirically-Based Intervals. The second type of prediction interval is based on empirical analyses of errors from past projections (e.g., Keyfitz, 1981; Smith & Sincich, 1988; Stoto, 1983; Tayman et al., 1998). Keyfitz (1981) took some 1,100 national projections made between 1939 and 1968 and, for each one, calculated the difference between the projected annual growth rate and the rate that actually occurred over time. He found that this difference was largely independent of the length of horizon over which the projections were made. He calculated that the RMSE for the entire sample was approximately 0.4 percentage points and developed 67% prediction intervals by applying that error to the growth rates projected for each country. For example, if growth were projected at 2% per year for the next 20 years, the probability would be approximately 67% that the actual growth rate would be somewhere between 1.6% and 2.4%.

Keyfitz refined his analysis by separating countries according to their population growth rates; he found a RMSE of 0.60 for rapidly growing countries, 0.48 for moderately growing countries, and 0.29 for slowly growing countries. He illustrated this refinement by applying the 0.29 RMSE to the U.S. growth rate of 0.79% per year projected by the Census Bureau, yielding annual growth rates of

0.50% and 1.08%. Applying those growth rates to the 1980 population implied a population of 260 million in 2000, with a prediction interval of 245–275 million. He concluded that the odds were about 2 to 1 that this interval would contain the U.S. population in that year.

Stoto (1983) followed a similar approach but analyzed projections that contained more temporal and geographic diversity. Like Keyfitz, he calculated forecast error as the difference between the projected annual growth rate and the rate actually realized over time. He differentiated between two components of error, one related to the launch year of the projection and the other to seemingly random events (the residual). He calculated the standard deviations for these two components of error and constructed prediction intervals in a manner similar to that used by Keyfitz. He applied those intervals to projections of the U.S. population and estimated that there was about a 67% probability that an interval of 241–280 million would contain the actual population in 2000. He compared his results to projections produced by the Census Bureau and concluded that the Census Bureau's low and high series were very similar to a 67% prediction interval. Keyfitz (1981) had reached the same conclusion.

Smith and Sincich (1988) also used the distribution of past forecast errors to construct prediction intervals but followed a different approach. They modified a technique developed by Williams and Goodman (1971), in which the predicted distribution of future forecast errors was based directly on the distribution of past forecast errors. An important characteristic of this technique is that it can accommodate any error distribution, including the asymmetric and truncated distributions typically found for absolute percent errors.

Using population data for states from 1900 to 1980, Smith and Sincich used four simple extrapolation methods to make a series of projections covering 10- and 20-year horizons. They calculated absolute percent errors for each target year by comparing projections with census counts and focused on the 90PE for each set of projections (i.e., the absolute percent error larger than exactly 90% of all absolute percent errors). They investigated two approaches to constructing 90% prediction intervals, one using the 90PE from the previous set of projections and the other using the 90PE from all other sets of projections. They found that both approaches provided relatively accurate prediction intervals. For most individual target years, 88–94% of state forecast errors fell within the predicted 90% interval. Summing over all target years, 91% of all forecast errors fell within the predicted 90% interval. They concluded that stability in the distribution of absolute percent errors over time made it possible to construct useful prediction intervals for state projections.

To our knowledge, Tayman et al. (1998) are the only analysts to develop statistical prediction intervals for subcounty population forecasts. They started with projections of population for *grid cells* in San Diego County; grid cells are geographic areas 2,000 ft. × 2,000 ft. defined for the most densely populated parts of the county. The projections used 1980 as a launch year and 1990 as a target year. Using repeated sampling techniques and randomly selected grid cells, they developed projections for a large number of areas that varied in population size from 500 to 50,000. Forecast errors were calculated by comparing the 1990 projections with 1990 census counts.

Rather than constructing prediction intervals for the population forecasts *per se*, Tayman and his colleagues developed predictions for the mean errors implied by those forecasts. Empirical prediction intervals for MAPEs and MALPEs were developed by using an approach similar to that used by and Smith and Sincich (1988). For areas that had 500 residents, they found a 95% prediction interval of 67.4–80.3% for the MAPE. For areas with 50,000 or more residents, the interval was 9.7–11.5%. For MALPE, the intervals were wider but centered closer to zero.

Evaluating the Evidence. Under formal definitions, probability statements about the accuracy of population projections cannot be made because the distribution of future forecast errors is unknown (and unknowable) at the time the projections are made. However, it is possible to construct prediction intervals based on specific models of population change or on the distribution of errors from past projections. If current projection methods are similar to those used in the past and if the degree of uncertainty is about the same in the future as it was in the past, then we can assume that future forecast errors will be drawn from the same distribution as past forecast errors (Keyfitz, 1981). If this is true, prediction intervals will provide a reasonable (albeit imperfect) view of the uncertainty surrounding current population forecasts.

The critical question, of course, is whether the distribution of forecast errors does indeed remain stable over time. Smith and Sincich (1988) showed that the distribution of absolute percent errors for states remained relatively stable over the decades of the twentieth century. Smith (1987) reported a similar result for county projections for the 1960s to the 1970s. Stoto (1983) divided forecast errors into two components and found that the distributions of both components remained fairly stable for national projections made between the 1940s and the 1970s. Some empirical evidence, then, supports the notion that the distribution of population forecast errors remains relatively stable over time.

More research is needed on constructing and interpreting prediction intervals for population forecasts. Which approach is best? What are the effects of differences in projection method, launch year, geographic region, and length of horizon? How can intervals be developed for demographic subgroups (e.g., age, sex, race) that are consistent with each other and with intervals for the entire population? Much remains to be done, but the potential payoff is high. Although we may never be able to forecast a specific *population* with a high degree of accuracy, we may be able to develop relatively accurate forecasts of the *distribution of errors* surrounding our population forecasts. Providing a realistic estimate of the uncertainty inherent in population forecasts may be the most useful service that the producers of population projections can provide to their users.

CONCLUSIONS

To close this chapter it may be helpful to summarize the empirical evidence on population forecast accuracy and bias. Forecast accuracy generally increases with population size. It tends to be greatest for places that have slow but positive growth rates and declines as growth rates deviate in either direction from those levels. Average errors vary from one launch year to another, especially for measures of bias. Average absolute errors grow steadily (and about linearly) with the projection horizon. For long-range projections based on simple extrapolation methods, 10 years of base data are generally necessary (and sufficient) to maximize forecast accuracy. For projections of total population, no single method is consistently more accurate than any other. These results have been found in so many circumstances that we believe they can be accepted as general characteristics of population forecast errors.

No general conclusions can be drawn regarding the direction of forecast errors, however. Some individual projections have large positive errors, others have large negative errors. Some sets of projections exhibit a substantial upward bias, others exhibit a substantial downward bias. There is no way to know in advance whether a particular projection (or set of projections) will be too high or too low. Over time, positive and negative errors seem to be roughly in balance. In this sense, we believe that most population projection methods are unbiased.

What level of error might a data user expect from a set of population forecasts? Using the evidence cited in this chapter and assuming that MAPEs grow about linearly with the projection horizon, we have developed a set of "typical" MAPEs by level of geography and length of horizon (Table 13.7). For states, MAPEs grow from 3% for five-year horizons to 18% for 30-year horizons; for counties, they grow from 6% to 36%; and for census tracts, they grow from 9% to 54%. Errors for any specific set of projections will be affected by factors such as projection method, population size, growth rate, and launch year, of course, but we believe that these numbers provide reasonable ballpark estimates of likely forecast errors.

These errors illustrate the high degree of uncertainty inherent in population projections, especially for small areas and for long projection horizons. Data users should be aware of these errors before making decisions based on population

Table 13.7. "Typical" MAPEs
for Population Projections by Level
of Geography and Length of Horizon

Level of geography	Length of horizon (years)					
	5	10	15	20	25	30
State	3	6	9	12	15	18
County	6	12	18	24	30	36
Census tract	9	18	27	36	45	54

projections. Projections that extend very far into the future simply cannot provide highly accurate forecasts. This may be disheartening news for the users of population projections, but it is a realistic portrayal of forecast accuracy, given the current state of the art.

Given this high degree of uncertainty, why should the analyst even bother making small-area projections? There are several reasons. First, the projection process itself is educational, teaching a great deal about changes in geographic boundaries over time, demographic and socioeconomic trends, special population subgroups, the occurrence of unusual events, and the potential impact of growth constraints. Second, projections are useful for analyzing the impact of alternative scenarios and combinations of assumptions, regardless of their forecast accuracy. Finally—and perhaps most important—there is really no alternative to making population projections. If people are not willing to make projections, they must either ignore potential change or assume that no change will occur. Neither of these options is particularly attractive. Ignoring potential change is not likely to be helpful; ignorance generally is *not* bliss. Furthermore, the assumption that no change will occur is itself a projection, albeit naïve and perhaps ill-founded. Projections based on no-change assumptions often lead to less accurate forecasts than can be obtained using other projection methods (e.g., Tayman, 1996a). Although population forecasts are almost always in error—sometimes by a wide margin—they represent our best hope of planning intelligently for the future.

Forecast accuracy is an important characteristic of population projections. However, it is not the only criterion upon which projections should be judged. In the final analysis, projections can best be judged according to their overall "utility," or their value in improving the quality of information upon which decisions are based (Swanson & Tayman, 1995). Even though they cannot perfectly predict future population trends, projections can point to potential growth constraints, highlight areas that are likely to lose population or grow very rapidly, show the implications of alternative public policy or land use decisions, and play other useful roles (Tayman, 1996a,b).

Do projections provide a stronger basis for decision making than the alternative, which is *not* to make projections? If so, are the gains large enough to offset the costs of making those projections? If these two questions can be answered affirmatively, population projections can play an important role in planning and analysis despite their sometimes less than stellar performance as forecasts.

CHAPTER 14

A Practical Guide to Small-Area Projections

We have provided a great deal of information on the construction and evaluation of population projections in this book. We have discussed commonly used projection methods; techniques and assumptions that can be used when applying those methods; potential sources of data; criteria for evaluating projections; and characteristics of forecast errors. This abundance of information may have left the reader feeling somewhat overwhelmed. How can a demographer, planner, market researcher, public official, or other analyst proceed when called upon to construct a set of population projections?

In this chapter, we present a set of guidelines that we hope will alleviate some of this confusion and anxiety (see Box 14.1). These guidelines are intended as a summary of the material presented throughout the book and as a road map to guide the analyst through the projection process. They will not answer every possible question, of course, but at least they will provide a checklist to highlight the issues that must be considered and the choices that must be made. We believe they will help the analyst make reasonable, defensible choices and—perhaps more important—avoid potentially disastrous pitfalls.

These guidelines focus on projections for small areas (i.e., counties and subcounty areas) for two reasons. First, the demand for small-area projections is large and growing rapidly. They are used for planning when and where to build new schools, roads, hospitals, and shopping centers; whether to expand the capacity of an electric power plant or a public transportation system; how to tailor a marketing plan to fit the needs of a specific client; and how to evaluate the environmental consequences of population growth. We believe that this demand will continue to grow.

Second, population size and data availability create methodological problems for small-area projections that do not exist (or are much less severe) for

---

Box 14.1

How to Make Small-Area Projections

1. Determine what is needed
   - Demographic characteristics
   - Geographic areas
   - Length of horizon and projection interval
   - Time and budget constraints
   - Other considerations
2. Construct the projections
   - Select computer software
   - Choose projection method(s)
   - Collect and evaluate data
   - Adjust for special events
   - Control for consistency
   - Account for uncertainty
3. Review and document results
   - Internal review
   - External review
   - Documentation

---

projections of larger areas. Consequently, some issues that must be considered when making projections for small areas are not present when making projections for large areas. In spite of this focus on small areas, however, much of the discussion in this chapter is applicable to state (or national) projections as well.

DETERMINE WHAT IS NEEDED

The first step in the projection process is to determine exactly what is needed. This may seem almost too obvious to mention, but many times clients (i.e., the persons, agencies, or organizations that request a set of projections) are not completely sure what they really need. Consequently, it is essential to discuss all aspects of a project at the very beginning and to cover relevant details clearly and completely. It is also helpful to discuss the purposes for which the projections are to be used. This will help the analyst determine the type of projections needed and choose the most appropriate data sources and projection methods.

The process will be different for general-purpose projections from that for customized projections. General-purpose projections are those produced without reference to a specific use or data user. Customized projections are those produced for a single data user or for a particular purpose. Decision-making criteria can be more clearly defined for customized projections than for general-purpose projections. Consequently, decisions that underlie general-purpose projections must be based on the analyst's expectations regarding the primary needs of the majority of data users. Time and budget constraints will play an important role in these decisions because the production of greater geographic and demographic detail requires greater resources.

Failing to clearly define all aspects of a project at the very beginning is an invitation to disaster, leading to wasted resources, unsatisfactory results, and unhappy clients. The following checklist will help the analyst identify issues that must be resolved before starting the production process.

Demographic Characteristics

Should the projections refer simply to total population size or also to characteristics such as age, sex, race, and ethnicity? If age-group projections are needed, what are the relevant age categories? What race or ethnic categories are needed? What types of cross-tabulations are required? The client will generally be able to answer these questions but may not realize their importance unless prompted by the analyst.

Age, sex, race, and ethnicity are the characteristics most commonly included in population projections. For some purposes, however, projections of other characteristics or population subgroups may be needed. Examples include persons with disabilities, the institutionalized population, the seasonal population, persons in the labor force, school enrollment, and the number of commuters. The model may also be used for projections of population-related variables such as households, average household size, births, deaths, and migration flows. Knowing the purpose for which projections are to be used helps determine which population characteristics are required.

Geographic Areas

Population projections are often made for well-defined areas such as states or counties. The geographic boundaries for these areas are easy to determine, match up clearly with the boundaries used for tabulating various types of data, and remain stable over time. For subcounty areas, however, the situation may be very different.

Boundaries for subcounty areas are subject to sudden and dramatic changes. It is not uncommon for cities to annex adjoining areas, for census tracts to be subdivided, for ZIP codes to be reconfigured, for service areas to be redefined, and for new school districts to be formed. The analyst must determine whether the boundaries of the area to be projected have changed during the period for which historical base data have been collected. If the boundaries have changed, the data must be adjusted to reflect a geographic area that remains constant over time; otherwise, historical changes in the base data will mix the effects of boundary changes and population changes, confound the projection process, and (probably) reduce forecast accuracy. If boundary changes have occurred but no relevant data are available, the analyst may be forced to make subjective adjustments based on anecdotal evidence or to use projection methods that require data from only one point in time (e.g., the constant-share method).

Sometimes projections must be made for areas that lack well-defined boundaries. For example, a client may have only a rough idea of the geographic boundaries that define the service area for a hospital, automobile dealership, or retail establishment. In these instances, the analyst and client will have to work together to establish a clear set of boundaries. Customer records or a sample survey may help determine the relevant area. When delineating the boundaries of a geographic area, it is helpful to match boundaries with those used for tabulating historical base data (e.g., counties, cities, census tracts, or block groups). This will make it easier to collect reliable data.

There may be circumstances in which the population to be projected is not defined geographically (e.g., the number of veterans from the armed forces or the number of retirees from a company). In these circumstances, of course, geographic boundaries are irrelevant (unless one is also concerned with the geographic distribution of that population, such as the number of veterans residing in New Jersey).

Length of Horizon and Projection Interval

What time span and projection intervals are needed? Should the projections extend for five years, 10 years, 20 years, or longer? Should they be made in one-year, five-year, or 10-year intervals?

Data availability plays an important role in the choice of the projection interval. Annual data are often unavailable for subcounty areas, which makes projections in one-year intervals difficult. Although data can be adjusted to match different time intervals, this process is subject to numerous problems (see the discussion of migration adjustments in Chapter 6). We believe that it is generally advisable to match projection intervals with the time periods for which base data are available. For example, when using gross migration data from the decennial census, it is better to make projections using five-year intervals than try to convert five-year data into one-year data. If annual projections are needed, they can be made by interpolating between target years (see Chapter 11).

Time and Budget Constraints

It is essential to start with a clear understanding of the time, money, and other resources available for developing the projections. This is true regardless of whether the client is an outside party or someone from within the analyst's own agency or organization. Indeed, many of the decisions made throughout the projection process will depend on deadlines and available resources. In general, the greater the amount of demographic and geographic detail required and the greater the attention paid to an area's unique characteristics and special populations, the greater the time and other resources needed to construct the projections. An inadequate budget or too short a time frame will lead to low levels of quality (for the projections) and high levels of stress (for the analyst).

It is much more time-consuming to construct a set of projections for the first time than it is to repeat the process a second, third, or fourth time. Developing the projection model from scratch and collecting, analyzing, and adjusting input data are time-consuming tasks. Updating a set of projections, on the other hand, is much simpler. The best advice for first-time producers of population projections is the same as for people about to remodel their kitchens: allow twice as much time as you think the project should take.

Other Considerations

A variety of other factors must also be considered. Will the projections be used solely as forecasts or will they play other roles as well (e.g., trace out the effects of different assumptions or scenarios)? Do they need to be tied to an independent projection of a larger area (e.g., census tract projections controlled to an independent county projection)? Will they be subject to some type of review process? If so, which parties will participate in this process? Will those parties have a strictly advisory role or will they have the power to require changes in the methodology, assumptions, or results? If there are disagreements, how will they be resolved?

Political considerations may be particularly important. In some instances, the analyst has complete freedom to select methods and assumptions; in other instances, outside parties are involved (e.g., clients, government agencies, groups of data users). In these instances, it may be helpful to put together an advisory panel made up of representatives from each party. Members of this panel can provide input regarding geographic areas, input data, techniques, assumptions, projection intervals, and other factors. This input is likely to improve both the quality and the political acceptability of the projections. Conflicts can be resolved more easily when dealt with early in the production process rather than at the end. Even when not required by law, achieving consensus among major stakeholders may raise the credibility of the projections.

As discussed in Chapter 12, political considerations can have either a positive or negative impact on the quality of projections, depending on the specific circumstances involved. Independent analysts have the option of saying "no" to a project if they believe political influences will harm their professional reputations or compromise the integrity of the projections. Staff members who work within a government agency or private business may not have this option; refusing to participate in a project or to accept nontechnical input may be tantamount to handing in one's resignation. Clearly, political considerations can present the analyst with difficult ethical dilemmas.

CONSTRUCT THE PROJECTIONS

Select Computer Software

The first step in constructing a set of projections is to select the computer software that will be used to organize and manipulate the input data, to develop the projection algorithms, and to present the projection results. Three broad types of software can be used: electronic spreadsheets (e.g., Excel, Quattro Pro), statistical analysis packages (e.g., SAS, SPSS), and customized routines written in computer programming languages (e.g., FORTRAN, C++, Visual Basic). Several critical functions must be considered when selecting the appropriate software:

1. *How does the software handle input data?* The most useful software will be adept at handling data from multiple platforms (Unix, PC, mainframe) and a variety of distribution media. Although many data files are now distributed on CDs, some historical data series are available only on mainframe cartridges. File structures have become more complex as programmers seek to reduce the amount of space required by any given file. Hierarchical structures have become commonplace (e.g., PUMS data) and a growing number of files are distributed in a packed binary format (e.g., the Census Bureau's County-to-County Migration files). The normalized data structure used by relational database managers to store and retrieve information from large data sets poses additional challenges. The software must be flexible and powerful enough to deal with an enormous variation in input data sources.

2. *How adept is the software at formulating and modifying the algorithms that will be used in constructing the projections?* How does it handle the sequence of data transformations and modeling steps required by the projection model? Statistical analysis packages and formal programming languages perform these functions very well and can easily accommodate changes in data and statistical routines. Spreadsheets are somewhat unwieldy because changes in the number of observations or operations cannot be handled seamlessly. Rather, formulas or worksheets must be duplicated, adding to the workload and creating additional

opportunities for error. Although the use of macros can substantially reduce these problems, the drawbacks of spreadsheets become more acute as the number of areas and the level of demographic detail increase.

3. *How does the software present its output?* Projection results are typically released as reports, tables, charts, and maps, in both hard-copy and computer-readable formats. Headers, footers, and appended notes must be programmed from scratch when a customized routine is used, but those functions are already included in most statistical packages and spreadsheets, along with relatively easy to use graphing and table-generating capabilities. Some software packages produce web-enabled files and generate a multipurpose data warehouse that can accept queries from web browsers. This is very useful for disseminating projections quickly and easily.

At present, there are no widely used off-the-shelf software packages for most population projection methods. The one exception is urban systems models, for which several software packages are available (but expensive). For trend extrapolation and cohort-component projections, however, this is not the case. The few packages currently available are either hard to find, difficult to use, or not flexible enough to handle a variety of settings and special circumstances.

Analysts who prepare population projections will generally have to write their own computer programs, using a spreadsheet, statistical analysis package, and/or programming language. We believe that spreadsheets are useful when the projections do not involve too many computations, but that statistical analysis packages or programming languages are more useful when the projections cover a large number of areas or require complex programming. It should be noted, however, that most of the computations, tables, and figures shown in this book were done by using a spreadsheet.

Choose Projection Method(s)

The next step is to choose the method(s) for constructing projections. This choice will depend on the purposes for which the projections will be used; the level of geographic and demographic detail needed; the amount of time, money, and other resources available; and the availability of relevant input data.

A variety of methods can be used to project total population. Simple extrapolation methods such as linear, exponential, shift-share, and share-of-growth will often be sufficient, especially for short-to-medium time horizons (e.g., 20 years or less). The reader is reminded, however, of the potential problems of using the exponential method for places that grew rapidly during the base period or using the shift-share method for places that declined (see Chapter 13).

Simple extrapolation methods can also be used for projections of race or ethnic groups, but projections by age will require some type of cohort approach.

For state and county projections, cohort-component models based on gross migration data are generally preferable to models based on net migration data (see Chapter 6 for an explanation). Simple two-region models are sufficient for most purposes, but projections of specific place-to-place migration flows will require the use of a full-blown multiregional model.

Net migration models can also be used, of course. Indeed, they offer several advantages over gross migration models, such as smaller data requirements and greater ease of application. Furthermore, projections based on net migration models are often similar to projections based on gross migration models, especially for relatively short projection horizons (e.g., five or 10 years). For rapidly growing areas, however, projections based on net migration tend to have an upward bias and larger-than-average forecast errors, especially for long projection horizons.

Because of problems of availability and reliability of gross migration data, net migration models are particularly useful for subcounty projections. Even the simple Hamilton–Perry method will often be sufficient for short- and medium-range projections. To avoid an upward bias, however, it is advisable to control projections based on the Hamilton–Perry method to independent projections of total population.

Other considerations may also come into play. Projections for simulation or policy analysis generally require the use of a cohort-component model and perhaps a structural model. Projections of special populations such as seasonal residents or public school enrollees require models that specifically account for those populations.

The critical point to remember when choosing a projection method is that no single model or technique is better than all others for all purposes. Rather, each has its own strengths and weaknesses and must be evaluated according to its face validity, timeliness, cost, data requirements, ease of application, and other characteristics. Some of these characteristics are complementary (e.g., low costs, low data requirements, and ease of application typically go together), but others conflict with each other (e.g., greater geographic and demographic detail imply greater time and money costs).

In the final analysis, the choice of the projection method will be determined by the analyst's judgment regarding the optimal combination of these characteristics. As a general rule, it is best to use the simplest method that can accomplish the task at hand. This allows the analyst to spend more time on activities that are likely to have an impact on the quality of the projections (e.g., checking the quality of the input data, adjusting for special populations and unique events, and reviewing the projection results) and spend less time on activities that are not likely to matter very much (e.g., developing an unnecessarily complex data base or projection model). The reader is also reminded of the potential benefits of combining projections from a variety of methods (see Chapter 13). We believe that combining is particularly valuable for small-area projections, where the potential for large errors is the greatest.

Collect and Evaluate Data

The availability and quality of input data has a major impact on the choice of the projection method. Simple trend extrapolation and ratio methods require only total population data for two points in time (one point for the constant share method). More complex extrapolation methods require data for numerous points. The Hamilton–Perry method requires data by age and sex from two points in time. More complex cohort-component models require data on births, deaths, migration, and the demographic make-up of the population. Structural models require data on explanatory variables as well as dependent variables. Urban systems models require data on vacant land, zoning restrictions, employment, transportation systems, and similar variables.

The lack of appropriate data may cause the analyst to choose a method different from the one that would have been chosen under ideal circumstances. Total population data are typically available for at least two time points (e.g., the two most recent censuses). Annual time series data are often unavailable for subcounty areas, however, which makes it difficult or impossible to use complex extrapolation methods. Similarly, the data required by structural models may be difficult or expensive to obtain. Information on demographic characteristics can generally be obtained from the decennial census, but a complete set of fertility, mortality, and migration rates is seldom available for subcounty areas. For age projections of subcounty areas, the analyst may have to use the Hamilton–Perry method or model schedules (i.e., fertility, mortality, and migration rates from another source considered representative of the projection area).

Whichever methods are chosen, efforts must be made to obtain the most recent data available. For demographic and socioeconomic characteristics of the population, the most recent data will often be from the most recent census. For total population, postcensal estimates may be available. If not, they can be constructed using electric customer, building permit, or similar types of data (e.g., Rynerson & Tayman, 1998; Smith & Cody, 1994). Using recent data is particularly important for small-area projections because growth trends change more rapidly and dramatically in small areas than in large areas.

It is also important to evaluate the quality of the input data. Although it is the closest thing to a "gold standard" for demographic data in the United States, the decennial census is not error-free. Sometimes these errors are corrected within a year or two after the census, but they often go uncorrected until the following census (or even longer). Many census errors cancel out at higher levels of geography (except for the well-known undercount problem), but they can be substantial for small areas, especially for particular subgroups of the population (e.g., specific age and race categories). Postcensal population estimates are subject to even larger errors than decennial census data. Data sources used for particular types of projections—such as birth and death statistics, IRS migration records, group quarters data, and employment forecasts—are also subject to error. It is essential to examine all input data, note potential errors, and make corrections or adjustments when necessary. This is a time-consuming task but can have a substantial payoff in improving the quality of the projections.

How long a base period is needed? As noted in Chapter 13, there is little uniformity among practitioners. Choices vary according to the projection method, the length of the projection horizon, and the availability of historical data. For simple extrapolation methods, only a few years of base data are needed for projections that extend only few years into the future. Given the impact of reporting errors and random fluctuations in historical data series, however, we believe that it is risky to use a single year of base data for *any* projection, even one that covers only a single year. For projections that extend beyond five years, we believe that approximately 10 years of base data are needed to achieve the greatest possible forecast accuracy. Longer base periods can be used, but they will not necessarily raise forecast accuracy and may even cause it to decline. One exception is EXPO projections for rapidly growing areas and SHIFT projections for either rapidly growing or declining areas, where increasing the base period to 20 years improves forecast accuracy by moderating the impact of extreme growth rates.

Very little research has considered this question for other projection methods. Some analysts believe that at least 50 observations are necessary to provide the best parameter estimates for time series models (e.g., McCleary & Hay, 1980), but this does not imply that 50 observations are needed to achieve acceptable forecast accuracy. In fact, some analysts have used fewer than 15 observations, seemingly with reasonable results (Campbell, 1996; Voss & Kale, 1985). Cohort-component models often use the most recent data on migration and fertility rates without attempting to account for changes over time. Structural models vary tremendously in their use of base data.

Other than for simple extrapolation methods, there are currently no general guidelines for the amount of base data that should be used in constructing population projections. In making this decision, analysts must rely on their professional judgment, informed by theoretical considerations and empirical evidence. In some instances, this decision will be determined primarily by the availability of historical data.

Adjust for Special Events

Before making the projections, it is important to adjust the base data for the effects of any special events that may have occurred. For example, large state prisons were built in a number of small, slowly growing Florida counties during the 1990s. The populations of these counties grew by 5, 10, or even 15% over a period of just one or two years. If projections were made using the 1990s as a base period and without accounting for these events, they in essence would be projecting the construction of similar prisons during each future time period. In these instances, adjustments should be made by taking special populations out of the base data, making projections from the remaining data, and adding them back as a final step.

Other special events may have substantial, one-time-only effects on population growth. Examples include the opening or closing of a military base, college, or retirement home; the construction of a large housing development; and the addition or loss of a major employer. The effects of such events on population change are particularly great for areas that have very small populations. For areas with larger populations, the effects of these types of events tend to cancel each other out and can usually be ignored.

Adjusting for special events requires intimate knowledge of the area to be projected. It also requires the application of professional judgment. Which events should be accounted for and which should be ignored? Should expectations of future events be considered, or only events that have already occurred? Are there any spin-off effects from these events that might affect other aspects of population growth (e.g., what are the employment implications of building a new prison)? The analyst must answer these and similar questions before constructing the projections.

In addition to accounting for special events, the analyst must consider the potential effects of constraints that may restrict future population growth. Constraints may be physical (e.g., swamps, lakes, and flood plains) or political (e.g., zoning restrictions, land use plans, and building moratoria). Such constraints are rarely important at the state or even the county level, but may be critical for subcounty areas. Some projection methods account for such constraints explicitly (e.g., urban systems models), but their impact should be considered when using other methods as well.

Control for Consistency

In some instances, it is useful to control small-area projections to projections of larger areas. For example, county projections might be controlled to state projections, or census tract projections controlled to county projections. Although there is no evidence that controlling improves forecast accuracy (Isserman, 1977; Voss & Kale, 1985), it makes small-area projections consistent with each other and with projections of larger areas. This will be particularly important when larger area projections are "official" projections whose use is mandated by law. Controlling can be achieved by using simple raking procedures or more complex adjustment techniques (see Chapter 11).

Account for Uncertainty

We can be pretty sure that the sun will set tonight and rise again tomorrow morning. The earth has been rotating on its axis for a long time. The chances of that pattern changing anytime soon are so slim that—for all practical purposes— we can accept its continuation as a certainty. Most future events, however, are subject to some degree of uncertainty. Will it rain tomorrow? Will the stock market go up or down next year? Will the Cubs ever win another World Series?

The course of future population change is also uncertain. The degree of uncertainty grows as the projection horizon becomes longer and as the size of the population becomes smaller. We can be much more confident of a one-year forecast than of a 20-year forecast. Similarly, we can be more certain of the size of the U.S. population in 20 years than of the size of the population of Portland, Maine. The errors reported in Chapter 13 illustrate the uncertainty inherent in population forecasts.

We believe that it is important to provide data users with some indication of this uncertainty. This can be done in several ways. One is to construct a range of projections based on two or more methods or different specifications of a particular method. For example, projections might be made using several trend extrapolation methods and/or different base periods for each one. A more common approach is to produce several sets of cohort-component projections based on different combinations of assumptions.

The primary benefit of producing a range of projections is that it shows the populations that stem from different but reasonable models, techniques, or sets of assumptions. The primary limitation is that it does not provide an explicit measure of uncertainty. How likely is it that the future population will fall within the range suggested by two alternative projections? This question cannot be answered simply by producing a range.

A second way to indicate uncertainty is to construct prediction intervals based on statistical models of population growth or empirical analyses of past forecast errors. Model-based prediction intervals are difficult to produce and are subject to a variety of specification errors. Empirically based prediction intervals require collecting a large amount of historical data. Both are valid only to the extent to which future error distributions are similar to past or simulated distributions. In spite of these problems, prediction intervals offer one major advantage over a range of projections: They provide an explicit estimate of the uncertainty of future population growth.

A third way to provide some indication of uncertainty is to construct tables that summarize errors from previous forecasts. Although this approach does not provide an explicit range or prediction interval, it does assess past performance and— by extension—provides a basis for predicting future performance. The reader is reminded, of course, that past performance is no guarantee of future performance.

In some instances, a data user may be better off using a high or low projection rather than a medium or "most likely" projection. For example, suppose that the cost of building too large a sewer system is relatively small, but the cost of building too small a system is very large. To reduce risk, it may be advisable to use a projection from the high series or at the high end of the prediction interval for planning the size of the system. Estimates of the cost of being wrong may play an important role in the choice of the projection used for a particular purpose. Measures of uncertainty help the data user make these choices.

REVIEW AND DOCUMENT THE RESULTS

The analyst may believe that the job is finished once the steps just described have been completed. That would be a mistake. At this point, the projections should be viewed as strictly preliminary. Before they are finalized, they must be thoroughly reviewed and evaluated. Are the results plausible? Do they make sense given historical population trends in the area and projected trends in other areas? Are they consistent with the area's demographic characteristics and economic conditions? It is possible that there were flaws in the original projection methodology, that errors were made in data entry or programming, or that something important was overlooked. A review and evaluation of the results often uncovers such problems. As the final step in the projection process, the entire methodology should be documented by describing data sources, models, techniques, and assumptions.

Internal Review

By *internal review*, we mean an examination of the results by the person or agency that produced the projections. This can be done in a variety of ways. We suggest several that we have found particularly helpful.

Suppose that we are reviewing a set of county projections. It is useful to observe historical population trends and to compare past changes with projected changes. Table 14.1 shows average annual population changes for several past and future time periods for selected counties in Florida. Some counties (Baker, Clay) exhibit fairly stable trends, whereas others (Miami-Dade, Gulf) exhibit a high degree of instability. A high degree of instability in historical trends is a red flag that warns the analyst to investigate the accuracy of the input data. If no errors are found, potential causes of the instability should be considered. In this example there was a logical explanation. Gulf County experienced a substantial increase in its prison population during the 1990s, and further increases were planned for 2000 and 2001. In Miami-Dade County, Hurricane Andrew reduced population growth dramatically in 1992–1993, but its impact was short-lived. Figuring out the causes of instability will help the analyst evaluate the accuracy of the input data and the validity of the assumptions used to create the projections.

Table 14.1. Average Annual Population
Change for Selected Counties in Florida,
1980–2020

Years	Baker	Clay	Miami-Dade	Gulf
1980–1985	327	3,791	32,760	75
1985–1990	312	3,995	29,557	95
1990–1995	358	2,982	15,345	353
1995–2000	373	4,138	24,744	333
2000–2005	319	3,439	22,937	401
2005–2010	295	3,246	21,373	150
2010–2015	309	3,398	22,192	156
2015–2020	321	3,555	23,040	164

Source: S. Smith and J. Nogle (1999). Population projections by
age, sex, and race for Florida and its counties, 1998–2010, *Florida Population Studies*, Bulletin 124, Bureau of Economic and
Business Research, University of Florida, and unpublished tables.

Analyzing historical trends also gives the analyst a basis for evaluating the
plausibility of the projections. Are projected population changes consistent with
historical changes? If not, why not? If a logical explanation cannot be found, it
may be a tip-off that an error was made somewhere in the projection process.
Inspecting patterns of population change in each projection interval also provides
clues regarding the plausibility of the projections. Do the projected changes
become larger, smaller, or remain about the same as the horizon becomes longer?
Is there a logical explanation for this pattern? If not, that is another red flag that
must be investigated.

Similar tables could be constructed showing percent changes rather than absolute changes, or showing county population as a share of state population. These
alternative forms provide different perspectives for viewing population change
and judging plausibility. The specific form doesn't matter. What matters is having
a set of numbers that provides a basis for comparing projected changes with
historical changes and for comparing changes in one area with changes in another.

It is also useful to construct summary tables for population characteristics
such as age, sex, race, and ethnicity. Table 14.2 provides an example of historical
data and projections for broad age groups for the United States, Florida, and
selected counties in Florida. Several patterns stand out from this table. One is the
huge differences in age structure found among the three counties. Alachua County
is the home of the University of Florida and has a very young population. Citrus
County is a magnet for retirees and has a very old population. Gilchrist County
falls in the middle, with an age structure similar to the nation as a whole. Florida's
population, of course, is considerably older than that of the nation.

Table 14.2. Percent Distribution by Age for the United States, Florida, and for Selected Counties in Florida, 1980–2020[a]

Region	Age	1970	1980	1990	2000	2010	2020
United States	0–20	39.6	33.9	30.5	30.2	28.9	28.1
	21–44	30.0	35.2	38.5	35.0	31.4	30.8
	45–64	20.6	19.6	18.6	22.2	26.5	24.6
	65+	9.8	11.3	12.5	12.6	13.2	16.5
Florida	0–20	35.8	29.2	26.3	26.2	24.8	23.5
	21–44	28.1	31.8	35.7	32.4	28.4	27.3
	45–64	21.6	21.6	19.7	23.1	27.8	25.9
	65+	14.5	17.3	18.2	18.3	18.9	23.3
Alachua	0–20	42.7	35.7	32.4	32.2	31.5	30.7
	21–44	36.0	43.8	43.8	42.1	39.5	38.6
	45–64	15.1	13.5	14.5	16.6	19.6	18.6
	65+	6.2	7.1	9.2	9.1	9.4	12.0
Citrus	0–20	27.6	22.1	20.1	18.7	16.7	15.0
	21–44	18.4	21.0	24.4	21.6	18.0	16.4
	45–64	28.0	28.0	24.3	26.6	31.4	28.6
	65+	26.0	29.0	31.2	33.1	33.9	39.9
Gilchrist	0–20	39.5	36.3	31.6	30.2	28.9	28.2
	21–44	27.2	30.8	34.0	34.1	30.9	29.3
	45–64	22.2	21.9	20.8	22.3	26.0	24.6
	65+	11.1	11.0	13.6	13.4	14.1	18.0

[a]Data for 1970, 1980, and 1990 are based on census counts; data for 2000, 2010, and 2020 are based on projections. Percentages may not add to 100 due to rounding.
Sources: J. Day (1996a). Population projections of the United States by age, sex, race, and Hispanic origin: 1995–2050, *Current Population Reports* P-25, No. 1130, Washington, DC: U.S. Bureau of the Census; S. Smith & J. Nogle (1999). Population projections by age, sex, and race for Florida and its counties, 1998–2010. *Florida Population Studies*, Bulletin 124, Bureau of Economic and Business Research, University of Florida, and unpublished tables.

Another pattern that stands out from this table is the aging of the population over time. The proportion of the population aged 20 and younger declines steadily between 1970 and 2020 for the United States, Florida, and all three counties. Similarly, the proportion aged 65+ grows steadily. The aging of the baby boom is also apparent. The proportion of the population aged 20 and younger is greatest in 1970, the proportion aged 21–44 is greatest in 1990, the proportion aged 45–64 is greatest in 2010, and the proportion 65+ is greatest in 2020. This pattern is found for the United States, Florida, and all three counties, despite their vast demographic differences.

Similar tables could be made showing the proportion female, proportion black, proportion Hispanic, and so forth. All provide data for judging the plausibility of the projections. Although inspecting historical and projected population changes is a tedious and time-consuming process, it is worth the effort. It improves the analyst's understanding of the population dynamics of the areas being projected, uncovers methodological flaws and data errors, and highlights the impact of special populations and growth constraints. Sometimes it points to the omission of factors that should have been included. We believe that internal review is useful for projections at all levels of geography but is particularly helpful for small areas because of their greater potential for error.

External Review

By *external review*, we mean an examination of the results by clients, public officials, advisory boards, and various groups of data users. In some circumstances, there is no formal external review. Once the results have been reviewed internally, the process is complete. This is the case for general-purpose projections produced by the Census Bureau, by some state and local government agencies, and by most private data companies.

Even in these circumstances, however, projections may be subject to a great deal of *informal* external review, as data users communicate their views (sometimes quite forcefully) regarding projection methodology and results. These comments may refer to the validity of the data, techniques, and assumptions used in constructing the projections; to the level of geographic or demographic detail provided; or to the plausibility of the results. Feedback from data users can have a positive impact on the usefulness of future projections.

In other circumstances, a formal external review is a central part of the projection process. This is often the case for customized projections produced for a particular client or for a particular purpose (e.g., local transportation planning). The client should always be given the opportunity to review and comment on the results before the projections are finalized; indeed, such an opportunity may be legally required. Given their knowledge of the projection area, clients may be able to spot irregularities that the analysts missed. Those observations may lead to improvements in the quality of the projections.

A formal review gives the analyst a chance to describe the projection methodology to the client(s), explain why particular techniques and assumptions were used, and answer any questions that might arise. The review process itself may help the analyst convince the client of the validity of the projections, which may be critical when the projections must have the client's approval before they can be finalized. The review also gives the client an opportunity to make suggestions regarding the optimal format in which to display the projections. Issues regarding geographic areas, age categories, projection horizons, and similar details, of course, should be resolved before work on the projections begins.

If the reviewers are dissatisfied with the results, the external review may be the most difficult part of the entire projection process. Dissatisfaction may be based on purely technical grounds, such as the choice of data sources, techniques, or assumptions. It is more likely, however, that dissatisfaction will be based on political or economic considerations. Population projections are used for distributing government funds, allocating various types of permits, regulating the expansion of businesses, and planning the development of infrastructure and public facilities. They often determine the winners and losers in high-stakes games that allocate political power and economic resources, and may even affect the salaries of public officials. It is no wonder that population projections are sometimes so controversial.

In some circumstances, external reviewers play only an advisory role; in other circumstances, they have the power to require that changes be made. This may put the analyst in a precarious position, attempting to balance technical competency with political expediency. At its best, an external review presents new insights, provides a final opportunity to catch errors, and promotes public support and acceptance. At its worst, it destroys the integrity of the process and the credibility of the projections. The successful analyst may have to be as skilled in the art of political persuasion as in the technical aspects of population forecasting.

Documentation

A written description of the methodology should accompany the projections. This report should cover data sources, projection methods, assumptions, special adjustments, and any other factor considered in constructing the projections. The reasons for choosing the forecast or "most likely" projection should be spelled out clearly. It is also helpful to discuss the range of projections, including the expected degree of forecast accuracy of the "most likely" projection, given its past performance or some other measure of uncertainty. A comparison of projected trends with past trends and with projections in other areas provides a context for considering the implications of the projections.

The methodological description should be comprehensive but clear. It should cover all aspects of the projection process in terms that can be understood by the data user. It is sometimes helpful to put the most technical material in an appendix or a separate report. Alternatively, a general description of the methodology and results can be put in an executive summary and the detailed description in the main body of the report. A clear description of the methodology helps data users understand and evaluate the projections and decide how they might best be used for their purposes.

Writing up the methodological description also helps the analyst review the entire projection process and note any parts that require further consideration. Are some data sources of dubious quality? Are some assumptions particularly questionable? Have all the relevant factors been taken into account? What improvements might be made next time around? Documenting all the steps in the projection process helps the analyst uncover weak spots in the methodology and develop a strategy for dealing with them.

Thorough documentation is essential for replicating projections in the future. It is much easier to repeat or revise a methodology used successfully in the past than to create a new methodology from scratch. Written documentation is particularly important when staff turnover takes away the analyst(s) who has direct knowledge of the methodological details of an earlier set of projections.

CONCLUSIONS

The guidelines presented in this chapter cannot answer every question and solve every problem that might be encountered when making small-area population projections. Every set of circumstances is unique in one way or another, with a variety of factors that must be considered during the projection process. In addition, not every guideline will be relevant in every situation. However, there are a number of commonalities shared by all small-area projections that provide a basis for developing a general set of guidelines that are applicable in a wide variety of settings.

Following these guidelines will not guarantee the accuracy of population forecasts, of course. Even the most brilliant analyst—armed with high-quality data, sophisticated models, and extensive knowledge of the area—may produce forecasts that turn out to be wildly inaccurate in some instances. We offer no crystal balls, no magic potions, and no guarantees. However, we believe that these guidelines can help the analyst focus on the relevant issues, make reasonable choices, and avoid common mistakes. These are modest accomplishments, perhaps, but they provide some degree of comfort to the analyst facing the sometimes daunting task of constructing a set of small-area population projections.

CHAPTER 15

New Directions in Population Projection Research

Using the past to project the future has been likened to driving a car by looking into the rearview mirror (Beck, 1996). There is certainly some truth in this analogy. All of the projection methods discussed in this book were based—in one way or another—on extrapolating past trends. But what is the alternative? If the windshield is covered with mud but the back window is clear, what offers the greater chance for keeping the car on the road (or at least on the shoulder), staring into the blank windshield in front of us or drawing inferences from the curves, hills, and potholes observed in the road behind us?

In reality, all objective projection methods—including cohort-component and structural models as well as trend extrapolations—are based on historical observations. Likewise, subjective projections based on intuition, experience, or expert judgment are distilled from the analyst's knowledge and understanding of past experiences. Only visions of the future inspired by dreams, tea leaves, or the stars may be truly free of the past. Even for these, who knows what part of a vision was actually inspired by historical events?

Projection methods differ in the specific variables, data sources, and time periods considered by the analyst; the ways in which those variables and data sources are related to each other; and assumptions regarding future changes in variables and their interrelationships. Therefore, future changes in the field of population projections will stem from changes in the availability of historical data, the tools for organizing and manipulating those data, our understanding of the way different variables interact to determine population change, and our ability to build new models or develop new methods based on these insights. The inspired analyst will incorporate factors not previously considered or will combine them in creative new ways. That inspiration, however, will still be firmly rooted in the past.

Frank and Ernest

© 1998 Thaves/Reprinted with permission.

What recent developments might change the way we make population projections? Where is current research headed? Are any "paradigm shifts" imminent? In this chapter, we describe several promising new developments in the field of state and local population projections, make some suggestions regarding areas that need further research, and offer several predictions regarding future developments. We also discuss several recent changes in the scope of projections and some of the challenges we see looming on the horizon.

We distinguish between two types of development. *Technological developments* are those that affect the availability of input data and the computing tools used to organize and manipulate those data. *Methodological developments* are those that affect the models used to formulate relationships among input data and to project those relationships into the future. Put another way, technological developments affect the availability of resources, whereas methodological developments reflect new ways of using those resources. Although this distinction is not always clear-cut, we believe that it helps clarify the developments discussed in this chapter.

TECHNOLOGICAL DEVELOPMENTS

Three technological developments—greater data availability, expanding computing power, and the growth of geographic information systems (GIS)—are transforming the way we make population projections. These developments have already had a substantial impact and promise to have an even greater impact in the coming years.

Data Availability

Sources of mortality, fertility, and migration data have evolved gradually during the past several decades. Perhaps the most important improvement was the development of annual migration estimates based on IRS records, which first became available during the late 1970s. These data were more timely than migration data from the decennial census, and they made it possible to apply projection methods based on annual data series (e.g., ARIMA time series models). Other than that, data on the components of population growth have changed very little during the last half century.

That may be about to change. As mentioned in Chapter 2, the American Community Survey (ACS) was begun on a limited basis in 1996 and has been expanded every year since that time. It is expected to cover some 3 million households each year by 2003. The ACS will provide annual estimates of demographic, economic, social, and housing characteristics for every state, as well as for cities and counties whose populations are 65,000 or more. For areas smaller than 65,000, it will take two to five years to collect enough data to provide reasonably accurate estimates. By 2008, the Census Bureau plans to issue annual estimates (based on multiyear averages) down to the block-group level for every county in the United States.

These data will be extremely useful for small-area population estimates and projections. Annual population characteristics will become available for a wide variety of geographic areas in which they were not previously available. The migration data will be particularly useful because there are few alternative sources. If carried out as planned, the ACS will provide a tremendously valuable new source of demographic data.

Although there have been few recent changes in the *sources* of data used for population projections, there have been major changes in the *formats* in which those data are available. For centuries, hard copy (i.e., printed reports) was the only available format. This format was slow, cumbersome, and expensive. Recent decades have seen the development of computer tapes, diskettes, CDs, and the Internet. Many types of data are now available instantly and free, simply at the click of a mouse. Information technology has revolutionized our ability to access and use data for population projections and a variety of other purposes. It takes no great leap of faith to anticipate substantial further improvements along these lines.

Population projections are often based on postcensal estimates. These estimates are typically based on symptomatic indicators of population change such as births, deaths, tax records, building permits, electric customers, school enrollment, and Medicare enrollees. Changes in information technology will increase the number and variety of data series that might be used as symptomatic indicators and will make them more readily available to a larger number of data users. In addition, the expansion of GIS will make it possible to tabulate these data at lower and lower levels of geography. These changes will have an important (albeit indirect) impact on population projections, especially for small areas.

The national Master Address File (MAF) assembled by the Census Bureau in preparation for the 2000 Census is another data product that is likely to have an impact on the construction of population projections. This is intended to be a "live" file, updated continuously in conjunction with the Census Bureau's geographically referenced TIGER file. In previous decades, the Census Bureau would create an address file just in time for each census but would not maintain an updated file between censuses. The development of a continuously updated MAF/ TIGER file will have a dramatic impact on the production of small-area population estimates, particularly if combined with ACS data (Hammer, Voss, & Blakely, 1999). By making it possible to construct accurate estimates annually, it will also have an important impact on the production of small-area projections.

Computing Capabilities

Before 1950, most population projections were generated manually by using pencils and paper or rudimentary mechanical devices for performing arithmetic operations. The dawn of the computer age made it possible to produce population projections much more quickly and easily for a wider range of geographic areas and demographic categories. It also facilitated the development of entirely new projection methods. For example, the urban systems models discussed in Chapter 10 are "computer-intensive"; they could not be implemented without a substantial amount of computing power. Increases in computing capabilities will undoubtedly influence future developments in the field of population projections, just as they have in other fields where projections are built on complex models driven by masses of data (e.g., meteorology).

In the early years of the electronic era, automated projection programs were written using a high-level programming language (usually FORTRAN) and were run through a centralized computer. These programs required hundreds of punch cards and were slow, inefficient, and often frustrating. The advent and widespread use of personal computers, spreadsheet programs, and statistical analysis packages has revolutionized the production, analysis, and evaluation of population projections. Spreadsheets and statistical analysis packages are simpler to use, more tolerant of errors, and require less training than formal programming languages. They have many built-in functions and macros to perform repetitive tasks and facilitate creating, displaying, printing, and graphing data (e.g., Klosterman, Brail, & Bossard, 1993). Greater computing power has also made possible the development of powerful desktop GIS applications. These developments have provided the technical infrastructure needed to support ever more complicated and data-intensive projection methods.

Distributed computing environments have made a comeback in recent years, along with new software and programming applications. Many analysts now work in networked computer environments, which make it easy to share information and obtain access to centralized sources of information. Relational database management systems, modern computing platforms, and user-friendly interfaces are being used more and more frequently for population projections (e.g., San

Diego Association of Governments, 1998). These new technologies greatly facilitate the development, analysis, and distribution of population projections.

Relational database storage and retrieval systems make it easier to manage, maintain, document, and verify information. Modern computing platforms contain structured and modular programming algorithms that are easy to maintain, can handle a wide range of computational algorithms, and permit detailed documentation of computer code. Finally, a well-designed, user-friendly graphical interface can tie together the entire projection system (e.g., data, assumptions, results, and computer programs). This frees the analyst from having to perform a number of tedious, time-consuming chores and makes it easier to explain projection methods to data users and to train new analysts.

The Internet, too, is beginning to influence the production of population projections. We believe that "projection tool kits" are likely to become available on the Internet in the years ahead. These tool kits will contain a variety of computational algorithms and have sophisticated data management and output capabilities that will enable analysts to apply their own data and assumptions and to develop a wide variety of reports, charts, graphs, and data files. This will substantially reduce the time required to put together the appropriate software for constructing population projections.

The Internet has already made it quicker, easier, and less costly to obtain the data needed for population projections. However, this is still mostly a "manual" process that requires locating information site by site. Looking ahead, we expect this process to become automated, perhaps through a common, easy-to-use interface that locates and integrates all of the relevant data sources, regardless of their file structure, format, or location on the Internet.

Geographic Information Systems

Most demographic analyses have a strong geographic component. For example, the population projections discussed in this book referred to specific geographic areas such as states, counties, or census tracts. Other data-intensive fields (e.g., urban planning, marketing, epidemiology, environmental science, sociology) also have close relationships with geography. It is not surprising, then, that a computer-based methodology has come into widespread use for the geographic display and analysis of data.

An *information system* is a chain of operations that involves observing and collecting data, storing and analyzing data, and using the derived information for some type of decision making (Star & Estes, 1990). A *geographic information system* (GIS) is designed specifically to work with data referenced by geographic or spatial coordinates. According to Star and Estes, a GIS has five essential elements: data acquisition, preprocessing (i.e., putting the data into a consistent format), data management, data manipulation and analysis, and product generation. GIS provides the tools for linking spatial data with nonspatial data and allows the analyst to organize data from one or more sources into a variety of geographic areas (e.g., census tracts, ZIP codes, market areas) and helps data users visualize spatial relationships.

GIS has been used for displaying data and conducting spatial analyses in a number of fields (e.g., Marks, Thrall, & Arno, 1992; Thrall, 1998). In recent years, it has gained acceptance for a variety of demographic activities as well (Pol & Thomas, 1997). In the coming years, we believe that it will play an increasingly important role in the production of population projections. A study by Swanson, Hough, Rodriguez, and Clemans (1998) illustrates how this might be done.

A group of consultants was hired by a school board to help them consolidate several independent school districts, to determine whether any new schools were needed, and to develop attendance zones for each school. The consultants approached this project by combining GIS tools with expert judgment and conventional population projection methods. They began by organizing school enrollment and population data by traffic analysis zone (TAZ). Projections were made for each TAZ. GIS was used to produce a computer-generated display that showed a variety of school attendance zones, each based on a different combination of TAZs. Decision makers were able to see projections for each of these potential attendance zones, make suggestions for reconfiguring those zones, and immediately see the reconfigured results. After a number of sessions (including public meetings), the school board adopted a final set of attendance zones. In this project, GIS provided a methodology for quickly and efficiently organizing and reorganizing a large amount of data, including population projections.

For the near future, the main value of GIS is likely to be in developing databases rather than constructing population projections. We provided a glimpse of this in Chapter 10 in our discussion of urban systems models. In recent years, more and more databases have included geographic identifiers. One company (MapInfo, 1998) estimated that 85% of all databases now contain some type of geographic information. The ability to geocode data such as building permits, electric customers, tax records, births, deaths, school enrollment, and other symptomatic indicators of population change will greatly enhance our ability to make estimates for very small geographic areas.

GIS makes it possible to analyze historical demographic trends for very small geographic areas. This capability will facilitate the refinement of current estimation and projection methods and the development of new ones. GIS may be particularly useful when combined with remote sensing to develop small-area population and housing estimates (e.g., Cowan & Jensen, 1998; Lo, 1995; Wicks, Swanson, Vincent, & De Almeida, 1999). Another potential use is the evaluation of projections: Maps and other visual formats are often more useful than printed output for identifying unreasonable projections.

GIS has a bright future in the field of small-area population projections. It is a powerful and flexible tool that has many potential benefits. However, GIS-based projection methods are complex and expensive. They require a substantial investment of time and effort. Clearly, the benefits of GIS must be balanced against the costs.

METHODOLOGICAL DEVELOPMENTS

The technological developments discussed before help improve the quality of the data and tools available for constructing projections. Will these developments lead to better projections? That depends primarily on improvements in the way those tools are used. Methodological advances are likely in at least six areas: (1) microsimulation models, (2) spatial diffusion models, (3) artificial neural networks, (4) integrating expert judgment, (5) measuring uncertainty, and (6) combining projections. All of these developments have the potential to increase the overall utility of population projections; some of them may improve forecast accuracy as well.

Microsimulation Models

For many years, population projections were made primarily at the national and state levels. In recent decades, projections have been carried out at progressively lower levels of geography. Now, projections are routinely made for subcounty areas such as census tracts, block groups, and traffic analysis zones. In Chapter 10, we discussed projection models designed for very small areas such as blocks and grid cells. We also noted a growing demand for projections for even smaller areas such as tax assessor parcels, block faces, and street segments. Taking this trend to its logical conclusion implies the development of projections for individual addresses, households, and people.

Making projections at this level may seem like science fiction, but several models for projecting individual households and people have already been constructed. These are often referred to as *microsimulation models*. The Transportation Analysis Simulation System (TRANSIMS) model funded the U.S. Department of Transportation is an example of a microsimulation model (University of California, 1999).

TRANSIMS was developed in response to federal transportation and clean-air legislation. It represents an attempt to evaluate topics such as traffic congestion, alternative development patterns, motor vehicle emissions, and intelligent transportation systems. TRANSIMS is a radical departure from traditional modeling approaches. Its major technical feature is its ability to identify individual travelers and to provide a detailed, second-by-second history of every traveler in the transportation system during a 24-hour day. These travelers come from synthetic households comprised of one or more individuals; each has an associated set of demographic characteristics (e.g., age, sex, income, labor force status). The current and future locations of these households and persons are determined from a variety of sources, including the decennial census, sample surveys, activity locations, land uses, the TIGER file, and detailed population and housing forecasts by block group (Beckman, Baggerley, & McKay, 1996).

TRANSIMS has been tested only on a very limited scale, but plans are currently underway to conduct large-scale tests during the next several years. These tests will attempt to demonstrate the model's capability for analyzing a wide range of polices, develop a base of experienced model users, identify problems and issues associated with implementing the model, and find solutions that permit widespread application of the model. Whether TRANSIMS becomes the standard methodology for transportation modeling remains to be seen, but it offers an example of the kinds of methods and data sets required by microsimulation projection models. We believe that projections for very small areas—including projections of individuals and households—will become increasingly common.

Spatial Diffusion Models

Population growth often moves progressively outward from (relatively) densely populated areas. For example, a city initially may grow rapidly within a small, central geographic area. As it becomes more densely populated, growth spills outward from the central city to nearby, less densely populated areas. As the nearby areas become more densely populated, growth spills over into areas even farther away. This is a widespread (albeit not universal) pattern that has characterized historical population growth at the national and local levels. Figure 15.1 illustrates this pattern in San Diego County, California, from 1940 to 2000.

Geographers refer to this growth pattern as *spatial diffusion*, which is defined as the spread of a particular phenomenon over space and time, starting from one specific time and place. It is a process in which behavior or characteristics in one area change as a result of what happened *earlier* in some *other* area (Morrill, Gaile, & Thrall, 1988). Spatial diffusion can refer to the spread of an idea, an attitude, a technological innovation, a religion, a language, a disease, and many other types of phenomena. It can also refer to the spread of population.

Spatial diffusion models offer a promising alternative to traditional projection models, especially for very small areas. Whereas traditional models are based on historical trends within a particular geographic area, spatial diffusion models are based on historical trends in neighboring areas. Because growth often spreads from one area to another over time, historical trends in neighboring areas may have a greater impact on an area's future growth than historical trends in the area itself. These models have the potential to predict when a stagnant or slowly growing area will suddenly start to grow more rapidly and when growth in a rapidly growing area will start to taper off.

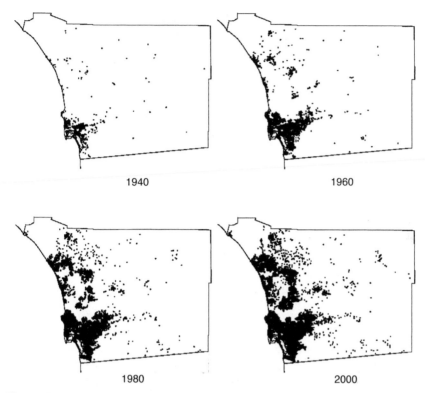

Figure 15.1. Spatial distribution of housing units, San Diego County, 1940–2000. The "year built" code in the 1990 Census allows one to estimate the count of units in each 1990 block group at various historical times. This is only an approximate estimate because it does not account for units built before 1990 and demolished. For 2000, we used census block estimates prepared by the San Diego Association of Governments. GIS techniques were used to randomly distribute the units to specific points within a block or block group. (*Sources*: Census of Population and Housing, 1990 Summary Tape File 3 [California]. U.S. Bureau of the Census, Washington, DC, 1991; San Diego Association of Governments, Population and Housing Estimates, 2000.)

The spatial diffusion of a phenomenon *within* an area has often been modeled using S-shaped curves, particularly the logistic curve (e.g., Casetti, 1969; O'Neil, 1981). Although these curves may be helpful for predicting when an area's growth may start to slow down, they are not very useful for predicting when growth will accelerate. For this purpose, models must be developed that incorporate historical growth rates in nearby areas. This can be done in a variety of ways, such as by developing regression models in which an area's future growth is determined by historical growth in adjacent areas and other nearby areas (Thrall, Sidman, Thrall,

& Fik, 1993). Further research is needed on the best ways to model population growth simultaneously over space and time.

Spatial diffusion models hold a great deal of promise for population projections of small areas such as census tracts, block groups, and individual blocks. We believe that research in this area will provide interesting new perspectives on the dynamics of small-area population growth and may lead to significant improvements in forecast accuracy as well.

Artificial Neural Networks

Artificial neural networks (sometimes simply called *neural networks*) are mathematical models inspired by the organization and functioning of biological neurons (Hill, Marquez, O'Connor, & Remus, 1994). These mathematical models are *not* based on the analyst's specification of relationships among independent and dependent variables; rather, artificial neural network (ANN) models develop these relationships on their own as they analyze input data by following procedures set forth in the model. Through trial and error, ANN models literally teach themselves how to perform a specific task. Many different ANN models have been proposed in recent years (e.g., Lippmann, 1987; Wasserman, 1989; Schocken & Ariav, 1994).

ANN models are based on the way the human brain is believed to operate. They are composed of a number of interconnected processing elements called *neurons* or *nodes*. Each neuron receives input signals from other neurons, processes those signals, and produces output signals going to other neurons. The links between neurons are expressed as weights that measure the impact on one neuron on another. Although each neuron implements its functions rather slowly and imperfectly, the network as a whole is capable of performing a number of tasks quite efficiently (Zhang, Patuwo, & Hu, 1998).

ANN models are specified by network topology, node characteristics, and learning rules. They learn from past examples and can capture both linear and nonlinear relationships. They have been widely used in areas of pattern recognition, pattern classification, signal processing, speech processing, and image recognition. They have also been used to forecast business failures, foreign exchange rates, stock prices, air traffic, rainfall, water demand, electricity consumption, and a number of other variables. They lend themselves well to many forecasting situations because they are data-driven, self-adapting methods that involve few a priori assumptions. A number of studies have found that ANN models produce relatively accurate forecasts, at least under some circumstances (e.g., Adya & Collopy, 1998; Zhang et al., 1998).

Forecasting models based on artificial neural networks have several shortcomings, however. First, there is no standard procedure for identifying and applying ANN models; consequently, they require a substantial amount of trial-and-error experimentation. Second, they are black-box models that have no explicit forms or relationships, which makes them very difficult to explain to data users. Finally, they produce satisfactory results only if based on a substantial amount of input data.

To our knowledge, ANN models have not yet been used for population projections. However—given the explosive growth of computing power and the increasing use of ANN models in other fields—we believe that they eventually will be used for population projections. Because they are nonlinear models, their greatest value may be for short-range projections that pick up seasonal and cyclical fluctuations. Perhaps they can be trained to spot unreasonable projections and develop their own corrections. Perhaps they can be used to develop "real time" probabilistic measures of uncertainty. Their ability to identify a variety of patterns from large data sets makes them intriguing candidates for further research.

Integrating Expert Judgment

This book has focused primarily on objective projection methods that can be quantified and replicated by other researchers. However—as we have emphasized throughout the book—every population projection method requires that the analyst apply expert judgment. Cohort-component projections require assumptions regarding future mortality, fertility, and migration rates. Structural models require decisions on independent variables and structural forms. Even simple trend extrapolation methods require choices regarding the length of the base period and adjustments for special events.

How are these decisions made? Compared to the mechanical application of extrapolative methods, does the application of expert judgment tend to raise, lower, or leave forecast accuracy unchanged? Are there some circumstances in which expert judgment improves accuracy but other circumstances in which it does not? How can expert judgment best be incorporated into the projection process? Can the application of judgment itself be quantified?

There are currently no definitive answers to these questions. Very few population researchers have addressed these issues, and the general forecasting literature is inconclusive. Some researchers have concluded that the application of expert judgment generally leads to more accurate forecasts than can be achieved by strictly extrapolative methods (e.g., Armstrong, 1983) but others have concluded that it has no consistent effect (e.g., Grove & Meehl, 1996). We believe that expert judgment is likely to improve forecast accuracy in some circumstances, but further research is needed before we can form clear conclusions.

How can expert judgment be incorporated directly into the projection process? One approach is to use the Delphi method, developed during the 1950s by researchers working on defense projects at the Rand Corporation (e.g., Dalkey & Helmer, 1963). Essentially, the Delphi method is a set of procedures for eliciting and refining the opinions of a group of experts. Although there are a number of potential variations, the process works something like this:

1. Experts are chosen who are knowledgeable about the topic under consideration. Ideally, they come from a variety of backgrounds and have different perspectives and areas of expertise.
2. Information and opinions on the topic are collected from each expert and distributed anonymously to the entire group.
3. After reviewing the information and opinions provided by the other experts, each expert revises or refines his/her original response.
4. The set of revised responses is sent to the entire group for further review and revision.

These steps are repeated until stability in the responses is achieved. The final responses may (or may not) reflect a general consensus of the group. The number of rounds is variable, but seldom goes beyond two complete iterations (Rowe & Wright, 1999).

The Delphi method is designed to maximize the positive aspects of interactions among a group of experts (e.g., collecting opinions from a variety of knowledgeable sources, synthesizing information) and to minimize the negative aspects (e.g., political or personality conflicts, domination by particular group members). It is generally not intended to replace statistical or model-based procedures, but rather to be used in situations in which those procedures are impractical or impossible because of a lack of data. It has been used for studies in a number of fields—including health care, education, transportation, engineering, and national defense—and often leads to more accurate judgments than can be obtained from individual experts or from other ways of eliciting expert opinion (Rowe & Wright, 1999).

The Delphi method has seldom been used for constructing population projections. In fact, methods that explicitly incorporate the views of a number of experts have not been frequently used. This may be changing now. One recent study used a panel of experts to develop a probabilistic range of assumptions regarding projections of mortality, fertility, and migration rates (Lutz, Sanderson, & Scherbov, 1999). Another discussed a variety of issues related to the use of experts in population forecasting and suggested an approach based on the validity of the arguments used to support specific assumptions (Lutz, Saariluoma, Sanderson, & Scherbov, 2000). We believe that the next few years will bring renewed interest in subjective projection methods and—more generally—in the role played by expert judgment in producing population projections and forecasts.

Measuring Uncertainty

Population projections cannot provide perfectly accurate forecasts of future population growth. The uncertainty inherent in population projections has traditionally been accounted for by developing a range of projections (e.g., Campbell, 1996; Day, 1996a). An alternative approach is to develop statistical prediction intervals to accompany population forecasts (e.g., Ahlo & Spencer, 1997; Cohen, 1986; Lee & Tuljapurkar, 1994). Whereas a range provides only a vague indication of uncertainty, prediction intervals provide an explicit statement of the level of error expected to accompany a population forecast.

We believe that future research will focus increasingly on the measurement of uncertainty in population forecasts. Although such research may not directly improve forecast accuracy, it will enhance our understanding of the uncertainty inherent in population forecasts, particularly how uncertainty varies by population size, geographic region, launch year, length of horizon, projection method, and perhaps other factors as well. Before long, private companies and federal, state, and local government agencies are likely to start producing explicit prediction intervals to accompany their population forecasts. This change will imply a shift from "population projections" to "population forecasts." We believe that prediction intervals will provide extremely valuable information to data users and will improve the quality of decision making based on population forecasts.

Combining Projections

Combining projections often improves forecast accuracy and reduces the variability of individual projections (e.g., Armstrong, 2001c; Clemen, 1989; Webby & O'Connor, 1996). This result has been found for many types of forecasts, including gross national product, inflation, corporate earnings, stock prices, exchange rates, rainfall, and psychiatric diagnoses. Combining can be done in a number of ways, including weighted averages, simple averages, trimmed means, and composite approaches (see Chapter 13 for further discussion).

We believe that combining offers the same potential benefits to population projections that have been achieved in other fields. Several recent studies found that combining population projections from different projection methods led to more accurate forecasts than could be achieved using a single method by itself (e.g., Ahlburg, 1999; Sanderson, 1999). Further research on this topic is likely to improve forecast accuracy and lead to the more frequent use of combining for population projections. It might also lead to new methods for constructing population prediction intervals.

SCOPE OF PROJECTIONS

Not only are changes occurring in the technology and methodology of population projections, but in their scope as well. In particular, projections are being made for smaller and more varied units of geography and for a broader array of demographic characteristics. This trend is driven by market demands for such projections and by technological changes that make such projections feasible.

The demand for small-area population projections is increasing. In the public sector, projections are used for constructing budgets, developing transportation systems, planning for future changes in school enrollment, and determining the optimal location of public facilities. Private sector uses include financial planning, site analyses, sales forecasting, target marketing, and new product introduction. Many of these uses require projections for very small levels of geography and very detailed population characteristics.

Recent trends in marketing illustrate the growing demand for geographic and demographic detail (Pol & Thomas, 1997). In the 1960s, mass marketing was the order of the day. The market was seen as a homogeneous mass of consumers, each one pretty much like any other. Over time, this concept gave way to the notion of target marketing in which specific products or brands were directed toward specific types of consumers. In response to the development of even more refined demographic clusters, target marketing is now giving way to *micromarketing*, which focuses on characteristics at the household level. Micromarketing attempts to pinpoint very small groups of potential customers and focus on their buying patterns. Armed with information from an ever-growing number of consumer databases, micromarketers are targeting customers down to the block or even the household level.

We believe that the demand for detailed projections for very small areas will continue to grow. Technological changes such as improvements in GIS and the availability of geocoded databases will allow practitioners to meet that demand more easily than in the past. The development of microsimulation models and spatial diffusion models will be particularly useful for these purposes. Greater geographic and demographic detail in population projections is clearly the wave of the future.

SOME CHALLENGES

Analysts who make population projections face several new challenges. One is an outgrowth of the same technological developments that have added to the analyst's projection tool kit. Because of the tremendous increase in the speed and ease with which information can be exchanged worldwide, more and more Americans are working at home rather than at traditional places of business. One's choice of a place to live is not as closely tied to the choice of a place to work as it once was. Similarly, the way businesses operate has changed dramatically. Work once done on-site can now be done off-site. For example, a firm in New York City can have its bookkeeping done in Omaha or Austin (or Bombay, São Paulo, or Singapore, for that matter).

What will these changes mean for future migration patterns? Will amenities such as a temperate climate, beaches, mountains, clean air, and cultural or recreational opportunities start to replace the economic factors that have traditionally had the greatest impact on long-distance migration decisions? Will long-standing migration trends evolve in new and heretofore unseen ways? Will new patterns emerge? Just as technological changes in the nineteenth and twentieth centuries led many people from the farm to the city and from the agricultural South to the industrial North, twenty-first century technological changes may shift migration flows to new locations. Foreseeing and accounting for these changes will be a major challenge for population forecasters.

Another challenge lies in the projection of a population's race and ethnic characteristics. Race and ethnicity are ambiguous concepts that are difficult to define, much less to measure. They are affected by a variety of social, cultural, linguistic, and psychological factors as well as by ancestry and genetics. Viewpoints regarding classification by race and ethnic identity vary from person to person and change over time; as a result, data on race and ethnic characteristics are not necessarily consistent across data sources and over time. This problem was exacerbated in the 2000 Census, when respondents were allowed to list themselves in more than one race category. Developing meaningful race and ethnic categories and constructing consistent historical data series is a major challenge facing population forecasters.

Earlier in this chapter, we mentioned that more and more databases are being developed, containing more and more information about more and more people. Although these databases are extremely useful for many purposes, their growth and widespread availability raise questions regarding the confidentiality of data. A backlash against collecting and using data—based on privacy concerns—could have serious repercussions for all data users. In particular, it could threaten the development of linked data systems based on geographic or personal identifiers. This would be a major blow for the production of population estimates and projections. Developing new data sources and maintaining access to current sources may become an increasingly difficult challenge for population forecasters.

The widespread availability of demographic data and software will make it possible for more and more people to make population projections and to make them faster, cheaper, with greater demographic detail, and for a wider variety of geographic areas than ever before. This trend will be generally beneficial for data users and will raise the overall usefulness of population projections. However, it will also lead to an increase in the number of projections based on ill-founded assumptions, inadequate attention to detail, vested political or economic interests, and a poor understanding of the causes of population growth and demographic change. This proliferation of projections will almost certainly be confusing. Data users will face a broader array of options than ever before and will have to do more homework to make the best choices. As elsewhere in a market-driven economy, *caveat emptor* (let the buyer beware) will be the order of the day.

Glossary

Administrative records.   Records kept by agencies of federal, state, and local governments for purposes of registration, licensing, and program administration. These records provide information on demographic events and population changes and are frequently used for constructing population estimates.

Age-specific rate.   A statistical measure that relates the number of demographic events (e.g., births, deaths) for a specific age group to the corresponding at-risk population. Age-specific rates are typically calculated by dividing the annual number of events by the midyear population in each age group.

Age structure (*see* **Population composition**).

American Community Survey (ACS).   A monthly household survey conducted by the Census Bureau designed to provide socioeconomic and demographic information for states, counties, places, and other small areas. According to current plans, it will eventually cover approximately 3 million households each year and will replace the long form of the decennial census in 2010.

Autoregressive Integrated Moving Average (ARIMA) model.   A model that bases projections of future values in a time series on the patterns of change in its historical values. Historical values are typically expressed as differences and future values are expressed as a function of previous values and previous forecast errors.

At-risk population.   The set of people to whom a demographic event (e.g., birth, death, migration) might potentially occur. Ideally, the at-risk population is measured by the total person-years lived during the relevant time interval. In reality, it is often approximated by the midyear population.

Base period.   The interval between the base year and launch year of a projection.

Base year.   The year of the earliest data used to make a projection.

Block.   A small geographic area bounded on all sides by identifiable features (e.g., roads, rivers, property lines, city limits). A block is the lowest geographic level for which decennial census data are tabulated.

Block group.   A cluster of blocks, typically containing between 250 and 550 housing units at the time of a decennial census.

Block numbering area.   Block numbering areas are equivalent to census tracts in nonmetropolitan counties for which census tracts have not been defined. This geographic concept was used only for the 1990 Census; for 2000, all counties were divided into census tracts.

Bottom-up model.   In a hierarchically nested geographic system, a model in which a projection for a higher geographic level (e.g., a state) is computed by summing the projections for a lower geographic level (e.g., counties).

Census.   A count of the entire population of a specific geographic area at a specific time. The U.S. government has conducted a census every 10 years since 1790; that is, the United States has a *decennial* census. The Census Bureau and state and local governments occasionally conduct special censuses for particular cities, counties, and other small areas for years between decennial censuses.

Census survival rate.   A measure of the probability of surviving from one census year to another. For censuses conducted every 10 years, it is calculated by dividing the population age $t+10$ in one census by the population age t in the previous census. Also called a *cohort-change ratio*, this measure includes the effects of mortality, migration, and census enumeration errors.

Census tract.   A small, relatively permanent statistical subdivision of a county. Census tracts are designed to be homogenous with respect to population and economic characteristics and typically contain between 2,500 and 8,000 persons at the time of a decennial census. Although census tract boundaries are intended to remain constant over time, changes reflecting population growth or decline often occur.

Child-woman ratio.   The number of children (e.g., aged 0–4) divided by the number of women of childbearing age (e.g., aged 15–44) at a given point in time.

Cohort.   A group of people who experience the same significant event during a particular time interval. For example, all persons married in 2000 are the marriage cohort for that year and all persons born during the 1990s are the birth cohort for that decade.

Cohort perspective.   A longitudinal view of demographic events and other life experiences for a particular cohort as it progresses through time.

Cohort-change ratio (*see* **Census survival rate**)

Cohort-component method. A method in which the components of population change are projected separately for each age-sex group in the population.

Components of population change. The demographic events that determine population change: births, deaths, and migration. A population grows through the addition of births and in-migrants and declines through the subtraction of deaths and out-migrants.

Controlling. The process of adjusting a geographic or demographic distribution to an independently derived total. For example, county population projections can be controlled to an independent state projection and age groups from one projection can be controlled to the total population from another projection. *Raking* is a synonym.

Crude rate. A statistical measure in which the number of demographic events during a time interval is divided by the total population. For example, the crude birth rate can be calculated by dividing the annual number of births by the midyear population.

Current Population Survey (CPS). A sample survey of approximately 50,000 U.S. households conducted monthly by the Census Bureau. This survey collects a wide range of socioeconomic and demographic data for the nation, all states, and some metropolitan areas.

Curve fitting. The process of finding the mathematical formula that best describes a particular data set (typically measured over time).

Demographic balancing equation. A basic demographic formula in which population change is expressed as births minus deaths plus in-migrants minus out-migrants. An error term is sometimes included to account for measurement errors.

Econometric model. An equation (or set of equations) in which relationships between independent and dependent variables are estimated using statistical methods. For population projections, independent variables are typically economic variables and dependent variables are demographic variables.

Emigration (*see* **International migration**)

Error of closure. A term added to the demographic balancing equation to account for errors in population counts (or estimates) and errors in measures of the components of population change.

Estimate. A calculation of a current or past value of a variable (e.g., population). Estimates are often based on symptomatic indicators of changes in the variable's values but can also be based on extrapolation or interpolation methods.

Extrapolation. The process of using mathematical formulas or graphical procedures to determine values that fall beyond the last known value in a series of numbers.

Face validity. The extent to which a projection uses the best methods for a particular purpose, is based on reliable data and reasonable assumptions, and accounts for the effects of relevant factors.

Forecast. The projection selected as the one most likely to provide an accurate prediction of the future value of a variable (e.g., population).

Forecast error. The difference between forecasted and actual (or estimated) values of a variable. The magnitude of the difference—measured in either absolute or percent terms—is called *accuracy* and the direction of the difference is called *bias*.

General fertility rate. The number of births during a time interval (e.g., a year) divided by the number of women of childbearing age (e.g., 15–44).

Geocoding. The assignment of a specific geographic location (e.g., latitude and longitude) to a person, household, or other entity. These locations are often based on information regarding street addresses, intersections, or other clearly recognized landmarks.

Geographic Information System (GIS). A chain of operations involving the collection, storage, analysis, and manipulation of data referenced by geographic or spatial coordinates.

Gravity model. A model based on the hypothesis that movement (e.g., migration or commuting) between two geographic areas is directly related to the size of their populations and inversely related to the distance between the two.

Gross migration. The movement of migrants into or out of an area.

Hamilton–Perry method. An abbreviated cohort-component method in which projections of population change are based on cohort-change ratios.

Immigration (*see* **International migration**)

Internal (or domestic) migration. Migration from one place to another within the same country. People who enter an area are called *in-migrants* and people who leave are called *out-migrants*.

International (or foreign) migration. Migration from one country to another. People entering a country are called *immigrants* and people leaving a country are called *emigrants*.

Interpolation. The process of using mathematical formulas, graphical procedures, and/or values from a related data series to calculate intermediate values that fall between two known values in a series of numbers.

Labor force. The sum of the employed (full-time or part-time) and unemployed (without a job but actively seeking work) populations at a given point in time.

Labor force participation rate. The proportion of the population in the labor force. The labor force participation rate is typically calculated by dividing the labor force population by the total adult population (e.g., age 16 and older). Rates can also be calculated for specific subgroups of the population (e.g., age, sex, race).

Launch year. The year of the most recent data used to make a projection.

Life expectancy. The average number of years of life remaining to people who reach a given age, assuming the continuation of a particular set of age-specific survival rates.

Life table. A statistical table that shows measures of mortality, survival, and life expectancy for each age group in the population. *Period* life tables use mortality and age data from a single point in time, whereas *cohort* life tables use mortality data for a particular birth cohort as it ages over time. *Complete* life tables contain information by single year of age, whereas *abridged* life tables contain information for broader age groups (e.g., five or 10 years).

Life table survival rate. A statistical measure that shows the probability of surviving from one exact age (or age group) to another, given a particular set of age-specific death rates.

Long form. The decennial census questionnaire given to a sample of households (approximately one in six). It collects information on a wide range of socioeconomic, demographic, and housing characteristics.

Master Address File (MAF). A set of records maintained by the Census Bureau that contains the address of every housing unit in the United States. The MAF forms the basis of the decennial census and a number of sample surveys.

Metropolitan area. A geographic area represented by a large population nucleus and adjacent communities that have a high degree of economic and social integration with the nucleus. A metropolitan area consists of one or more counties and must contain either a place with a minimum population of 50,000 or a Census Bureau-defined urban area with a total population of at least 100,000.

Migrant. A person who changes his or her place of residence from one political or administrative area to another.

Migration. The process of changing one's place of residence from one political or administrative area to another.

Migration interval. The period of time over which migration is measured.

Mobility. The process of changing one's place of residence from one address (e.g., house or apartment) to another.

Model schedule. A set of demographic rates (birth, death, migration) derived from one population that can be used as a proxy for the rates in another population. Model schedules can be used for population estimates and projections in areas that lack detailed demographic data.

Mover. A person who changes his or her place of residence from one address (e.g., house or apartment) to another.

Multiregional model. A model in which migration is represented by a series of origin-destination-specific gross migration flows (e.g., a 51 × 51 matrix that depicts migration between each pair of states and the District of Columbia).

Natural increase (decrease). The excess of births (deaths) over deaths (births).

Net migration. The difference between the number of in-migrants and the number of out-migrants.

Nonrecursive model. A structural model that accounts for two-way interactions between independent and dependent variables. For example, the model may account for the effect of wage rates on migration and the effect of migration on wage rates.

Period perspective. A cross-sectional view of demographic events and other life experiences in which the combined events and experiences of all cohorts are measured at a given point in time.

Person-years lived. The total number of years lived by a population during a given time interval (e.g., five years). It is calculated by adding up the exact number of years (or fractions thereof) lived by each member of the population during the interval.

Place of residence. According to Census Bureau guidelines, this is the place where a person lives and sleeps most of the time. It is sometimes called the place of *usual* or *permanent* residence.

Plausibility. The extent to which a projection is consistent with historical trends, with the assumptions inherent in the model, and with projections for other areas.

Population composition. The classification of members of a population according to one or more characteristics such as age, sex, race, ethnicity, income, and educational attainment.

Population distribution. The spatial spread of a population among geographic areas such as states, counties, census tracts, and traffic analysis zones.

Prediction interval. An estimate of the probability that a given range of projections will contain the actual future value of a variable (e.g., population). Prediction intervals can be based on past forecast errors, statistical models, expert judgment, or a combination of these factors.

Projection. The numerical outcome of a particular set of assumptions regarding future values of a variable (e.g., population).

Projection horizon. The interval between the launch year and target year of a projection.

Projection interval. The increments in which projections are made. For population projections, one- and five-year intervals are the most common.

Public Use Microdata Sample (PUMS). A sample of responses selected from long-form questionnaires in the decennial census. PUMS data sets provide population and housing information for individual respondents. Two different samples are planned for the 2000 Census, one based on 5% of the respondents and another based on 1%.

Raking (*see* **Controlling**)

Recursive model. A structural model that accounts only for one-way interactions between independent and dependent variables. For example, the model may account for the effect of wage rates on migration but not for the effect of migration on wage rates.

Sex ratio. The number of males per 100 females in a population.

Short form. The decennial census questionnaire given to approximately five in six households. It collects information on a limited number of basic demographic and housing characteristics. All questions on the short form are also included on the long form, providing 100% coverage for these characteristics.

Special population. A group of persons who reside in an area because of an administrative or legislative action. Common types include college students, prison inmates, residents of nursing homes, and military personal and their dependents.

Structural model. A statistical model that relates changes in one variable (e.g., population or one of its components) to changes in one or more other variables (e.g., employment, income, land use, and the transportation system).

Survival rate (*see* **Census survival rate**; **Life table survival rate**)

Synthetic method. A method in which changes in demographic rates for one area are based on changes in those rates in another area. For example, age-specific death rates for a state could be projected to change at the same rate as age-specific death rates for the nation.

Target year. The year for which a variable is projected.

Time series. An ordered sequence in which the values of a variable are measured at equally spaced time intervals.

Top-down model. In a hierarchically nested geographic system, a model in which projections for a lower geographic level (e.g., counties) are adjusted so that they add to a projection for a higher geographic level (e.g., a state).

Topologically Integrated Geographic Encoding and Referencing System (TIGER). A digital database developed by the Census Bureau in which residential addresses, physical features (e.g., streets, rivers), political boundaries (e.g., cities, counties), and census statistical boundaries are assigned exact spatial locations.

Total fertility rate. The average number of children that a group of women would have during their lifetimes if their fertility behavior conformed to a given set of age-specific birth rates.

Traffic analysis zone (TAZ). A small geographic area designed for purposes of transportation modeling and planning. In most instances, a TAZ is smaller that a census tract.

Trend extrapolation method. A projection method in which projected values of a variable are based solely on its historical values.

Vital statistics. Data that reflect the registration of *vital events* such as births, deaths, marriages, divorces, and abortions.

ZIP code area. A postal delivery area delineated by the U.S. Postal Service. Although ZIP code areas do not always correspond to census geographic boundaries, they are widely used for small-area demographic analyses.

References

Adya, M., & Collopy, F. (1998). How effective are neural networks at forecasting and prediction? A review and evaluation. *Journal of Forecasting*, 17, 481–495.

Ahlburg, D. (1982). How accurate are the U.S. Bureau of the Census projections of total live births. *Journal of Forecasting*, 1, 365–374.

Ahlburg, D. (1983). Good times, bad times: A study of the future path of U.S. fertility. *Social Biology*, 30, 17–23.

Ahlburg, D. (1986). Forecasting regional births: An economic–demographic approach. In A. Isserman (Ed.), *Population change in the economy: Social science theory and models* (pp. 31–51). Boston, MA: Kluwer-Nijhoff.

Ahlburg, D. (1995). Simple versus complex models: Evaluation, accuracy, and combining. *Mathematical Population Studies*, 5, 281–290.

Ahlburg, D. (1999). Using economic information and combining to improve forecast accuracy in demography. Unpublished paper. Rochester, MN: Industrial Relations Center, University of Minnesota.

Ahlburg, D., & Vaupel, J. (1990). Alternative projections of the U.S. population. *Demography*, 27, 639–652.

Alho, J. (1997). Scenarios, uncertainty and conditional forecasts of the world population. *Journal of the Royal Statistical Society A*, 160, part 1, 71–85.

Alho, J., & Spencer, B. (1990). Error models for official mortality forecasts. *Journal of the American Statistical Association*, 85, 609–616.

Alho, J., & Spencer, B. (1997). The practical specification of the expected error of population forecasts. *Journal of Official Statistics*, 13, 203–225.

Alinghaus, D. (1994). *Practical handbook of curve fitting*. New York: CRC Press.

Alonso, W. (1964). *Location and land use*. Cambridge, MA: Harvard University Press.

American Statistical Association (1977). Report on the conference on economic demographic methods for projecting population. Washington, DC.

Anas, A. (1982). *Residential location models and urban transportation*. New York: Academic Press.

Anas, A. (1992). *NYSIM (the New York simulation model): A model for cost-benefit analysis of transportation projects*. New York: Regional Plan Association.

Anderson, M. (1988). *The American census: A social history*. New Haven, CT: Yale University Press.

Anderson, M., & Fienberg, S. (1999). *Who counts? The politics of census-taking in contemporary America*. New York: Russell Sage Foundation.

Armstrong, J. (1983). Relative accuracy of judgemental and extrapolative methods in forecasting annual earnings. *Journal of Forecasting*, 2, 437–447.

Armstrong, J. (1985). *Long range forecasting: From crystal ball to computer*. New York: Wiley.

Armstrong, J. (2001a). *Principles of forecasting: A handbook for researchers and practitioners*. Norwell, MA: Kluwer Academic.

Armstrong, J. (2001b). Evaluating forecasting methods. In J. Armstrong (Ed.), *Principles of forecasting: A handbook for researchers and practitioners* (pp. 443–473). Norwell, MA: Kluwer Academic.

Armstrong, J. (2001c). Combining forecasts. In J. Armstrong (Ed.), *Principles of forecasting: A handbook for researchers and practitioners* (pp. 417–439). Norwell, MA: Kluwer Academic.

Armstrong, J., & Collopy, F. (1992). Error measures for generalizing about forecasting methods: Empirical comparisons. *International Journal of Forecasting*, 8, 69–80.

Armstrong, J., & Fildes, R. (1995). Correspondence on the selection of error measures for comparisons among forecasting methods. *Journal of Forecasting*, 14, 67–71.

Ascher, W. (1978). *Forecasting: An appraisal for policy-makers and planners*. Baltimore: Johns Hopkins University Press.

Ascher, W. (1981). The forecasting potential of complex models. *Policy Sciences*, 13, 247–267.

Ashton, A., & Ashton, R. (1985). Aggregating subjective forecasts: Some empirical results. *Management Science*, 31, 1499–1508.

Astone, N., & McLanahan, S. (1994). Family structure, residential mobility and school report: A research note. *Demography*, 31, 575–584.

Bartel, A. (1979). The migration decision: What role does job mobility play? *American Economic Review*, 69, 775–786.

Batchelor, R., & Dua, P. (1990). Forecaster ideology, forecasting technique, and the accuracy of economic forecasts. *International Journal of Forecasting*, 6, 3–10.

Bates, D., & Watts, D. (1988). *Nonlinear regression analysis & its applications*. New York: Wiley.

Bates, J., & Granger, C. (1969). The combination of forecasts. *Operational Research Quarterly*, 20, 451–468.

Batty, M. (1976). *Urban modelling: Algorithms, calibrations, predictions*. Cambridge, UK: Cambridge University Press.

Batty, M. (1992). Urban modeling in computer-graphic and geographic information system environments. *Environment and Planning B*, 19, 689–708.

Batty, M. (1994). A chronicle of scientific planning: The Anglo-American modeling experience. *Journal of the American Planning Association*, 60, 7–16.

Batty, M., Cote, C., Howes, D., Pelligrini, P., & Zheng, X. (1995). Draft: Data requirements for land use modeling: First thoughts and a preliminary assessment. Proceedings from the Travel Model Improvement Program: Land Use Modeling Conference, DOT-T-96-09. Washington, DC: U.S. Department of Transportation.

Beaumont, P., & Isserman, A. (1987). Comment. *Journal of the American Statistical Association*, 82, 1004–1009.

Beck, A. (1996). Forecasting: Fiction and utility in jail construction planning. Internet site: http://www.justiceconcepts.com/forec.htm.

Becker, G. (1960). An economic analysis of fertility. In National Bureau of Economic Research (Ed.), *Demographic and economic change in developed countries* (pp. 209–240). Princeton, NJ: Princeton University Press.

Beckman, R., Baggerley, K., & McKay, M. (1996). Creating synthetic baseline populations. *Transportation Research A*, 30, 415–429.

Behr, M., & Gober, P. (1982). When a residence is not a house: Examining residence-based migration definitions. *Professional Geographer*, 34, 178–184.

Bell, F., Wade, A., & Goss, S. (1992). Life tables for the United States social security area 1900–2080. Actuarial Study No. 107. Washington, DC: Social Security Administration.

Bennett, N., & Olshansky, S. (1996). Forecasting U.S. age structure and the future of social security: The impact of adjustments to official mortality schedules. *Population and Development Review*, 22, 703–727.

Berger, M. (1985). The effect of cohort size on earnings growth: A reexamination of the evidence. *Journal of Political Economy*, 93, 561–573.

Berger, M. (1989). Demographic cycles, cohort size, and earnings. *Demography*, 26, 311–321.

Berry, W. (1984). *Nonrecursive causal models.* Beverly Hills, CA: Sage.

Bhrolchain, M. (1992). Period paramount? A critique of the cohort approach to fertility. *Population and Development Review*, 18, 599–629.

Blanco, C. (1963). The determinants of interstate population movements. *Journal of Regional Science*, 5, 77–84.

Bloom, D., & Trussell, J. (1984). What are the determinants of delayed childbearing and permanent childlessness in the United States? *Demography*, 21, 591–611.

Bogue, D. (1998). Techniques for indirect estimation of total, marital, and extra-marital fertility for small areas and special populations. Paper presented at the meeting of the Federal–State Cooperative Program for Population Projections. Chicago, IL.

Bogue, D., Hinze, K., & White, M. (1982). *Techniques for estimating net migration.* Chicago: University of Chicago Press.

Bolton, R. (1985). Regional economic models. *Journal of Regional Science*, 25, 495–520.

Bongaarts, J., & Feeney, G. (1998). On the quantum and tempo of fertility. *Population and Development Review*, 24, 271–291.

Borts, G., & Stein, J. (1964). *Economic growth in a free market.* New York: Columbia University Press.

Bowles, G., & Tarver, J. (1965). *Net migration of the population, 1950–1960, by age, sex and color.* Washington, DC: U.S. Department of Agriculture.

Bowles, G., Beale, C., & Lee, E. (1975). *Net migration of the population, 1960–1970, by age, sex and color.* Athens: University of Georgia.

Bowley, A. (1924). Births and population in Great Britain. *The Economic Journal*, 34, 188–192.

Box, G., & Jenkins, G. (1976). *Time series analysis: Forecasting and control.* San Francisco: Holden-Day.

Boyce, D. (1988). Renaissance of large scale models. *Papers of the Regional Science Association*, 65, 1–10.

Brass, W. (1974). Perspectives in population prediction: Illustrated by the statistics in England and Wales. *Journal of the Royal Statistical Society, A*, 137, 532–570.

Brodie, R., & DeKluyver, C. (1987). A comparison of the short term forecasting accuracy of econometric and naive extrapolation models of market share. *International Journal of Forecasting*, 3, 423–437.

Butz, W., & Ward, M. (1979). The emergence of countercyclical U.S. fertility. *American Economic Review*, 69, 318–328.

Caindec, E., & Prastacos, P. (1995). A description of POLIS: The projective optimization land use model. Working Paper 95– 1. Oakland, CA: Association of Bay Area Governments.

Caldwell, J. (1982). The failure of theories of social and economic change to explain demographic change: Puzzles of modernization or westernization. *Research in Population Economics*, 4, 217–232.

Campbell, P. (1996). Population projections for states by age, sex, race, and Hispanic origin: 1995 to 2050. Report PPL-47. Washington, DC: U.S. Bureau of the Census.

Cannan, E. (1895). The probability of a cessation of the growth of population in England and Wales during the next century. *The Economic Journal*, 5, 506–515.

Carbone, R., & Armstrong, J. (1982). Evaluation of extrapolative forecasting methods: Results of a survey of academicians and practitioners. *Journal of Forecasting*, 1, 215–217.

Carlino, G., & Mills, E. (1987). The determinants of county growth. *Journal of Regional Science*, 27, 39–53.

Carter, L., & Lee, R. (1986). Joint forecasts of U.S. marital fertility, nuptiality, births, and marriages using time series models. *Journal of the American Statistical Association*, 81, 902–911.

Carter, L., & Lee, R. (1992). Modeling and forecasting U.S. sex differentials in mortality. *International Journal of Forecasting*, 8, 393–411.

Casetti, E. (1969). Why do diffusion processes conform to logistic trends? *Geographic Analysis*, 1, 101–105.

Center for Immigration Studies (1995). Immigration-related statistics 1995. *Backgrounder*, 2–95. Washington, DC.

Center for the Continuing Study of the California Economy (1997). *California county projections, 1997*. Palo Alto, CA.

Chen, R., & Morgan, S. (1991). Recent trends in the timing of first births in the United States. *Demography*, 28, 513–533.

Choldin, H. (1994). *Looking for the last percent: The controversy over census undercounts.* New Brunswick, NJ: Rutgers University Press.

Clark, D., & Cosgrove, J. (1991). Amenities versus labor market opportunities: Choosing the optimal distance to move. *Journal of Regional Science*, 31, 311–328.

Clark, D., & Hunter, W. (1992). The impact of economic opportunity, amenities and fiscal factors on age-specific migration rates. *Journal of Regional Science*, 32, 349–365.

Clark, D., & Murphy, C. (1996). Countywide employment and population growth: An analysis of the 1980s. *Journal of Regional Science*, 36, 235–256.

Clark, D., Knapp, T., & White, N. (1996). Personal and location-specific characteristics and elderly interstate migration. *Growth and Change*, 27, 327–351.

Clemen, R. (1989). Combining forecasts: A review and annotated bibliography. *International Journal of Forecasting*, 5, 559–583.

Clemen, R., & Guerard, J. (1989). Econometric GNP forecasts: Incremental information relative to naive extrapolation. *International Journal of Forecasting*, 5, 417–426.

Cochran, G., & Orcutt, G. (1949). Application of least squares regression to relationships containing autocorrelated error terms. *Journal of the American Statistical Association*, 64, 32–61.

Cohen, J. (1986). Population forecasts and confidence intervals for Sweden: A comparison of model-based and empirical approaches. *Demography*, 23, 105–126.

Congdon, P. (1992). Multiregional demographic projections in practice: A metropolitan example. *Regional Studies*, 26, 177–191.

Conway, R. (1990). The Washington projection and simulation model. *International Regional Science Review*, 13, 141–165.

Coulter, L., Stow, D., Kiracofe, B., Langeuin, C., Chen, D., Daeschner, S., Service, D., & Kaiser, J. (1999). Deriving current land-use information for metropolitan transportation planning through integration of remotely sensed data and GIS. *Photogrammetric Engineering & Remote Sensing*, 65, 1293–1300.

Cowan, D., & Jensen, J. (1998). Extraction and modeling of urban attributes using remote sensing technology. In D. Liverman, E. Moran, R. Rindfuss, & P. Stern (Eds.), *People and pixels: Linking remote sensing and social science* (pp. 164–188). Washington, DC: National Academy Press.

Crecine, J. (1968). *A dynamic model of urban structure.* Santa Monica, CA: The Rand Corporation.

Crimmins, E., Saito, E., & Ingegneri, D. (1997). Trends in disability-free life expectancy in the United States, 1970–1990. *Population and Development Review*, 23, 555–572.

Cushing, B. (1987). A note on the specification of climate variables in models of population migration. *Journal of Regional Science*, 27, 641–649.

Dalkey, N., & Helmer, O. (1963). An experimental application of the Delphi method to the use of experts. *Management Science*, 9, 458–467.

Data Resources Incorporated. (1998). *Review of the U.S. Economy: Long-range focus, Winter 97–98.* Lexington, MA: Data Resources Incorporated.

DaVanzo, J. (1978). Does unemployment affect migration? Evidence from micro-data. *Review of Economics and Statistics*, 60, 504–514.

DaVanzo, J. (1983). Repeat migration in the United States: Who moves back and who moves on? *Review of Economics and Statistics*, 65, 552–559.

DaVanzo, J., & Morrison, P. (1978). *Dynamics of return migration: Descriptive findings from a longitudinal study*. Report P-5913. Santa Monica, CA: The Rand Corporation.

DaVanzo, J., & Morrison, P. (1981). Return and other sequences of migration in the United States. *Demography*, 18, 85–101.

Davis, H. (1995). *Demographic projection techniques for regions and smaller areas*. Vancouver, Canada: UBC Press.

Day, J. (1992). Population projections of the United States, by age, sex, race, and Hispanic origin: 1992 to 2050. *Current Population Reports*, P25, No. 1092. Washington DC: U.S. Bureau of the Census.

Day, J. (1996a). Population projections of the United States by age, sex, race, and Hispanic origin: 1995 to 2050. *Current Population Reports*, P25, No. 1130. Washington, DC: U.S. Bureau of the Census.

Day, J. (1996b). Projections of the number of households and families in the United States: 1995 to 2010. *Current Population Reports*, P25, No. 1129. Washington, DC: U.S. Bureau of the Census.

De Beer, J. (1993). Forecast intervals of net migration: The case of the Netherlands. *Journal of Forecasting*, 12, 585–599.

de la Barra, T. (1989). *Integrated land use and transport modelling*. Cambridge, UK: Cambridge University Press.

Deakin, E. (1995). Land use model conference keynote address. Proceedings from the Travel Model Improvement Program: Land Use Modeling Conference, DOT-T-96-09. Washington, DC: U.S. Department of Transportation.

Delaware Valley Regional Planning Commission (1996). Review of land use models and recommended model for DVRPC. Philadelphia: Delaware Valley Regional Planning Commission.

Deming, W. (1943). *Statistical adjustment of data*. New York: Dover.

Dennis, R. (1987). Forecasting errors: The importance of the decision-making context. *Climatic Change*, 11, 81–96.

Department of Rural Sociology (1998). Projections of the population of Texas and counties in Texas by age, sex, and race/ethnicity for 1999–2030. College Station: Texas A&M University.

Diebold, F., & Pauly, P. (1990). The use of prior information in forecast combination. *International Journal of Forecasting*, 6, 503–508.

Dorn, H. (1950). Pitfalls in population forecasts and projections. *Journal of the American Statistical Association*, 45, 311–334.

Easterlin, R. (1961). The American baby boom in historical perspective. *American Economic Review*, 51, 1–44.

Easterlin, R. (1978). What will 1984 be like? Socioeconomic implications of recent twists in age structure. *Demography*, 15, 397–421.

Easterlin, R. (1987). *Birth and fortune: The impact of numbers on personal welfare*. Chicago: University of Chicago Press.

Echenique, M. (1983). The use of planning models in developing countries. In L. Chatterji & P. Nijkamp (Eds.), *Urban and regional policy analysis in developing countries: Some case studies* (pp. 115–158). Aldershot, Hampshire, UK: Gower.

Echenique, M. (1994). Urban and regional studies at the Martin Centre: Its origins, its present, its future. *Environment and Planning B*, 21, 517–533.

Edmonston, B., & Passel, J. (1992). Immigration and immigrant generations in population projections. *International Journal of Forecasting*, 8, 459–476.

Edmonston, B., & Schultze, C. (1995). *Modernizing the U.S. census: Panel on census requirements in the year 2000 and beyond*. Washington, DC: National Academy Press.

Engels, R., & Healy, M. (1981). Measuring interstate migration flows: An origin-destination network based on Internal Revenue Service records. *Environment and Planning A*, 13, 1345–1360.

Evans, A. (1990). The assumption of equilibrium in the analysis of migration and interregional differences: A review of some recent research. *Journal of Regional Science*, 30, 515–531.

Evans, M. (1986). American fertility patterns: A comparison of white and nonwhite cohorts born 1903–56. *Population and Development Review*, 12, 267–293.

Faber, C. (2000). Geographical mobility, March 1997 to March 1998. *Current Population Reports*, P20–520. Washington, DC: U.S. Census Bureau.

Fildes, R. (1985). Quantitative forecasting—The state of the art: Econometric models. *Journal of the Operational Research Society*, 36, 549–580.

Fildes, R. (1992). The evaluation of extrapolative forecasting methods. *International Journal of Forecasting*, 8, 81–98.

Fischer, M., & Nijkamp, P. (1987). Spatial labor market analysis: Labor and scope. In M. Fischer & P. Nijkamp (Eds.), *Regional labor markets: Analytical comparisons and cross-national comparisons* (pp. 1–33). Amsterdam: North Holland.

Flotow, M., & Burson, R. (1996). Allocation errors of birth and death records to subcounty geography. Paper presented at the meeting of the Population Association of America. New Orleans, LA.

Fogel, R., & Costa, D. (1997). A theory of the technophysio evolution, with some implications for forecasting population, health care costs, and pension costs. *Demography*, 34, 49–66.

Fonseca, L., & Tayman, J. (1989). Postcensal estimates of household income distribution. *Demography*, 26, 149–160.

Foot, D. (1981). *Operational urban models*. London: Methuen.

Foot, D., & Milne, W. (1989). Multiregional estimation of gross internal migration flows. *International Regional Science Review*, 12, 29–43.

Foster, A. (1990). Cohort analysis and demographic translation: A comparative study of recent trends in age specific fertility rates from Europe and North America. *Population Studies*, 44, 287–315.

Fotheringham, S. (1984). *Gravity and spatial interaction models*. Beverly Hills, CA: Sage.

Freedman, R. (1975). *The sociology of human fertility*. New York: Irvington.

Friedman, D., Hechter, M., & Kanazawa, S. (1994). A theory of the value of children. *Demography*, 31, 375–401.

Fries, J. (1980). Aging, natural death, and the compression of morbidity. *The New England Journal of Medicine*, 303, 130–135.

Fries, J. (1989). The compression of morbidity: Near or far? *The Milbank Quarterly*, 67, 208–232.

Fuguitt, G., & Brown, D. (1990). Residential preferences and population redistribution. *Demography*, 27, 589–600.

Fullerton, H. (1997). Labor Force 2006: Slowing down and changing composition. *Monthly Labor Review*, November. Washington, DC: U.S. Bureau of Labor Statistics.

Gabbour, I. (1993). SPOP: Small area population projection. In R. Klosterman, R. Brail, & E. Bossard (Eds.), *Spreadsheet models for urban and regional analysis* (pp. 69–84). New Brunswick, NJ: Center for Urban Policy Research, Rutgers University.

Garin, R. (1966). A matrix formulation of the Lowry model for inter-metropolitan activity location. *Journal of the American Institute of Planners*, 32, 361–364.

George, M. (1998). Personal communication.

Gibbs, R. (1994). The information effects of origin on migrants' job search behavior. *Journal of Regional Science*, 34, 163–178.

Gober, P. (1993). Americans on the move. *Population Bulletin*, 48(3). Washington, DC: Population Reference Bureau.

Golini, A. (1998). How low can fertility be? An empirical exploration. *Population and Development Review*, 24, 59–73.

Gordon, I. (1985). The cyclical relationship between regional migration, employment and unemployment: A time series analysis for Scotland. *Scottish Journal of the Political Economy*, 32, 135–158.

Gouldner, W. (1980). Critique and evaluation of the ABAG modeling system. Working Paper. Oakland, CA: Bay Area Association of Governments.

Gouldner, W., Rosenthal, S., & Meredith, J. (1972). *Projective land use model-PLUM: Theory and practice*. Berkeley: Institute for Transportation and Traffic Engineering, University of California.

Granger, C. (1989). Invited review: Combining forecasts—twenty years later. *Journal of Forecasting*, 8, 167–173.

Graves, P. (1979). A life-cycle empirical analysis of migration and climate by race. *Journal of Urban Economics*, 6, 135–147.

Graves, P. (1980). Migration and climate. *Journal of Regional Science*, 20, 227–237.

Graves, P. (1983). Migration with a composite amenity. *Journal of Regional Science*, 23, 541–546.

Graves, P., & Knapp, T. (1988). Mobility behavior of the elderly. *Journal of Urban Economics*, 24, 1–8.

Graves, P., & Linneman, P. (1979). Household migration: Theoretical and empirical results. *Journal of Urban Economics*, 6, 383–404.

Graves, P., & Mueser, P. (1993). The role of equilibrium and disequilibrium in modeling growth and decline. *Journal of Regional Science*, 33, 69–84.

Greenberg, M., Krueckeberg, D., & Michaelson, C. (1978). *Local population and employment projection techniques*. New Brunswick, NJ: Center for Urban Policy and Research, Rutgers University.

Greenwood, M. (1975). Simultaneity bias in migration models: An empirical investigation. *Demography*, 12, 519–536.

Greenwood, M. (1981). *Migration and economic growth in the United States: National, regional and metropolitan perspectives*. New York: Academic Press.

Greenwood, M. (1985). Human migration: Theory, models, and empirical studies. *Journal of Regional Science*, 25, 521–544.

Greenwood, M. (1997). Internal migration in developed countries. In M. Rosenzweig & O. Stark (Eds.), *Handbook of population and family economics* (pp. 647–720). Amsterdam: Elsevier Science.

Greenwood, M., & Hunt, G. (1984). Migration and interregional employment distribution in the United States. *American Economic Review*, 74, 957–969.

Greenwood, M., & Hunt, G. (1989). Jobs versus amenities in the analysis of metropolitan migration. *Journal of Urban Economics*, 25, 1–16.

Greenwood, M., & Hunt, G. (1991). Forecasting state and local population growth with limited data: The use of employment migration relationships and trends in vital rates. *Environment and Planning A*, 23, 987–1005.

Greenwood, M., Hunt, G., & McDowell, J. (1986). Migration and employment change: Empirical evidence on the spatial and temporal dimensions of the linkages. *Journal of Regional Science*, 26, 223–234.

Greenwood, M., Hunt, G., Rickman, D., & Treyz, G. (1991). Migration, regional equilibrium, and the estimation of compensating differentials. *American Economic Review*, 81, 1382–1390.

Grove, W., & Meehl, P. (1996). Comparative efficiency of informal (subjective, impressionistic) and formal (mechanical, algorithmic) prediction procedures: The clinical-statistical controversy. *Psychology, Public Policy, and Law*, 2, 293–323.

Hagerty, M. (1987). Conditions under which econometric models will outperform naive models. *International Journal of Forecasting*, 3, 457–460.

Hahn, R., Mulinare, J., & Teutsch, S. (1992). Inconsistencies in coding of race and ethnicity between births and deaths in U.S. infants: A new look at infant mortality, 1983 through 1985. *Journal of the American Medical Association*, 267, 259–263.

Hamilton, C. (1965). Practical and mathematical considerations in the formulation and selection of migration rates. *Demography*, 2, 429–443.

Hamilton, C., & Perry, J. (1962). A short method for projecting population by age from one decennial census to another. *Social Forces*, 41, 163–170.

Hammer, R., Voss, P., & Blakely, R. (1999). Spatially arrayed growth forces and small area population estimates methodology. Paper presented at the Population Estimates Conference, U.S. Census Bureau, Suitland, MD.

Hansen, K. (1993). Selected place of work migration statistics for states, 1990. 1990 Census of Population, CPH-L-121. Washington, DC: U.S. Bureau of the Census.

Harrigan, F., & McGregor, P. (1993). Equilibrium and disequilibrium perspectives on regional labor migration. *Journal of Regional Science*, 33, 49–67.

Harris, B. (1965). New tools for planning. *Journal of the American Institute of Planners*, 30, 90–95.

Harris, B. (1994). The real issues concerning Lee's requiem. *Journal of the American Planning Association*, 60, 31–34.

Haurin, D., & Haurin, J. (1988). Net migration, unemployment and the business cycle. *Journal of Regional Science*, 28, 239–254.

Henderson, J. (1982). Evaluating consumer amenities and interregional welfare differences. *Journal of Urban Economics*, 11, 32–59.

Herbert, J., & Stevens, B. (1960). A model for the distribution of residential activity in urban areas. *Journal of Regional Science*, 2, 21–36.

Hill, T., Marquez, L., O'Conner, M., & Remus, W. (1994). Artificial neural network models for forecasting and decision making. *International Journal of Forecasting*, 10, 5–15.

Horiuchi, S., & Wilmoth, J. (1998). Deceleration in the age pattern of mortality at older ages. *Demography*, 35, 391–412.

Hunt, D. (1994a). Calibrating the Naples land-use and transport model. *Environment and Planning B*, 21, 569–590.

Hunt, D. (1994b). A description of the MEPLAN framework for land use and transportation interaction modelling. Paper presented at the meeting of the Transportation Research Board, Washington, DC.

Hunt, D., & Simmons, D. (1993). Theory and application of an integrated land-use and transport modelling framework. *Environment and Planning B*, 20, 221–244.

Hunt, G. (1993). Equilibrium and disequilibrium in migration modelling. *Regional Studies*, 27, 341–349.

Irwin, R. (1977). Guide for local area population projections. Technical Paper # 39. Washington, DC: U.S. Bureau of the Census.

Isserman, A. (1977). The accuracy of population projections for subcounty areas. *Journal of the American Institute of Planners*, 43, 247–259.

Isserman, A. (1984). Projection, forecast, and plan: On the future of population forecasting. *Journal of the American Planning Association*, 50, 208–221.

Isserman, A. (1985). Economic–demographic modeling with endogenously determined birth and migration rates: Theory and prospects. *Environment and Planning A*, 17, 25–45.

Isserman, A. (1993). The right people, the right rates: Making population estimates and forecasts with an interregional cohort-component model. *Journal of the American Planning Association*, 59, 45–64.

Isserman, A., & Fisher, P. (1984). Population forecasting and local economic planning: The limits on community control over uncertainty. *Population Research and Policy Review*, 3, 27–50.

Isserman, A., Plane, D., & McMillen, D. (1982). Internal migration in the United States: An evaluation of federal data. *Review of Public Data Use*, 10, 285–311.

Isserman, A., Plane, D., Rogerson, P., & Beaumont, P. (1985). Forecasting interstate migration with limited data: A demographic-economic approach. *Journal of the American Statistical Association*, 80, 277–285.

Johnston, R., & de la Barra, T. (1997). Comprehensive regional modeling for long-range planning: Integrated urban models and geographic information systems. Paper presented at the meeting of the Transportation Research Board, Washington, DC.

Joun, R., & Conway, R. (1983). Regional economic-demographic forecasting models: A case study of the Washington and Hawaii models. *Socio-Economic Planning Sciences*, 17, 345–353.

Judson, D. (1997). FSCP member survey. Reno: Nevada State Demographer's Office.

Kale, B., Voss, P., Palit, C., & Krebs, H. (1981). On the question of errors in population projections. Paper presented at the meeting of the Population Association of America, Washington, DC.

Kale, B., Voss, P., & Krebs, H. (1985). Small area population projections: The Wisconsin experience.

Paper presented at the meeting of the Federal State Cooperative Program for Population Projections, Boston, MA.

Keilman, N. (1990). *Uncertainty in national population forecasting*. Amsterdam: Swets and Zeitlinger.

Keilman, N. (1999). How accurate are the United Nations world population projections? In W. Lutz, J. Vaupel, & D. Ahlburg (Eds.), *Frontiers of population forecasting* (pp. 15–41). New York: The Population Council (A supplement to *Population and Development Review*, 24).

Keyfitz, N. (1968). *An introduction to the mathematics of population*. Reading, MA: Addison-Wesley.

Keyfitz, N. (1972). On future population. *Journal of the American Statistical Association*, 67, 347–362.

Keyfitz, N. (1981). The limits of population forecasting. *Population and Development Review*, 7, 579–593.

Keyfitz, N. (1982). Can knowledge improve forecasts? *Population and Development Review*, 8, 729–751.

Kim, T. (1989). *Integrated urban systems modeling: Theory and applications*. Cambridge, MA: Harvard University Press.

Kintner, H., & Swanson, D. (1994). Estimating vital rates from corporate databases: How long will GM's salaried retirees live? In H. Kintner, T. Merrick, P. Morrison, & P. Voss (Eds.), *Demographics: A Casebook for business and government* (pp. 265–295). Boulder, CO: Westview Press.

Klosterman, R. (1994). Large scale urban models: Retrospect and prospects. *Journal of the American Planning Association*, 60, 3–6.

Klosterman, R. (1999). Operational urban models: A report on the state of the art. Paper presented at the meeting of the Association of Collegiate Schools of Planning, Chicago, IL.

Klosterman, R., Brail, R., & Bossard, E. (1993). *Spreadsheet models for urban and regional analysis*. New Brunswick, NJ: Center for Urban Policy Research, Rutgers University.

Kmenta, J. (1971). *Elements of econometrics*. New York: Macmillan.

Krieg, R., & Bohara, A. (1999). A simultaneous profit model of earnings, migration, job change with wage heterogeneity. *The Annals of Regional Science*, 33, 453–467.

Kriesberg, E., & Vining, D. (1978). On the contribution of outmigration to changes in net migration: A time series analysis of Beale's cross-sectional results. *The Annals of Regional Science*, 12, 1–11.

Kulkarni, M., & Pol, L. (1994). Migration expectancy revisited: Results for the 1970s, 1980s and 1990s. *Population Research and Policy Review*, 13, 195–202.

Land, K. (1986). Methods for national population forecasts: A review. *Journal of the American Statistical Association*, 81, 888–901.

Landis, J. (1994). The California urban futures model: A new generation of metropolitan simulation models. *Environment and Planning B*, 31, 399–420.

Landis, J. (1995). Imagining land use futures: Applying the California urban futures model. *Journal of the American Planning Association*, 61, 438–457.

Landis, J., & Zhang, M. (1998a). The second generation of the California urban futures model: Part I: Model Logic. *Environment and Planning B*, 25, 657–666.

Landis, J., & Zhang, M. (1998b). The second generation of the California urban futures model: Part 2: Specification and calibration results from the land use change submodel. *Environment and Planning B*, 25, 795–824.

Landis, J., Zhang, M., & Zook, M. (1998). *CUFII: The second generation of the California urban futures model*. Berkeley: UC Transportation Center, University of California.

Leach, D. (1981). Re-evaluation of the logistic curve for human populations. *Journal of the Royal Statistical Society A*, 144, 94–103.

Lee, D. (1973). Requiem for large-scale models. *Journal of the American Institute of Planners*, 39, 163–178.

Lee, E. (1966). A theory of migration. *Demography*, 3, 47–57.

Lee, E., Miller, A., Brainerd, C., & Easterlin, R. (1957). Methodological considerations and reference tables. In S. Kuznets & D. Thomas (Eds.), *Population redistribution and economic growth: United States, 1870–1950* (pp. 15–56). Philadelphia: The American Philosophical Society.

Lee, J., & Hong, W. (1974). *Regional demographic projections: 1960–1985*. Report 72–R-1. Washington, DC: National Planning Association.

Lee, R. (1974). Forecasting births in post-transition populations: Stochastic renewal with serially correlated fertility. *Journal of the American Statistical Association*, 69, 607–617.

Lee, R. (1976). Demographic forecasting and the Easterlin hypothesis. *Population and Development Review*, 2, 459–468.

Lee, R. (1992). Stochastic demographic forecasting. *International Journal of Forecasting*, 8, 315–327.

Lee, R. (1993). Modeling and forecasting the time series of U.S. fertility: Age distribution, range, and ultimate level. *International Journal of Forecasting*, 9, 187–202.

Lee, R., & Carter, L. (1992). Modeling and forecasting U.S. mortality. *Journal of the American Statistical Association*, 87, 659–675.

Lee, R., & Tuljapurkar, S. (1994). Stochastic population forecasts for the United States: Beyond high, medium, and low. *Journal of the American Statistical Association*, 89, 1175–1189.

Leitch, G., & Tanner, J. (1995). Professional economic forecasts: Are they worth their costs? *Journal of Forecasting*, 14, 143–157.

LeSage, J. (1990). Forecasting metropolitan employment using an export-base error-correction model. *Journal of Regional Science*, 30, 307–323.

Lesthaeghe, R. (1983). A century of demographic and cultural change in Western Europe: An exploration of underlying dimensions. *Population and Development Review*, 9, 411–435.

Lesthaeghe, R., & Surkyn, J. (1988). Cultural dynamics and economic theories of fertility change. *Population and Development Review*, 14, 1–45.

Lesthaeghe, R., & Willems, P. (1999). Is low fertility a temporary phenomenon in the European Union? *Population and Development Review*, 25, 211–228.

Levernier, W., & Cushing, B. (1994). A new look at the determinants of intrametropolitan distribution of population and employment. *Urban Studies*, 31, 1391–1405.

Lichter, D., & DeJong, G. (1990). The United States. In C. Nam, W. Serow, & D. Sly (Eds.), *International handbook on internal migration* (pp. 391–417). New York: Greenwood Press.

Lippmann, R. (1987). An introduction to computing with neural nets. *ISEE ASSP Magazine*, 4, 4–22.

Lo, C. (1995). Automated population dwelling unit estimation from high-resolution satellite images: A GIS approach. *International Journal of Remote Sensing*, 16, 17–34.

London, D. (1988). *Survival models and their estimation*. New Britain, CT: ACTEX.

Long, J. (1989). The relative effects of fertility, mortality, and immigration on projected population age structure. Paper presented at the meeting of the Population Association of America. Baltimore, MD.

Long, J. (1995). Complexity, accuracy, and utility of official population projections. *Mathematical Population Studies*, 5, 203–216.

Long, J., & Boertlein, C. (1990). Comparing migration measures having different intervals. *Current Population Reports*, P-23, No. 166. Washington, DC: U.S. Bureau of the Census.

Long, J., & McMillen, D. (1987). A survey of Census Bureau population projection methods. *Climatic Change*, 11, 141–177.

Long, L. (1988). *Migration and residential mobility in the United States*. New York: Russell Sage Foundation.

Long, L. (1991). Residential mobility differences among developed countries. *International Regional Science Review*, 14, 133–147.

Long, L., & Hansen, K. (1979). Reasons for interstate migration. *Current Population Reports*, P-23, No. 81. Washington, DC: U.S. Bureau of the Census.

Lowry, I. (1964). *A model of metropolis*. Report RM 4125-RC. Santa Monica, CA: The Rand Corporation.

Lutz, W., Sanderson, W., & Scherbov, S. (1999). Expert-based probabilistic population projections. In W. Lutz, J. Vaupel, & D. Ahlburg (Eds.), *Frontiers of population forecasting* (pp. 139–155). New York: The Population Council (A supplement to *Population and Development Review*, 24).

Lutz, W., Saariluoma, P., Sanderson, W., & Scherbov, S. (2000). *New developments in the methodology of expert- and argument-based probabilistic population projection.* Interim Report 1R-00–020. Luxenburg, Austria: International Institute for Applied Statistical Analysis.

MacDonald, M., & Rindfuss, R. (1978). Relative economic status and fertility: Evidence from a cross-section. *Research in Population Economics*, 1, 291–307.

Mahmoud, E. (1984). Accuracy in forecasting: A survey. *Journal of Forecasting*, 3, 139–159.

Mahmoud, E. (1987). The evaluation of forecasts. In S. Makridakis & S. Wheelwright (Eds.), *The handbook of forecasting* (pp. 504–522). New York: Wiley.

Makridakis, S. (1986). The art and science of forecasting: An assessment and future directions. *International Journal of Forecasting*, 2, 15–39.

Makridakis, S. (1993). Accuracy measures: Theoretical and practical concerns. *International Journal of Forecasting*, 9, 527–529.

Makridakis, S., & Winkler, R. (1983). Averages of forecasts: Some empirical results. *Management Science*, 29, 987–996.

Makridakis, S., Andersen, A., Carbone, R., Fildes, R., Hibon, M., Lewandowski, R., Newton, J., Parzen, E., & Winkler, R. (1982). The accuracy of extrapolation (time series) methods: Results of a forecasting competition. *Journal of Forecasting*, 1, 111–153.

Makridakis, S., Hibon, M., Lusk, E., & Belhadjali, M. (1987). Confidence intervals: An empirical investigation of the series in the M-competition. *International Journal of Forecasting*, 3, 489–508.

Makridakis, S., Wheelwright, S., & Hyndman, R. (1998). *Forecasting: Methods and applications.* New York: Wiley.

Makridakis, S., Chatfield, C., Hibon, M., Lawrence, M., Mills, T., Ord, K., & Simmons, L. (1993). The M2–competition: A real time judgementally based forecasting study. *International Journal of Forecasting*, 9, 5–22.

Manton, K., Patrick, C., & Stallard, E. (1980). Mortality model based on delays in progression of chronic diseases: Alternative to cause elimination model. *Public Health Reports*, 95, 580–588.

Manton, K., Stallard, E., & Tolley, H. (1991). Limits to human life expectancy: Evidence, prospects, and implications. *Population and Development Review*, 17, 603–637.

MapInfo. (1998). *MapInfo professional user's guide V.5.0.* Troy, NY: MapInfo Corporation.

Marchetti, C., Meyer, P., & Ausubel, J. (1996). Human population dynamics revisited with the logistic model: How much can be modeled and predicted. *Technological Forecasting and Social Change*, 52, 1–30.

Maricopa Association of Governments. (1996). GIS analysis and data enhancement study. Phoenix, AZ.

Marks, A., Thrall, G., & Arno, M. (1992). Siting hospitals to provide cost-effective health care. *Geo Info Systems*, 2, 58–66.

Martin, P., & Midgley, E. (1999). Immigration to the United States. *Population Bulletin*, 54. Washington, DC: Population Reference Bureau.

Mason, K. (1997). Explaining fertility transitions. *Demography*, 34, 443–454.

Massey, D., Alarcon, R., Durand, J., & Gonzalez, H. (1987). *Return to Aztlan: The social process of international migration from western Mexico.* Berkeley: University of California Press.

McCleary, R., & Hay, R. (1980). *Applied time series analysis for the social sciences.* Beverly Hills, CA: Sage.

McDonald, J. (1979). A time series approach to forecasting Australian total live-births. *Demography*, 16, 575–601.

McDonald, J. (1981). Modeling demographic relationships: An analysis of forecast functions for Australian births. *Journal of the American Statistical Association*, 76, 782–801.

McHugh, K. (1985). Reasons for migrating or not. *Sociology and Social Research*, 69, 585–588.

McKibben, J. (1996). The impact of policy changes on forecasting for school districts. *Population Research and Policy Review*, 15, 527–536.

McNees, S. (1992). The uses and abuses of 'consensus' forecasts. *Journal of Forecasting*, 11, 703–710.

McNown, R., & Rogers, A. (1989). Forecasting mortality: A parameterized time series approach. *Demography*, 26, 645–660.

McNown, R., & Rogers, A. (1992). Forecasting cause-specific mortality using time series methods. *International Journal of Forecasting*, 8, 413–432.

McNown, R., Rogers, A., & Little, J. (1995). Simplicity and complexity in extrapolative population forecasting models. *Mathematical Population Studies*, 5, 235–257.

Mentzer, J., & Kahn, K. (1995). Forecasting technique familiarity, satisfaction, usage, and application. *Journal of Forecasting*, 14, 465–476.

Miller, A. (1967). The migration of employed persons to and from metropolitan areas of the United States. *Journal of the American Statistical Association*, 62, 1418–1432.

Mills, E. (1967). An aggregative model of resource allocation in metropolitan areas. *American Economic Review*, 57, 187–210.

Mills, E., & Lubuele, L. (1995). Projecting growth in metropolitan areas. *Journal of Urban Economics*, 37, 344–360.

Mincer, J. (1978). Family migration decisions. *Journal of Political Economy*, 86, 749–773.

Moen, E. (1984). Voodoo forecasting: Technical, political and ethical issues regarding the projection of local population growth. *Population Research and Policy Review*, 3, 1–25.

Molho, I. (1986). Theories of migration: A review. *Scottish Journal of Political Economy*, 33, 396–419.

Morrill, R., Gaile, G., & Thrall, G. (1988). *Spatial diffusion*. Newbury Park, CA: Sage.

Morrison, P. (1971). Chronic movers and the future redistribution of population: A longitudinal analysis. *Demography*, 8, 171–184.

Morrison, P. (1977). The functions and dynamics of the migration process. In A. Brown & E. Neuberger (Eds.), *Internal migration: A comparative perspective* (pp. 61–72). New York: Academic Press.

Mueser, P., & White, M. (1989). Explaining the association between rates of in-migration and out-migration. *Papers of the Regional Science Association*, 67, 121–134.

Murdock, S., & Ellis, D. (1991). *Applied demography: An introduction of basic, concepts, methods, and data*. Boulder, CO: Westview Press.

Murdock, S., & Leistritz, F. (1980). Selecting socioeconomic assessment models: A discussion of criteria and selected models. *Journal of Environmental Management*, 10, 1–12.

Murdock, S., Leistritz, F., Hamm, R., Hwang, S., & Parpia, B. (1984). An assessment of the accuracy of a regional economic-demographic projection model. *Demography*, 21, 383–404.

Murdock, S., Jones, L., Hamm, R., & Leistritz, F. (1987). *The Texas assessment modeling system (TAMS): Users guide*. College Station: Texas Agricultural Experiment Station, Texas A&M University.

Murdock, S., Hamm, R., Voss, P., Fannin, D., & Pecotte, B. (1991). Evaluating small area population projections. *Journal of the American Planning Association*, 57, 432–443.

Muth, R. (1971). Migration: Chicken or egg. *Southern Economic Journal*, 37, 295–306.

Nakosteen, R. (1989). Detailed population projections for small areas: The Massachusetts experience. *Socio-Economic Planning Sciences*, 23, 125–138.

Namboodiri, K. (1972). On the ratio-correlation and related methods of subnational population estimation. *Demography*, 9, 443–454.

National Center for Health Statistics. (1997). U.S. decennial life tables for 1989–91. Vol. 1, No. 1. Washington, DC.

National Center for Health Statistics. (1999). Births and deaths; Preliminary data for 1998, *National Vital Statistics Reports* Vol. 47. Washington, DC.

Nelson, C. (1973). *Applied time series analysis for managerial forecasting*. San Francisco: Holden-Day.

Newbold, P., & Granger, C. (1974). Experience with forecasting univariate time series and the combination of forecasts. *Journal of the Royal Statistical Society*, 137, 131–165.

Newell, C. (1988). *Methods and models in demography*. New York: Guilford.

O'Hare, W. (1976). Report on a multiple regression method for making population estimates. *Demography*, 13, 369–380.

O'Neil, W. (1981). Estimation of a logistic growth and diffusion model describing neighborhood change. *Geographical Analysis*, 13, 391–397.

Olshansky, S. (1987). Simultaneous/multiple cause-delay (SIMCAD): An epidemiological approach to projecting mortality. *Journal of Gerontology*, 42, 358–365.

Olshansky, S. (1988). On forecasting mortality. *The Milbank Quarterly*, 66, 482–530.

Olshansky, S., Carnes, B., & Cassel, C. (1990). In search of Methuselah: Estimating the upper limits to human longevity. *Science*, 250, 634–640.

Opitz, W., & Nelson, H. (1996). Short-term population-based forecasting in the public sector: A dynamic caseload simulation model. *Population Research and Policy Review*, 15, 549–563.

Pant, P., & Starbuck, W. (1990). Innocents in the forest: Forecasting and research methods. *Journal of Management*, 16, 433–460.

Pearl, R., & Reed, L. (1920). On the rate of growth of the population of the United States since 1790 and its mathematical representation. *Proceedings of the National Academy of Science*, 6, 275–287.

Pflaumer, P. (1988). Confidence intervals for population projections based on Monte Carlo methods. *International Journal of Forecasting*, 4, 135–142.

Pflaumer, P. (1992). Forecasting U.S. population totals with the Box-Jenkins approach. *International Journal of Forecasting*, 8, 329–338.

Pignataro, L., & Epling, J. (2000). The TELUS story: Information tool for transportation planning makes its debut. *Transportation Research News*, 210, 9–12, 25.

Pittenger, D. (1976). *Projecting state and local populations*. Cambridge, MA: Ballinger.

Pittenger, D. (1980). Some problems in forecasting population for government planning purposes. *The American Statistician*, 34, 135–139.

Plane, D. (1989). Population migration and economic restructuring in the United States. *International Regional Science Review*, 12, 263–280.

Plane, D. (1993). Demographic influences on migration. *Regional Studies*, 27, 375–383.

Plane, D., Rogerson, P., & Rosen, A. (1984). The cross-regional variation of in-migration and out-migration. *Geographical Analysis*, 16, 162–175.

Pol, L., & Thomas, R. (1997). *Demography for business decision making*. Westport, CT: Quorum Books.

Population Reference Bureau. (1999). World population data sheet. Washington, DC.

Prastacos, P. (1986a). An integrated land use and transportation model for the San Francisco region: Design and mathematical structure. *Environment and Planning A*, 18, 307–322.

Prastacos, P. (1986b). An Integrated land use and transportation model for the San Francisco region: Empirical results and estimation. *Environment and Planning A*, 18, 511–528.

Prastacos, P. (1991). Integrating GIS technology in urban transportation planning and modeling. *Transportation Research Record*, 1305, 123–130.

Pritchett, H. (1891). A formula for predicting the population of the United States. *Publications of the American Statistical Association*, 14, 278–296.

Putman, S. (1979). *Urban residential location models*. Boston: Martinus Nijhoff.

Putman, S. (1983). *Integrated urban models*. London: Pion.

Putman, S. (1991). *Integrated urban models: II*. London: Pion.

Putman, S. (1994). Integrated transportation and land use models: An overview of progress with DRAM and EMPAL, with suggestions for further research. Paper presented at the meeting of the Transportation Research Board, Washington, DC.

Putman, S. (1995). EMPAL and DRAM location and land use models: An overview. Paper presented at the Land Use Modeling Conference. Dallas, TX.

Rainford, P., & Masser, I. (1987). Population forecasting and urban planning practice. *Environment and Planning A*, 19, 1463–1475.

Ravenstein, E. (1889). The laws of migration. *The Journal of the Royal Statistical Society*, LII, 241–301.

Raymondo, J. (1992). *Population estimation and projection: Methods for marketing, demographic, and planning personnel*. New York: Quorum Books.

Rees, P. (1977). The measurement of migration from census data and other sources. *Environment and Planning A*, 9, 247–272.

Rees, P., & Kupiszewski, M. (1999). Internal migration: What data are available in Europe? *Journal of Official Statistics*, 15, 551–586.

Reeve, T., & Perlich, P. (1995). State of Utah demographic and economic projection modeling system. Salt Lake City, UT: Governor's Office of Planning and Budget.

Rives, N., Serow, W., Lee, A., Goldsmith, H., & Voss, P. (1995). *Basic methods for preparing small-area population estimates*. Madison, WI: Applied Population Laboratory.

Rogers, A. (1967). A regression analysis of interregional migration in California. *Review of Economics and Statistics*, 49, 262–267.

Rogers, A. (1985). *Regional population projection models*. Beverly Hills, CA: Sage.

Rogers, A. (1990). Requiem for the net migrant. *Geographical Analysis*, 22, 283–300.

Rogers, A. (1992). Elderly migration and population redistribution in the United States. In A. Rogers (Ed.), *Elderly migration and population redistribution* (pp. 226–248). London: Belhaven Press.

Rogers, A. (1995a). *Multiregional demography: Principles, methods and extensions*. Chichester, UK: Wiley.

Rogers, A. (1995b). Population forecasting: Do simple models outperform complex models? *Mathematical Population Studies*, 5, 187–202.

Rogers, A., & Castro, L. (1984). Model migration schedules. In A. Rogers (Ed.), *Migration, urbanization and spatial population dynamics* (pp. 41–88). Boulder, CO: Westview Press.

Rogers, A., & Raymer, J. (1999). *Origin dependence, return migration, and the spatial redistribution of the U.S.-born population*, Working Paper, WP-99–6. Boulder, CO: Population Program, University of Colorado.

Rogers, A., & Williams, P. (1986). Multistate demoeconomic modeling and projection. In A. Isserman (Ed.), *Population change and the economy: Social science theory and methods* (pp. 177–202). Boston, MA: Kluwer-Nijhoff.

Rogers, A., & Woodward, J. (1991). Assessing state population projections with transparent multi-regional demographic models. *Population Research and Policy Review*, 10, 1–26.

Rogers, R. (1995). Sociodemographic characteristics of long-lived and healthy individuals. *Population and Development Review*, 21, 33–58.

Romaniuc, A. (1990). Population projection as prediction, simulation and prospective analysis. *Population Bulletin of the United Nations*, 29, 16–31.

Rothenberg, J. (1977). On the microeconomics of internal migration. In A. Brown & E. Neuberger (Eds.), *Internal migration: A comparative perspective* (pp. 183–205). New York: Academic Press.

Rowe, G., & Wright, G. (1999). The Delphi technique as a forecasting tool: Issues and analysis. *International Journal of Forecasting*, 15, 353–375.

Ryder, N. (1965). The cohort as a concept in the study of social change. *American Sociological Review*, 30, 843–861.

Ryder, N. (1986). Observations on the history of cohort fertility in the United States. *Population and Development Review*, 12, 617–643.

Ryder, N. (1990). What is going to happen to American fertility? *Population and Development Review*, 16, 433–453.

Rynerson, C., & Tayman, J. (1997). An integrated system for estimating sub-county household income distributions. Paper presented at the meeting of the Population Association of America, Washington, DC.

Rynerson, C., & Tayman, J. (1998). An evaluation of address-level administrative records used to prepare small area population estimates. Paper presented at the meeting of the Population Association of America, Chicago, IL.

Saboia, J. (1974). Modeling and forecasting populations by time series: The Swedish case. *Demography*, 11, 483–492.

San Diego Association of Governments. (1998). Urban development model, Vol. 2: Technical description. San Diego, CA.

San Diego Association of Governments. (1999). Demographic and economic forecasting model: Vol. 2: Technical description. San Diego, CA.

Sanderson, W. (1999). Knowledge can improve forecasts: A review of selected socioeconomic population projection models. In W. Lutz, J. Vaupel, & D. Ahlburg (Eds.), *Frontiers of population forecasting* (pp. 88–117). New York: The Population Council (A supplement to *Population and Development Review*, 24).

Schachter, J., & Althaus, P. (1989). An equilibrium model of gross migration. *Journal of Regional Science*, 29, 134–159.

Schmidley, A., & Gibson, C. (1999). Profile of the foreign born population in the United States: 1997. *Current Population Reports*, P-23, No. 195. Washington, DC: U.S. Census Bureau.

Schmidt, R., Barr, C., & Swanson, D. (1997). Socioeconomic impacts of the proposed federal gaming tax. *International Journal of Public Administration*, 20, 1675–1698.

Schmitt, R., & Crosetti, A. (1951). Accuracy of the ratio method for forecasting city population. *Land Economics*, 27, 346–348.

Schnaars, S. (1986). A comparison of extrapolation models on yearly sales forecasts. *International Journal of Forecasting*, 2, 71–85.

Schocken, S., & Ariav, G. (1994). Neural networks for decision support: Problems and opportunities. *Decision Support Systems*, 11, 393–414.

Schoen, R., Kim, Y., Nathanson, C., Fields, J., & Astone, N. (1997). Why do Americans want children? *Population and Development Review*, 23, 333–358.

Schroeder, E. (1987). Testing labor force and unemployment projections. *Demography*, 24, 649–661.

Shao, G., & Treyz, G. (1993). Building U.S. national and regional economic models. *Economic Systems Research*, 5, 63–77.

Shryock, H., & Siegel, J. (1973). *The methods and materials of demography*. Washington, DC: U.S. Government Printing Office.

Siegel, J. (1972). Development and accuracy of projections of population and households in the United States. *Demography*, 9, 51–68.

Siegel, J., & Davidson, M. (1984). Demographic and socioeconomic aspects of aging in the United States. *Current Population Reports*, P-23, No. 138. Washington, DC: U.S. Bureau of the Census.

Sink, L. (1997). Race and ethnicity classification consistency between the Census Bureau and the National Center for Health Statistics. Population Division Working Paper Series No. 17. Washington, DC: U.S. Bureau of the Census.

Sjaastad, L. (1960). The relationship between income and migration in the United States. *Papers and Proceedings of the Regional Science Association*, 6, 37–64.

Sjaastad, L. (1962). The costs and returns of human migration. *Journal of Political Economy*, 70, 80–93.

Smith, D. (1992). *Formal demography*. New York: Plenum Press.

Smith, J., & Welch, F. (1981). No time to be young: The economic prospects for large cohorts in the United States. *Population and Development Review*, 7, 71–83.

Smith, S. (1986). Accounting for migration in cohort-component projections of state and local populations. *Demography*, 23, 127–135.

Smith, S. (1987). Tests of forecast accuracy and bias for county population projections. *Journal of the American Statistical Association*, 82, 991–1003.

Smith, S. (1989). Toward a methodology for estimating temporary residents. *Journal of the American Statistical Association*, 84, 430–436.

Smith, S. (1995). Population growth and demographic change. In J. Scoggins and A. Pierce (Eds.), *The economy of Florida*. Gainesville: Bureau of Economic and Business Research, University of Florida.

Smith, S., & Ahmed, B. (1990). A demographic analysis of the population growth of states, 1950–1980. *Journal of Regional Science*, 30, 209–227.

Smith, S., & Cody, S. (1994). Evaluating the housing unit method: A case study of 1990 population estimates in Florida. *Journal of the American Planning Association*, 60, 209–221.

Smith, S., & Fishkind, H. (1985). Elderly migration in rapidly growing areas: A time series approach. *The Review of Regional Studies*, 15, 11–20.

Smith, S., & Nogle, J. (1999). Population projections by age, sex and race for Florida and its counties, 1998–2010. *Florida Population Studies*, Bulletin 124. Gainesville: Bureau of Economic and Business Research, University of Florida.

Smith, S., & Nogle, J. (2000). Projections of Florida's population by county, 1999–2030. *Florida Population Studies*, Bulletin 126. Gainesville: Bureau of Economic and Business Research, University of Florida.

Smith, S., & Shahidullah, M. (1995). An evaluation of population projection errors for census tracts. *Journal of the American Statistical Association*, 90, 64–71.

Smith, S., & Sincich, T. (1988). Stability over time in the distribution of population forecast errors. *Demography*, 25, 461–474.

Smith, S., & Sincich, T. (1990). The relationship between the length of base period and population forecast errors. *Journal of the American Statistical Association*, 85, 367–375.

Smith, S., & Sincich, T. (1991). An empirical analysis of the effect of the length of forecast horizon on population forecast errors. *Demography*, 28, 261–274.

Smith, S., & Sincich, T. (1992). Evaluating the forecast accuracy and bias of alternative population projections for states. *International Journal of Forecasting*, 8, 495–508.

Smith, S., & Swanson, D. (1998). In defense of the net migrant. *Journal of Economic and Social Measurement*, 24, 249–264.

Southworth, F. (1995). A technical review of urban land use-transportation models as tools for evaluating vehicle travel reduction strategies. Report ORNL-6881. Oak Ridge, TN: Oak Ridge National Laboratory.

Spencer, G. (1989). Projection of the population of the United States by age, sex, and race: 1988 to 2008. *Current Population Reports*, P-25, No. 1018. Washington, DC.: U.S. Bureau of the Census.

Star, J., & Estes, J. (1990). *Geographic information systems: An introduction*. Englewood Cliffs, NJ: Prentice-Hall.

Steinnes, D. (1982). "Do people follow jobs" or "do jobs follow people"? A causality issue in urban economics. *Urban Studies*, 19, 187–192.

Stolnitz, G. (1964). The demographic transition: From high to low birth and death rates. In R. Freedman (Ed.), *Population: The vital revolution*. Garden City, NY: Anchor Books.

Stone, L. (1971). On the correlation between metropolitan area in- and out-migration by occupation. *Journal of the American Statistical Association*, 66, 693–701.

Stoto, M. (1983). The accuracy of population projections. *Journal of the American Statistical Association*, 78, 13–20.

Swanson, D. (1980). Improving accuracy in multiple regression estimates of population using principles from causal modeling. *Demography*, 17, 413–427.

Swanson, D., & Beck, D. (1994). A new short-term county population projection method. *Journal of Economic and Social Measurement*, 20, 25–50.

Swanson, D., & Tayman, J. (1995). Between a rock and a hard place: The evaluation of demographic forecasts. *Population Research and Policy Review*, 14, 233–249.

Swanson, D., Vaidya, E., Yehya, B., Bennett, B., & Prevost, R. (1989). Impact of census error adjustments on state population projections: The case of Ohio. *The Ohio Journal of Science*, 89, 26–32.

Swanson, D., Hough, G., Rodriguez, J., & Clemans, C. (1998). K-12 enrollment forecasting: Merging methods and judgement. *ERS Spectrum*, 16, 24–31.

Tabuchi, T. (1985). Time series modeling of gross migration and dynamic equilibrium. *Journal of Regional Science*, 25, 65–83.

Tayman, J. (1994). Estimating population, housing, and employment for micro-geographic areas. In K. Rao & J. Wicks (Eds.), *Studies in applied demography* (pp. 101–108). Bowling Green, OH: Bowling Green State University.

Tayman, J. (1996a). The accuracy of small-area population forecasts based on a spatial interaction land-use modeling system. *Journal of the American Planning Association*, 62, 85–98.

Tayman, J. (1996b). Forecasting, growth management, and public policy decision making. *Population Research and Policy Review*, 15, 491–508.

Tayman, J., & Swanson, D. (1996). On the utility of population forecasts. *Demography*, 33, 523–528.

Tayman, J., & Swanson, D. (1999). On the validity of MAPE as a measure of population forecast accuracy. *Population Research and Policy Review*, 18, 299–322.

Tayman, J., Parrott, B., & Carnevale, S. (1994). Locating fire station sites: The response time component. In H. Kintner, P. Voss, P. Morrison, & T. Merrick (Eds.), *Applied demographics: A casebook for business and government* (pp. 203–217). Boulder, CO: Westview Press.

Tayman, J., Schafer, E., & Carter, L. (1998). The role of population size in the determination and prediction of population forecast errors: An evaluation using confidence intervals for subcounty areas. *Population Research and Policy Review*, 17, 1–20.

Texas Water Development Board. (1997). Water for Texas: A consensus-based update to the state water plan: Vol. II, technical planning appendix. Doc # GF-6-2. Austin, TX.

Theil, H. (1966). *Applied economic forecasting*. Amsterdam: North-Holland.

Thomas, R. (1994). Using demographic analysis in health services planning: A case study in obstetrical services. In H. Kintner, T. Merrick, P. Morrison, & P. Voss (Eds.), *Demographics: A casebook for business and government* (pp. 159–179). Boulder, CO: Westview Press.

Thomlinson, R. (1962). The determination of a base population for computing migration rates. *Milbank Memorial Fund Quarterly*, 40, 356–366.

Thompson, W., & Whelpton, P. (1933). *Population trends in the United States*. New York: McGraw-Hill.

Thrall, G. (1998). Common geographic errors of real estate analysis. *Journal of Real Estate Literature*, 6, 45–54.

Thrall, G., Sidman, C., Thrall, S., & Fik, T. (1993). The Cascade GIS diffusion model for measuring housing absorption by small area with a case study of St. Lucie County, Florida. *The Journal of Real Estate Research*, 8, 401–420.

Treadway, R. (1997). Population projections for the state and counties of Illinois. Springfield: State of Illinois.

Treyz, G. (1993). *Regional economic modeling: A systematic approach to economic forecasting and policy analysis*. Boston, MA: Kluwer Academic.

Treyz, G. (1995). Policy analysis: Application of REMI economic forecasting and simulation models. *International Journal of Public Administration*, 18, 13–42.

Treyz, G., Rickman, D., & Shao, G. (1992). The REMI economic-demographic forecasting and simulation model. *International Regional Science Review*, 14, 221–253.

Treyz, G., Rickman, D., Hunt, G., & Greenwood, M. (1993). The dynamics of U.S. internal migration. *Review of Economics and Statistics*, 75, 209–214.

Twain, M. (1874). *Life on the Mississippi*. New York: Harper and Brothers.

U.S. Bureau of the Census. (1950). Illustrative projections of the population of the United States: 1950 to 1960, *Current Population Reports*, P-25, No. 43. Washington, DC.

U.S. Bureau of the Census. (1957). Illustrative projections of the population, by state, 1960, 1965, and 1970, *Current Population Reports*, P-25, No. 160. Washington, DC.

U.S. Bureau of the Census. (1966). Illustrative projections of the population of states: 1970 to 1985. *Current Population Reports*, P-25, No. 326. Washington, DC.

U.S. Bureau of the Census. (1972). Preliminary projections of the population of states: 1975–1990, *Current Population Reports*, P-25, No. 477. Washington DC.

U.S. Bureau of the Census. (1975). Projections of the population of the United States: 1975 to 2050, *Current Population Reports*, P-25, No. 601. Washington, DC.

U.S. Bureau of the Census. (1977). Projections of the population of the United States: 1977 to 2050, *Current Population Reports*, P-25, No. 704. Washington, DC.

U.S. Bureau of the Census. (1979). Illustrative projections of state populations by age, race, and sex: 1975 to 2000, *Current Population Reports*, P-25, No. 796. Washington, DC.

U.S. Bureau of the Census. (1992). Summary population and housing characteristics. *1990 Census of Population*, CPH-1-1. Washington, DC.

U.S. Bureau of the Census. (1993a). Social and economic characteristics, United States. *1990 Census of Population*, CP-2-1. Washington, DC.

U.S. Bureau of the Census. (1993b). Social and economic characteristics, Florida. *1990 Census of Population*, CP-2-11. Washington, DC.

U.S. Bureau of the Census. (1998). Population estimates by age, sex, race, and Hispanic origin. Internet site: http://www.census.gov, released December, 1998.

U.S. Bureau of Economic Analysis. (1985). *BEA Regional Projections*. Washington, DC.

U.S. Bureau of Economic Analysis. (1995). *BEA Regional Projections to 2045: Volume I: States*. Washington, DC.

U.S. Bureau of Economic Analysis. (1999). *Regional Economic Information System (REIS) 1969–97*. Released on CD in August, 1999.

U.S. Census Bureau. (1999a). About the American Community Survey. Internet site: http://www.census.gov/acs/www/acs.htm.

U.S. Census Bureau. (1999b). Population estimates. Internet site: http://www.census.gov/population/www/estimates/popest.html.

U.S. Immigration and Naturalization Service. (1999). *Statistical yearbook of the Immigration and Naturalization Service 1997*. Washington, DC.

University of California. (1999). TRansportation ANalysis SIMulation System (TRANSIMS), Version: TRANSIMS-LANL-1.0, Volume 0 - Overview. Report LA—UR-99-1658. Berkeley, CA.

Vijverberg, W. (1993). Labor market performance as a determinant of migration. *Economica*, 60, 143–160.

Voss, P., & Kale, B. (1985). Refinements to small-area population projection models: Results of a test based on 128 Wisconsin communities. Paper presented at the meeting of the Population Association of America, Boston, MA.

Wachter, K. (1975). A time-series fertility equation: The potential for a baby-boom in the 1980's. *International Economic Review*, 16, 609–624.

Waddell, P. (2000). A behavioral simulation model for metropolitan policy analysis and planning: Residential location and housing market components of UrbanSim. *Environment and Planning B*, 27, 247–263.

Wasserman, P. (1989). *Neural computing: Theory and practice*. New York: Van Nostrand Reinhold.

Watterson, T. (1990). Adapting and applying existing urban models: DRAM/EMPAL in the Seattle region. *Journal of the Urban and Regional Information System*, 2, 35–46.

Weatherby, C. (1995). Summary of workshop observations and recommendations. Proceedings from the Travel Model Improvement Program: Land Use Modeling Conference, DOT-T-96-09. Washington, DC: U.S. Department of Transportation.

Webby, R., & O'Connor, M. (1996). Judgemental and statistical time series forecasting: A review of the literature. *International Journal of Forecasting*, 12, 91–118.

Weeks, J. (1999). *Population*. Belmont, CA: Wadsworth.

Wegener, M. (1986). Transportation network equilibrium and regional deconcentration. *Environment and Planning A*, 18, 437–456.

Wegener, M. (1994). Operation urban models: State of the art. *Journal of the American Planning Association*, 60, 17–30.

Wegener, M. (1995). Current and future land use models. Proceedings from the Travel Model Improvement Program: Land Use Modeling Conference, DOT-T-96-09. Washington, DC: U.S. Department of Transportation.

Wegener, M., Mackett, R., & Simmons, D. (1991). One city, three models: Comparison of land use/ transport policy simulation models for Dortmund. *Transport Reviews*, 11, 107–129.

Welch, F. (1979). Effects of cohort size on earnings: The baby boom babies' financial bust. *Journal of Political Economy*, 87, 865–897.

Weller, B. (1990). Predicting small region sectoral responses to changes in aggregate economic activity: A time series approach. *Journal of Forecasting*, 9, 273–281.

West, C., & Fullerton, T. (1996). Assessing the historical accuracy of regional economic forecasts. *Journal of Forecasting*, 15, 19–36.

Westoff, C., & Ryder, N. (1977). The predictive validity of reproductive intentions. *Demography*, 14, 431–453.

Wetrogan, S. (1983). Provisional projections of the population of states. by age and sex: 1980 to 2000, *Current Population Reports*, P-25, No. 937. Washington, DC: U.S. Bureau of the Census.

Wetrogan, S. (1988). Projections of the population of states, by age, sex, and race: 1988–2010, *Current Population Reports*, P-25, No. 1017. Washington, DC: U.S. Bureau of the Census.

Wetrogan, S. (1990). Projections of the population of states by age, sex, and race: 1989–2010, *Current Population Reports*, P-25, No. 1053. Washington, DC: U.S. Bureau of the Census.

Wetrogan, S., & Long, J. (1990). Creating annual state-to-state migration flows with demographic data, *Current Population Reports*, P-23, No. 166. Washington, DC: U.S. Bureau of the Census.

Whelpton, P. (1928). Population of the United States, 1925 to 1975. *American Journal of Sociology*, 34, 253–270.

Whelpton, P., Eldridge, H., & Siegel, J. (1947). Forecasts of the population of the United States: 1945–1975. Washington, DC: U.S. Government Printing Office.

White, H. (1954). Empirical study of the accuracy of selected methods of projecting state populations. *Journal of the American Statistical Association*, 49, 480–498.

White, K., & Preston, S. (1996). How many Americans are alive because of twentieth-century improvements in mortality? *Population and Development Review*, 22, 415–429.

White, M., Mueser, P., & Tierney, J. (1987). Net migration of the population of the United States by age, race and sex: 1970–1980. Ann Arbor, MI: Inter-University Consortium for Political and Social Research.

Wicks, J., Swanson, D., Vincent, R., & De Almeida, J. (1999). Population estimates from remotely sensed data: A discussion of recent technological developments and future research plans. Paper presented at the Canadian Population Society Meetings, Lennoxville, Quebec.

Williams, W., & Goodman, M. (1971). A simple method for the construction of empirical confidence limits for economic forecasts. *Journal of the American Statistical Association*, 66, 752–754.

Wilson, A. (1974). *Urban and regional models in geography*. London: Wiley.

Wilson, A., Coelho, J., Macgill, S., & Williams, H. (1981). *Optimization in locational and transport analysis*. Chichester, UK: Wiley.

Wingo, L. (1961). *Transportation and urban land*. Baltimore: The Johns Hopkins University Press.

Yokum, J., & Armstrong, J. (1995). Beyond accuracy: Comparison of criteria used to select forecasting methods. *International Journal of Forecasting*, 11, 591–597.

Yu, J. (1982). *Transportation engineering: Introduction to planning design and operations*. New York: Elsevier.

Yule, G. (1925). The growth of population and the factors which control it. *Journal of the Royal Statistical Society*, 38, 1–58.

Zarnowitz, V. (1984). The accuracy of individual and group forecasts from business outlook surveys. *Journal of Forecasting*, 3, 11–26.

Zax, J. (1994). When is a move a migration? *Regional Science and Urban Economics*, 24, 341–360.

Zelinsky, W. (1971). The hypothesis of the mobility transition. *Geographical Review*, 61, 219–249.

Zelinsky, W. (1980). The impasse in migration theory: A sketch map for potential escapees. In P.

Morrison (Ed.), *Population movements: Their forms and functions in urbanization and development* (pp. 19–46). Liege, France: Orlina Editions.

Zhang, G., Patuwo, B., & Hu, M. (1998). Forecasting with artificial neural networks: The state of the art. *International Journal of Forecasting*, 14, 35–62.

Zhao, Y., Carlson, J., & Swanson, D. (1994). An evaluation of the demographic components of a proprietary economic forecasting and simulation system: The REMI model as used by SAIC, Inc. for the Yucca Mountain Project in Nevada. Paper presented at the International Conference on Applied Demography, Bowling Green, OH.

Author Index

Subject Index